Organelle Proteomics

T0189826

METHODS IN MOLECULAR BIOLOGY™

John M. Walker, SERIES EDITOR

METHODS IN MOLECULAR BIOLOGY™

Organelle Proteomics

Edited by

Delphine Pflieger

Laboratoire de Neurobiologie et Diversité Cellulaire, CNRS UMR 7637, Ecole Supérieure de Physique et de Chimie Industrielles, Paris, France and Laboratoire d'Analyse et de Modélisation pour la Biologie et l'Environnement, CNRS UMR 8587, Université d'Evry Val d'Essonne, Evry, France

and

Jean Rossier

Laboratoire de Neurobiologie et Diversité Cellulaire, CNRS UMR 7637, Ecole Supérieure de Physique et de Chimie Industrielles, Paris, France

Editor
Delphine Pflieger
Laboratoire de Neurobiologie et Diversité
Cellulaire, CNRS UMR 7637
Ecole Supérieure de Physique et de
Chimie Industrielles
Paris, France
and
Laboratoire d'Analyse et de Modélisation
pour la Biologie et l'Environnement
CNRS UMR 8587, Université d'Evry Val
d'Essonne, Evry, France

Jean Rossier
Laboratoire de Neurobiologie et Diversité
Cellulaire, CNRS UMR 7637
Ecole Supérieure de Physique et de
Chimie Industrielles
Paris, France

Series Editor
John M. Walker
School of Life Sciences
University of Hertfordshire
Hatfield, Herts., AL10 9AB
UK

ISBN: 978-1-61737-772-3   e-ISBN: 978-1-59745-028-7

ISSN: 1064-3745

Cover Illustration: Figure 1, Chapter 4, "Purification of *Saccharomyces cerevisiae* Mitochondria by Zone Electrophoresis in a Free Flow Device," by Hans Zischka, Norbert Kinkl, Ralf J. Braun, and Marius Ueffing

Printed on acid-free paper

9 8 7 6 5 4 3 2 1

springer.com

Preface

Human genome sequencing has identified about 25,000 genes, most being of unknown function. Localization of the final gene products—that is, the proteins—in a specific organelle is a key to deciphering the proteins' roles within the cell. Over the past 20 years, proteomic analyses have progressively proved to be an invaluable tool to obtain high-throughput protein identification from low-abundance, complex biological samples. These analyses boomed thanks to dramatic technological progresses in mass spectrometry instrumentation, optimization of its coupling to capillary liquid chromatography, and the development of software enabling processing of the vast amount of generated data. In the context of organelle study, such analyses have allowed greater depth in the characterization of the proteins constitutive of, or transiently present in, these large functional modules. For example, in 2002, we published the analysis of a total yeast mitochondrial protein extract by the coupling between capillary liquid chromatography and tandem mass spectrometry, known as liquid chromatography-tandem mass spectrometry (LC-MS/MS) *(1)*. We were then able to identify 179 proteins (http://mitochondria.cgm.cnrs-gif.fr/) out of about 500 expected mitochondrial constituents. Among these, 132 were already recorded as mitochondrial in the Yeast Protein Database, YPD *(2)*, 28 were described to be of unknown localization and function, and 19 were described to belong to other subcellular compartments. Among the 28 identified proteins that were uncharacterized in early 2002, eight were further functionally studied by other groups and demonstrated to play a role in mitochondria. For example, Ykr065cp was shown to be involved in the import of mitochondrial matrix proteins *(3)*, Ylr201cp in ubiquinone biosynthesis *(4)*, and Ynl177cp was proved to be a mitochondrial ribosomal protein *(5)*. Another 17 were identified in more recent proteomic analyses of yeast mitochondria (such as reference *[6]*) confirming our results. Little to no data are currently available in yeast databases to support mitochondrial localization of the three remaining proteins. This study, among many others, illustrates that proteomic analysis of a carefully prepared organelle sample reliably reveals new proteins constitutive of the cellular compartment and paves the way for their detailed functional characterization.

Cell proteomics is faced with the extraordinary chemical diversity of expressed proteins and the very large dynamic range of their cellular concentrations. In any given human cell, the most abundant protein is usually actin, present at above 10^8 molecules per cell, whereas cellular receptors, signaling proteins, and transcription factors may exist at a few hundreds of copies or even less. Yet, the understanding of the diverse cellular processes requires the study of the expression of those proteins—needles in the cellular haystack. To achieve that goal, it is indispensable to enrich such target proteins within a subcellular sample of reduced complexity. To characterize proteins of minor abundance in an organelle, this subcellular compartment can be divided into fractions, such as soluble and insoluble membrane proteins. In one last step of fractionation, the proteomic analysis can be focused on a given protein machinery, such as a single complex. *Organelle Proteomics* focuses on these three levels of subcellular organization by describing the preparation of samples and their proteomic analysis.

This book starts with a chapter by Dr. Edwin Romijn and Prof. John R. Yates III, who introduce the different analytical strategies developed and successfully utilized to study organelle proteomes and detail the use of multidimensional liquid chromatography coupled to tandem mass spectrometry for peptide sample analysis. This book is further composed of two main sections. First, detailed protocols are provided to perform the purification of the various organelles present in eukaryotic cells, as well as to prepare certain subfractions of organelles (Chapters 2–22). In all cases, the samples are aimed to be analyzed by a mass spectrometry technique. Although an exhaustive list of chapters covering all the proteomic analyses of organelles and organelle fractions was not conceivable, we nevertheless wanted to provide analysis examples reflecting the trend toward more specific purifications of organelle subfractions, which will allow reaching the more comprehensive and accurate characterization of the organelle. Most of the chapters cover the whole analytical procedure of organelle characterization, from its purification starting with whole cells up to protein identification using mass spectrometry. In some cases, the chapter may provide a detailed description of the purification process wherein less classical techniques appear, which are implemented by a minority of laboratories (e.g., free flow electrophoresis). Second, however optimized the organelle purification protocol—and skilled the operator—the sample of interest will never consist of the pure targeted organelle. Therefore, among the proteins identified, one has to separate the true from the intruders. The actual subcellular localization of some individual proteins newly attributed to the studied organelle can be evaluated by orthogonal assays, such as microscopy, by expressing the GFP-tagged version of the protein candidates. Yet, this approach is labor-intensive and is usually restricted to a few selected proteins. We devoted the second section of this book to methods enabling a global estimate of the reliability of the protein list assigned to an organelle. An average ratio of proteins wrongly attributed to the organelle of interest is provided by assessing sample purity (Chapter 23). To determine whether every identified protein is an actual component of the purified organelle, quantitative mass spectrometry methods can be employed (Chapters 24–26). In Chapter 26, Dr. Wei Yan et al. more specifically demonstrate the utility of quantitative approaches to scrutinize protein shuttling between organelles. The examples of quantitative mass spectrometry analysis of organelle fractions presented use a few commercially available isotope-tagged reagents, but many other chemicals, either commercial or prepared in-house, can be utilized. A larger variety of the existing polypeptide-labeling strategies can be found in another volume of this series entitled *Quantitative Proteomics*, edited by Dr. Salvatore Sechi. Finally, the last chapter of this book, by Dr. Wallace F. Marshall, addresses the use of transcriptomic data to identify genes potentially encoding organelle proteomes.

One should keep in mind that some, if not the majority, of peptide sequences assigned by software tools to raw mass spectra must be rejected. The degree of false-positive protein identification can be estimated by statistical interpretation of mass spectrometry results. This aspect, while of greatest importance to generate interlaboratory databases of proteins organized by localization and function, is beyond the scope of this book. It is dealt with in an alternative volume of this series, *Mass Spectrometry Data Analysis in Proteomics*, edited by Dr. Rune Matthiesen.

In terms of chapter format, each chapter begins by introducing the protocol to be described, with its goals and possible advantages over other techniques. In the Materials section, all the equipment and reagents necessary for performing the protocol are listed. The Methods section details the different steps of the protocol while the Notes collect remarks, tricks, and troubleshooting that are likely to help dealing with difficulties that might be encountered during the protocol.

Delphine Pflieger
Jean Rossier

References

1. Pflieger, D., Le Caer, J. P., Lemaire, C., Bernard, B. A., Dujardin, G., Rossier, J. (2002) Systematic identification of mitochondrial proteins by LC-MS/MS. *Anal. Chem.* **74**, 2400–2406.
2. Hodges, P. E., Payne, W. E., Garrels, J. I. (1997) The Yeast Protein Database (YPD): a curated proteome database for *Saccharomyces cerevisiae*. *Nucleic Acids Res.* **26**, 68–72.
3. van der Laan M., Chacinska, A., Lind, M., Perschil, I., Sickmann, A., Meyer, H., et al. (2005) Pam17 is required for architecture and translocation activity of the mitochondrial protein import motor. *Mol. Cell. Biol.* **25**, 7449–7458.

4. Johnson, A., Gin, P., Marbois, B. N., Hsieh, E. J., Wu, M., Barros, M. H., et al. (2005) COQ9, a new gene required for the biosynthesis of coenzyme Q in Saccharomyces cerevisiae. *J. Biol. Chem.* **280**, 31397–31404.

5. Gan X., Kitakawa, M., Yoshino, K., Oshiro, N., Yonezawa, K., Isono, K. (2002) Tag-mediated isolation of yeast mitochondrial ribosome and mass spectrometric identification of its new components. *Eur. J. Biochem.* **269**, 5203–5214.

6. Sickmann, A., Reinders, J., Wagner, Y., Joppich, C., Zahedi, R., Meyer, H. E., et al. (2003) The proteome of Saccharomyces cerevisiae mitochondria. *Proc. Natl. Acad. Sci. U.S.A.* **100**, 13207–13212.

Contents

Contributors

RUEDI AEBERSOLD • *Institute for Systems Biology, Seattle, WA; Institute of Molecular Systems Biology, ETH-Zurich, Zurich, Switzerland; and Faculty of Science, University of Zurich, Zurich, Switzerland*

JOHN D. AITCHISON • *Institute for Systems Biology, Seattle, WA*

ERIK ALEXANDERSSON • *Department of Biochemistry, Centre for Chemistry and Chemical Engineering, Lund University, Lund, Sweden*

PHILIP C. ANDREWS • *Department of Biological Chemistry, The University of Michigan, Ann Arbor, MI*

RICK BAGSHAW • *Research Institute, The Hospital for Sick Children, Toronto, Ontario; and Department of Laboratory Medicine and Pathobiology, University of Toronto, Toronto, Ontario, Canada*

KATJA BERNFUR • *Department of Biochemistry, Centre for Chemistry and Chemical Engineering, Lund University, Lund, Sweden*

ADELE R. BLACKLER • *Department of Pharmacology, University of Colorado School of Medicine, Aurora, CO*

MARIE-PIERRE BOUSQUET-DUBOUCH • *Institut de Pharmacologie et de Biologie Structurale, Centre National de la Recherche Scientifique, Toulouse, France*

RALF J. BRAUN • *GSF-National Research Center for Environment and Health, Institute of Human Genetics, Munich-Neuherberg, Germany*

ODILE BURLET-SCHILTZ • *Institut de Pharmacologie et de Biologie Structurale, Centre National de la Recherche Scientifique, Toulouse, France*

JOHN W. CALLAHAN • *Research Institute, The Hospital for Sick Children, Toronto, Ontario; and Departments of Biochemistry and Pediatrics, University of Toronto, Toronto, Ontario, Canada*

AGNÈS CHAPEL • *CEA, DSV, DRDC, Laboratoire de Chimie des Protéines, Grenoble, France; INSERM, Grenoble, France; and Université Joseph Fourier, Grenoble, France*

JULIE CHAROLLAIS • *Global Health Institute, Ecole Polytechnique Fédérale de Lausanne, Lausanne, Switzerland*

XUEQUN CHEN • *Department of Biological Chemistry, The University of Michigan, Ann Arbor, MI*

HOLLY D. COX • *Department of Pharmaceutical Sciences, University of Montana, Missoula, MT. Current address: Department of Pathology, Salt Lake City, UT*

SANDRINE DA CRUZ • *Department of Cell Biology, University of Geneva, Geneva, Switzerland*

MANUELLE DUCOUX-PETIT • *Institut de Pharmacologie et de Biologie Structurale, Centre National de la Recherche Scientifique, Toulouse, France*

GENEVIÈVE DUJARDIN • *CNRS, Centre de Génétique Moléculaire, Avenue de la Terrasse, Gif sur Yvette, France*

TOM P. J. DUNKLEY • *Cambridge Centre for Proteomics Department of Biochemistry, University of Cambridge, Cambridge, UK*

XIAOLIAN FAN • *Research Institute, The Hospital for Sick Children, Toronto, Ontario, Canada*

MYRIAM FERRO • *CEA, DSV, iRTSV, Laboratoire de l'Etude de la Dynamique des Protéomes, Grenoble, France; INSERM, Grenoble, France; and Université Joseph Fourier, Grenoble, France*

LAURENCE FLORENS • *The Stowers Institute for Medical Research, Kansas City, MO*

KIICHI FUKUI • *Laboratory of Dynamic Cell Biology, Department of Biotechnology, Graduate School of Engineering, Osaka University, Osaka, Japan*

KAZUHIRO FURUKAWA • *Department of Chemistry, Faculty of Science, Niigata University, Niigata, Japan*

JÉRÔME GARIN • *CEA, DSV, DRDC, Laboratoire de Chimie des Protéines, Grenoble, France; INSERM, Grenoble, France; and Université Joseph Fourier, Grenoble, France*

NIKLAS GUSTAVSSON • *Department of Biochemistry, Centre for Chemistry and Chemical Engineering, Lund University, Lund, Sweden*

TSUNEYOSHI HORIGOME • *Department of Chemistry, Faculty of Science, Niigata University, Niigata, Japan*

KATHRYN E. HOWELL • *Department of Cell and Developmental Biology, University of Colorado School of Medicine, Aurora, CO*

DAEHEE HWANG • *Institute for Systems Biology, Seattle, WA. Current address: I-Bio Program & Department of Chemical Engineering, Pohang University of Science and Technology, Kyungbuk, Republic of Korea*

ISABELLE IOST • *Laboratoire de Génétique Moléculaire, Ecole Normale Supérieure, Paris, France*

KOHEI ISHII • *Department of Chemistry, Faculty of Science, Niigata University, Niigata, Japan*

MARKUS ISLINGER • *Department of Anatomy and Cell Biology II, University of Heidelberg, Heidelberg, Germany*

AGNÈS JOURNET • *CEA, DSV, DRDC, Laboratoire de Chimie des Protéines, Grenoble, France; INSERM, Grenoble, France; and Université Joseph Fourier, Grenoble, France*

JACQUES JOYARD • *Laboratoire de Physiologie Cellulaire Végétale, Grenoble, France; Université Joseph Fourier/CNRS UMR-5168/INRA/CEA-Grenoble, Grenoble, France; and iRTSV, Grenoble, France*

ADINE KARLSSON • *Department of Biochemistry, Centre for Chemistry and Chemical Engineering, Lund University, Lund, Sweden*

LANI C. KELLER • *Department of Biochemistry & Biophysics, University of California, San Francisco, CA*

JACOB KENNEDY • *ZymoGenetics, Inc, Seattle, WA*

SYLVIE KIEFFER-JAQUINOD • *CEA, DSV, DRDC, Laboratoire de Chimie des Protéines, Grenoble, France; INSERM, Grenoble, France; and Université Joseph Fourier, Grenoble, France*

NORBERT KINKL • *Technical University Munich, Institute of Human Genetics, Munich, Germany; GSF-National Research Center for Environment and Health, Institute of Human Genetics, Munich-Neuherberg, Germany*

PER KJELLBOM • *Department of Biochemistry, Centre for Chemistry and Chemical Engineering, Lund University, Lund, Sweden*

NADIA KORFALI • *The Wellcome Trust Centre for Cell Biology and Institute of Cell Biology, University of Edinburgh, Edinburgh, UK*

CHRISTER LARSSON • *Department of Biochemistry, Centre for Chemistry and Chemical Engineering, Lund University, Lund, Sweden*

CLAIRE LEMAIRE • *CNRS, Centre de Génétique Moléculaire, Gif sur Yvette, France*

KATHRYN S. LILLEY • *Cambridge Centre for Proteomics Department of Biochemistry, University of Cambridge, Cambridge, UK*

MARK S. LOWENTHAL • *Department of Pharmacology, University of Colorado School of Medicine, Aurora, CO*

DON J. MAHURAN • *Research Institute, The Hospital for Sick Children, Toronto, Ontario; and Department of Laboratory Medicine and Pathobiology, University of Toronto, Toronto, Ontario, Canada*

MARCELLO MARELLI • *Homestead Clinical Corporation, Seattle, WA*

WALLACE F. MARSHALL • *Department of Biochemistry & Biophysics, University of California, San Francisco, CA*

JEAN-CLAUDE MARTINOU • *Department of Cell Biology, University of Geneva, Geneva, Switzerland*

MARIETTE MATONDO • *Institut de Pharmacologie et de Biologie Structurale, Centre National de la Recherche Scientifique, Toulouse, France*

BERNARD MONSARRAT • *Institut de Pharmacologie et de Biologie Structurale, Centre National de la Recherche Scientifique, Toulouse, France*

ALEXEY I. NESVIZHSKII • *Department of Pathology, University of Michigan, Ann Arbor, MI*

DELPHINE PFLIEGER • *Laboratoire Analyse et Modélisation pour la Biologie et l'Environnement, Université d'Evry Val d'Essonne, Evry, France*

THIERRY RABILLOUD • *CEA/DSV/IRTSV/BBSI, UMR CNRS 5092, Biologie et Biophysique des systèmes intégrés, CEA Grenoble, 17 rue des martyrs, Grenoble, France*

NORBERT ROLLAND • *Laboratoire de Physiologie Cellulaire Végétale, Grenoble, France; Université Joseph Fourier/CNRS UMR-5168/INRA/CEA-Grenoble, Grenoble, France; and iRTSV, Grenoble, France*

EDWIN P. ROMIJN • *Philips Research Eindhoven, High Tech Campus, Eindhoven, The Netherlands*

JEAN ROSSIER • *Laboratoire de Neurobiologie et Diversité Cellulaire, CNRS UMR 7637, Ecole Supérieure de Physique et de Chimie Industrielles, Paris, France*

DANIEL SALVI • *Laboratoire de Physiologie Cellulaire Végétale, Grenoble, France; Université Joseph Fourier/CNRS UMR-5168/INRA/CEA-Grenoble, Grenoble, France; iRTSV, Grenoble, France*

ERIC C. SCHIRMER • *The Wellcome Trust Centre for Cell Biology and Institute of Cell Biology, University of Edinburgh, Edinburgh, UK*

HIDEAKI TAKATA • *Laboratory of Dynamic Cell Biology, Department of Biotechnology, Graduate School of Engineering, Osaka University, Osaka, Japan*

CHARLES M. THOMPSON • *Department of Biomedical and Pharmaceutical Sciences, Missoula, MT*

SUSUMU UCHIYAMA • *Laboratory of Dynamic Cell Biology, Department of Biotechnology, Graduate School of Engineering, Osaka University, Osaka, Japan*

MARIUS UEFFING • *Technical University Munich, Institute of Human Genetics, Munich, Germany; GSF-National Research Center for Environment and Health, Institute of Human Genetics, Munich-Neuherberg, Germany*

SANDRINE UTTENWEILER-JOSEPH • *Institut de Pharmacologie et de Biologie Structurale, Centre National de la Recherche Scientifique, Toulouse, France*

JOËLLE VINH • *Laboratoire de Neurobiologie et Diversité Cellulaire, Ecole Supérieure de Physique et de Chimie Industrielles, Paris, France*

GERHARD WEBER • *BD Diagnostics – Preanalytical Systems, Planegg/Martinsried, Germany*

JACEK R. WIŚNIEWSKI • *Department of Proteomics and Signal Transduction, Max-Planck Institute of Biochemistry, Martinsried, Germany*

CHRISTINE C. WU • *Department of Pharmacology, University of Colorado School of Medicine, Aurora, CO*

WEI YAN • *Institute for Systems Biology, Seattle, WA; Department of Pathology, University of Washington, Seattle, WA*

JOHN R. YATES III • *The Scripps Research Institute, La Jolla, CA*

EUGENE C. YI • *ZymoGenetics, Inc, Seattle, WA*

HUIWEN ZHANG • *Research Institute, The Hospital for Sick Children, Toronto, Ontario, Canada*

HANS ZISCHKA • *GSF-National Research Center for Environment and Health, Institute of Toxicology, Munich-Neuherberg, Germany*

1

Analysis of Organelles by On-Line Two-Dimensional Liquid Chromatography–Tandem Mass Spectrometry

Edwin P. Romijn and John R. Yates III

Summary

The sequencing of the genomes of many different species has greatly helped our understanding of organelles. This information has driven the development of mass spectrometry (MS)-based methods for large-scale analysis of proteins. One of these methods uses two-dimensional liquid chromatography (2DLC) coupled on-line to tandem MS and is described here. In this method, proteins are first proteolytically digested, and then the peptides are separated in two dimensions. Typically, separation in the first dimension is based on charge interactions with a strong cation exchange (SCX) resin, whereas separation in the second dimension is based on hydrophobic interactions with a reversed-phase (RP) support. Peptides are eluted from the SCX resin using increasing concentrations of a salt and subsequently trapped on the RP resin. Next, the salt is washed from the system, and the peptides are eluted using an increasing gradient of a non-polar organic solvent. Eluting peptides are mass analyzed and fragmented to generate tandem mass spectra. These tandem mass spectra can be used to search sequence databases to identify peptides by matching amino-acid sequences to each spectrum.

Key Words: MudPIT; strong cation exchange; reversed-phase chromatography; tandem mass spectrometry; multidimensional chromatography; organelle.

1. Introduction

With the DNA sequences of more than 100 genomes completed, as well as a draft sequence of the human genome *(1,2)*, one of the major challenges in modern biology is to understand the expression, function, and regulation of the

From: *Methods in Molecular Biology, vol. 432: Organelle Proteomics*
Edited by: D. Pflieger and J. Rossier © Humana Press, Totowa, NJ

entire set of proteins that carry out defined functions in organelles. In the post-genomic era, large-scale proteomics technologies have made organelle-scale studies possible and have become powerful methods for studying organelles. One of the core components of (organelle) proteomics is the ability to system-atically quantify and identify every expressed protein. The technology for such proteome analysis involves *(1)* methods for the separation of proteins and peptides, *(2)* analytical science for the identification and quantification of these biomolecules, and *(3)* bioinformatics for data management. Systems biology relies on linking proteomics data to data obtained through other genome-wide approaches. Proteomics is a multidisciplinary research activity wherein separation science and mass spectrometry (MS) play pivotal roles. The most popular MS-based strategies rely on enzymatic digestion of proteins before introduction into the mass spectrometer. Digestion of proteins into peptides not only increases sample complexity but helps to overcome some inherent problems associated with intact proteins, for example, solubility and handling problems. The benefits of coupling LC on-line with MS include an enhanced ability to handle complex peptide mixtures, improved dynamic range, and increased sensitivity arising from the concentration of each mixture component into a small (nanoliter range) volume delivered from the chromatographic column to the mass spectrometer ion source. This chapter only describes the use of an electrospray ionization (ESI) ion source and will not deal with matrix-assisted laser desorption ionization (MALDI). The implementation of nanoflow liquid chromatography (LC)-MS offers unique opportunities for speed, sensi-tivity, and automation of proteomics research *(3–6)*. In addition, the LC-MS combination allows facile desalting of samples prior to MS analysis *(7)*. Nano-LC columns with 50–100 μm diameters are now routinely used for the analysis of protein digests and require the delivery of solvent gradients at a few hundreds of nanoliter per minute flow rates. This has been typically achieved using generic high-performance liquid chromatography (HPLC) pumping systems for the delivery of microliter per minute gradients that are either flow-split or sampled. At the end of the capillary, a tip is often pulled with an inner diameter of ~10 μm diameter.

In a one-dimensional (1D) mode, the total peptide mixture is typically first loaded onto a nanocolumn (usually 75 μm internal diameter) containing reversed phase C18 material and then eluted by using a gradient directly into a tandem mass spectrometer. The peptides are eluted at a low flow rate of typically 100–200 nL per minute, and the elution time of each peptide is ~10–30 s. The peptides elute off the column into the ionization source of the mass spectrometer. First, a survey scan is taken, in which a mass spectrum of the intact peptides is obtained. In the data-dependent acquisition mode, the instrument can then be set to automatically fragment and collect MS/MS data on any number of peptides observed in the original MS spectrum based on their intensity, m/z value, or charge state.

Using LC-MS/MS in proteomics, quite a few different strategies may be taken to analyze the protein sample of interest. In one approach, close to the conventional one, 2D gel electrophoresis is still employed for the separation of the intact proteins. Digests resulting from proteolysis of the cut-out spots are analyzed by LC-MS or LC-MS/MS. This approach may be complementary to a conventional peptide mapping approach, as different sets of peptides are identified when using either matrix-assisted laser desorption ionization (MALDI) or ESI. Such a parallel MALDI peptide mapping and LC-ESI-MS approach may therefore enhance the protein coverage *(8)*.

To overcome some of the inherent disadvantages of 2D gels, several 2D gel-independent LC-MS/MS approaches have been introduced. In one such approach, the proteins in the total proteome are separated and resolved by molecular weight using 1D gels. This 1D gel is cut into pieces, and all proteins in such a band are digested, and the mixture of peptides are analyzed by LC-MS and/or LC-MS/MS *(9–14)*. This approach provides an intermediate form between analyzing the very complex large peptide mixture obtained when digesting all proteins of a lysate and the single protein digested when using a 2D gel. As compared to the 2D gel, a 1D gel-based approach is less elaborate. Additionally, very large and basic proteins (known to be difficult to analyze using 2D gels) are easier to handle using single-dimensional gels.

In another approach, the entire whole cell lysate is digested chemically or by a protease and analyzed using multidimensional protein identification technology (MudPIT) *(5,15,16)*. This generates a very large and complex set of peptides that is beyond the separation capacity of 1D separation techniques. For the analysis of such complex mixtures, various multidimensional separation techniques have been developed *(17)*. In one approach, peptides are first separated using a strong cation exchange (SCX) column and a stepped salt gradient. They are then introduced into a reversed-phase (RP) column that is coupled to a tandem mass spectrometer for analysis. In an off-line system, the fractions from the cation exchange column are collected, and each fraction is analyzed by RP LC-MS/MS. The direct coupling of the two stages of separation is feasible and the pioneering work by the group of Yates demonstrated the power of this procedure *(3,15,17–20)*. They introduced the rapid and large-scale proteome analysis of total cell lysates by MudPIT *(5,15,16)*. This approach typically employs a biphasic column with SCX resin packed behind RP resin. Application of MudPIT technology to the proteome analysis of *Saccharomyces cerevisiae* as well as off-line SCX fractionation followed by LC-MS/MS *(21)* yielded one of the largest proteome analyses to date with respect to the total number of proteins identified/detected *(15)*. A total of 1500 proteins were detected and identified by these methods *(15,21)*. Categorization of these hits demonstrated the ability of this technology to detect and identify proteins rarely seen in proteome analysis, including low-abundance proteins like transcription factors and protein kinases.

On the other hand, an "off-line" approach may provide a more versatile approach. In an "off-line" study of the yeast proteome *(21)*, 1 mg of yeast lysate was separated into 80 fractions on a SCX column, and each of these fractions was individually analyzed by RP LC-MS/MS, resulting in the identification of 1504 proteins. This exceeds the number of proteins that can readily be identified using gel-based approaches using this amount of material, and additional methods such as the use of different/multiple proteases *(22,23)* and the introduction of new mass spectrometers *(24)* may further extend the identification of proteins.

Compared to the conventional 2D gel-based approach, multidimensional LC has also proven to increase detection and sequence coverage of membrane proteins. Membrane proteins are usually underrepresented on conventional 2D gels. MudPIT strategies identified 131 proteins in a yeast total extract with three or more predicted transmembrane domains *(3,16)*, and more recently, Schirmer et al. used MudPIT technology to identify 13 known nuclear envelope integral membrane proteins and 67 uncharacterized open reading frames with predicted membrane-spanning regions. Remarkably, of the eight proteins tested all targeted to the nuclear envelope, indicating that there are substantially more integral proteins of the nuclear envelope than previously thought *(25)*. Together with controlled proteolysis by non-specific proteases, MudPIT described topological information of membrane proteins *(26)*.

LC-LC-MS/MS or MudPIT has been proven to be very powerful for the analysis of complex peptide mixtures, but these mixtures often also contain salts that can interfere with binding of the peptides to the SCX resin. McDonald et al. *(7)* reported the use of a third dimension/phase consisting of an additional section of RP material behind the first RP and SCX to circumvent this potential problem. MudPIT analysis is also constrained by the amount of data generated in each experiment. Depending on the type of instrument used, a MudPIT analysis will result in over 3000 spectra per hour for a LCQ-Deca 3D ion trap mass spectrometer and over 13,000 spectra per hour for a LTQ linear ion trap mass spectrometer. The first step after data acquisition is the comparison of these spectra against theoretical spectra generated from database sequence information. Depending on the computing resources available, the size of the database being searched, the sample complexity, the enzyme specificity used to search the spectra, and taking into account potential post-translational modifications, this might take anywhere between a few hours and a few days. Once peptide "hits" are scored for the identification of proteins, manual validation to verify real protein identifications versus false positives is frequently necessary. While this process has been streamlined by the development of several computer programs that remove spectra of poor quality, filter and sort data, normalize cross-correlation scores reported by SEQUEST *(27)*, and aid in assembly of proteins from their peptide sequences, interpretation of proteomic data remains a possible bottleneck in the identification of proteins using MudPIT.

MudPIT has been used to identify proteins in organelles *(28–32)*. Organelles represent subcellular compartments within cells and often have constrained functions and a limited number of proteins. The unique potential of organellar proteomics to focus on structures below the magnitude of a cell may explain the fascination of this approach. Once specific compartments or protein complexes are isolated in high purity, it is thus possible to investigate them by MudPIT. The core in organellar proteomics is the isolation of the given target structure. Due to the high sensitivity of MudPIT, the success of the overall procedure depends on the initial purity of the organelle under investigation.

2. Materials

2.1. Sample Preparation (Reduction/Alkylation/Digestion)

1. Trichloroacetic acid (Sigma, St. Louis, MO, USA) for protein precipitation.
2. Resuspension buffer for digestion with Proteinase K: 200 mM Na_2CO_3 (J.T. Baker, Phillipsburg, NJ), pH 11.
3. Solid urea.
4. Resuspension buffer for digestion with Lys-C/trypsin: 100 mM Tris, pH 8.5, 8 M urea (both from Sigma).
5. Protein reducing agent: Tris(2-carboxyethyl)phosphine (TCEP) (Pierce, Rockford, IL).
6. Protein alkylating agent: iodoacetamide (Sigma).
7. Proteinase K (Roche Diagnostics, Indianapolis, IN).
8. Sequencing-grade endoproteinase Lys-C (Roche Diagnostics) and sequencing-grade trypsin (Promega, Madison, WI).
9. 50 mM ammonium bicarbonate (J.T. Baker), pH 8.5.
10. 100 mM $CaCl_2$ (Mallinckrodt Baker, Paris, KY).
11. Thermomixer (Brinkmann, Westbury, NY) for shaking protein tubes to be digested at 37°C.
12. To stop the digestion: formic acid (Sigma).

2.2. Preparation of Packed Capillary Columns and Sample Loading

1. For preparing chromatography columns: 75 µm × 365 µm, 250 µm × 365 µm, and 180 µm × 365 µm fused silica capillary tubing (Polymicro Technologies, Phoenix, AZ).

2. For frits: Kasil® 1624 (PQ Corporation, Southpoint, PA) and formamide (Sigma).
3. Ceramic capillary cutter (Upchurch Scientific, Oak Harbor, WA) for cutting fused silica.
4. For pulling electrospray tips to fused silica capillaries, use a CO_2 laser puller (P2000; Sutter Instruments, Novato, CA). Alternatively, use commercially prepared nanocolumns (New Objective, Woburn, MA and Nanoseparations, Nieuwkoop, The Netherlands).
5. Kimwipes for cleaning.
6. For slurry packing: a stainless steel pneumatic pressure bomb (The Scripps Research Institute, La Jolla, CA or Cytopea) and a helium tank for pressurizing.
7. Stationary phases: Aqua (Phenomenex, Torrance, CA) C18 5 μm, Aqua C18 3 μm, or Partisphere SCX resin (Whatman, Clifton, NJ).
8. HPLC grade methanol (Fisher Scientific, Fairlawn, NJ).
9. For equilibration and washing of columns: buffer A is H_2O/formic acid/HPLC grade acetonitrile, 94.9/0.1/5 (v/v/v). Acetonitrile is from Fisher Scientific.
10. For coupling of columns: a true zero dead volume union (Upchurch Scientific).

2.3. LC-MS/MS Analysis

1. For liquid chromatography: quaternary HPLC system (Agilent 1100 series LC; Agilent Technologies, Palo Alto, CA) or other system capable of multidimensional chromatography at 0.1–1 mL/min flow rates.
2. For measuring the flow through the column: 1–5 μL calibrated pipets (Drummond Scientific Company, Broomall, PA).
3. For mass spectrometry: any mass spectrometer capable of tandem MS with automated data acquisition can be used, for example, LCQ 3D-ion trap tandem mass spectrometer, LTQ linear ion trap mass spectrometer, LTQ-FT hybrid mass spectrometer, LTQ-Orbitrap hybrid mass spectrometer (all Thermo Electron Corporation, San Jose, CA), or a QqToF-type instrument (Waters Cooperation, Milford, MA or Applied Biosystems, Foster City, CA).
4. Peek MicroTees (Upchurch Scientific) for setting up the column.
5. For peptide chromatographic separation: buffer A (H_2O/formic acid/acetonitrile, 94.9/0.1/5, v/v/v), buffer B (H_2O/formic acid/acetonitrile, 20/0.1/79.9, v/v/v), and buffer C (500 mM ammonium acetate in H_2O/formic acid/acetonitrile, 94.9/0.1/5, v/v/v).

3. Methods

3.1. Sample Preparation (Reduction/Alkylation/Digestion)

1. Organelle protein samples are first precipitated with trichloroacetic acid in a microcentrifuge tube.
2. When performing Proteinase K digestions, the sample is resuspended in 150 μL of 200 mM Na_2CO_3 (*see* **Note 1**), pH 11, for 1 h on ice. After 1 h, the sample

solution is adjusted to 8 M urea by adding solid urea. When performing Lys-C/trypsin digestions, the sample is directly resuspended in 100 μL of 100 mM Tris, pH 8.5, and 8 M urea.

3. For reduction of the sample, TCEP is added to a final concentration of 5 mM and incubated for 20 min at room temperature.

4. To alkylate the sample, iodoacetamide is added to a final concentration of 10 mM and incubated in the dark for 0.5 h.

5. For a Proteinase K digestion, simply add Proteinase K in a ratio of 1:50 (enzyme : substrate, w/w) to the sample solution and incubate with shaking at 37°C for 5 h in the dark. For Lys-C/trypsin digestions, add Lys-C in a ratio of 1:100 (enzyme : sample, w/w) and incubate overnight in the dark with shaking at 37°C. Dilute the sample to 2 M urea using 50 mM ammonium bicarbonate, pH 8.5, and add $CaCl_2$ to a final concentration of 1 mM. Finally, add trypsin in a 1:100 (enzyme : sample, w/w) ratio and incubate overnight in the dark with shaking at 37°C.

6. After digestion, add formic acid to a final concentration of 5% (v/v) to acidify the sample and stop the enzymatic reaction. Centrifuge sample at maximum speed, typically 30,000 g, to remove insoluble material possibly present in the sample to prevent clogging of columns while loading the sample (*see* **Note 2**).

3.2. Preparation of packed capillary columns and sample loading

1. Porous ceramic frits are prepared in 180 μm or 250 μm ID undeactivated (*see* **Note 3**) fused silica capillaries (~20 cm long) to provide a frit for column packing. A mixture of 300 μL of the potassium silicate solution Kasil® 1624 and 100 μL of formamide are thoroughly mixed at room temperature. The mixture is deposited immediately inside the capillary for a few seconds by capillary action. The material is polymerized by heating in an oven for at least 4 h, but typically overnight at 100°C. The resulting porous frits are stored and used without any further treatment (e.g., deactivation), cut into 0.5–1 mm lengths with a fused silica precision cutter just prior to use.

2. Analytical columns are prepared in fused silica capillaries containing an electrospray/nanospray tip (no frit). Electrospray tips are prepared using a CO_2 laser puller. This is a critical step in the procedure as every column ID has its optimal flow rate, and every flow rate has its optimal tip opening, for example, a column with an internal diameter of 75 μm has an optimal flow rate of ~200 nL/min with an optimal tip opening of ~8 μm (*see* **Note 4**). Detailed documentation is available from the manufacturer (Sutter Instruments) and requires optimization as it may vary from instrument to instrument and may change slightly in time. Typical settings for pulling ~8 μm tips to 365 μm OD × 75 μm ID can be found in **Table 1**. Approximately 5 cm of the polyimide coating in the middle of a 50 cm long 365 μm OD × 75 μm ID is completely burned off using an alcohol burner and subsequently cleaned with a Kimwipe and ethanol. The capillary is placed in the CO_2 laser puller making sure that the clear section of the capillary is in the chamber of the puller. The experimentally determined program to pull the

Table 1
Sample Program for Pulling ~8-µm Tips to a 365 µm OD × 75 µm ID Fused Silica Capillary Using a CO_2 Laser Puller

	Heat	Filament	Velocity	Delay	Pull
Line 1	280	0	25	200	0
Line 2	270	0	15	200	0
Line 3	280	0	20	200	0
Line 4	270	0	25	200	0

    desired tip opening is chosen. In case a CO_2 laser puller is unavailable, pre-pulled columns are commercially available (New Objective and Nanoseparations).

3. Fused silica capillary columns are prepared by slurry packing using a stainless steel pneumatic pressure bomb (*see* **Note 5**). Slurries are prepared by adding ~1 mg Aqua C18 5 µm, Aqua C18 3 µm, or SCX resin in a microcentrifuge tube containing 1 mL of methanol. The microcentrifuge tube containing the slurry is placed in the stainless steel pneumatic pressure bomb. The bomb is closed and a pressure of ~600 psi is applied using a helium tank. After the pressure has been applied, the packing material should begin to fill the capillary (*see* **Note 6**). Using this method, the fritted loading column is packed with 3 cm of 5 µm C-18 resin and 3 cm of 5 µm SCX resin and the analytical column with 15 cm of 3 µm C-18 resin.

4. Sample is loaded onto the loading column containing 3 cm of 5 µm C-18 resin and 3 cm of 5 µm SCX resin. The loading column is first equilibrated using buffer A (94.9% H_2O, 0.1% formic acid, 5% acetonitrile, v:v), before the acidified sample (5% formic acid) is loaded using the pressure bomb (*see* **Note 7**). After sample loading, the column is equilibrated again using at least five column volumes of buffer A. Finally, the loading column is coupled through a true zero dead volume union to a 15 cm length, 75 µm ID analytical column packed with 3 µm C-18 resin.

3.3. LC-MS/MS Analysis

1. Multidimensional chromatography is performed using a setup as depicted in **Fig. 1**. The flow from the HPLC pump (300 µL/min) is split using a 365 µm OD × 50 µm ID fused silica capillary, resulting in a 200 nL/min flow through the column. The flow through the column is measured using a 1–5 µL calibrated pipet while running buffer A (*see* **Note 8**). To decrease flow through the column, a shorter split line may be used, and to increase the flow through the column, a longer split line may be used (*see* **Notes 9** and **10**). The voltage (1.8 kV for a LCQ 3D-ion trap mass spectrometer) is applied at the Peek MicroTee in the back using a 1 cm gold wire. The column setup (loading column coupled through a true zero dead

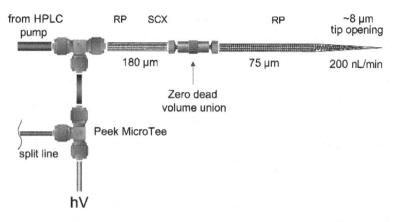

Fig. 1. Integrated multi-dimensional liquid chromatography setup. Please note that this picture is not on scale. All fused silica tubing is much smaller than depicted and is held in place by sleeves (not in picture).

volume union to the analytical column) is inserted in the Peek MicroTee in the front. The two Peek MicroTees are placed on a XYZ stage, allowing the tip of the column to be positioned exactly in front of the orifice of the mass spectrometer.

2. In general, the more complex the sample, the more salt steps should be used. For digests from organelles analyzed by a LCQ-Deca 3D-ion trap tandem mass spectrometer, 12 salt steps are used. The first step consists of 0% buffer C and the last step 100% buffer C. The exact percentages of salt in steps 2–11 may be optimized for each different sample (*see* **Note 11**). An example of a 12-step 2D-LC run can be found in **Fig. 2**.

3. The LCQ-Deca 3D-ion trap tandem mass spectrometer is set up to operate in positive ion mode and data-dependent mode. This means that the mass spectrometer acquires one full MS-scan over m/z 400–1400 Da, typically followed by 3 MS/MS scans from the top three most intense ions detected in the previous full MS-scan. Dynamic exclusion is applied to prevent an abundant ion from being continually sampled for MS/MS analysis; once selected and analyzed in MS/MS mode, the ion is excluded for a period of time (typically 1 min), making sampling of less intense ions more likely.

3.4. Database Searching and Bio-Informatics

Fragment ions produced in the MS/MS fragmentation process can be separated into two classes. One class retains the charge on the N-terminal region while cleavage is observed from the C-terminal region, resulting in a_n, b_b, and c_n type ions [nomenclature according to *(33)*]. The second class retains the charge on the C-terminal region while cleavage is observed from the N-terminal region,

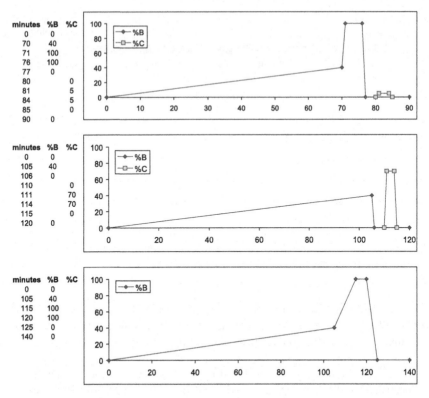

Fig. 2. Typical 12-step MudPIT HPLC gradient program. Top panel depicts the first step of the procedure and the bottom panel the last. The middle panel shows one of the steps 2–11 (e.g., step with 70% C). Note that the salt step is done at the end of the previous gradient to maximize the separation window and has all peptides from the same salt step in one data file.

resulting in x_n, y_b, and z_n type ions (*see* **Fig. 3**). As peptides fragment in general in this predictable manner, sequences from databases can be used to generate theoretical fragmentation spectra, which can be compared with the observed MS/MS spectra in the acquired data set. Several different software packages are commercially available to perform this database searching, and alternatively, web-based search tools are also available at http://prospector.ucsf.edu, http://www.matrixscience.com, and http://prowl.rockefeller.edu. We describe here the process by which proteins are identified from fragmentation spectra using the SEQUEST algorithm *(27)*. Collected MS/MS spectra are first processed using RawExtract *(34)* and then filtered using the PARC algorithm *(35)*. RawExtract is used to extract the peak lists from the raw data, and the PARC algorithm is used to filter out the worst spectra that basically only

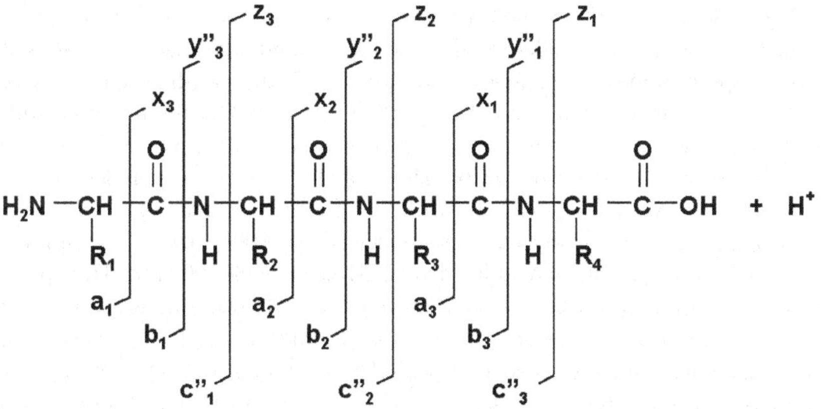

Fig. 3. Roepstorff nomenclature for peptide fragmentation. Fragmentation is observed predominantly at the amide bond. N-terminal fragments are named a_n, b_b, and c_n type ions and C-terminal fragments x_n, y_b, and z_n type ions.

have a very limited amount of peaks present. The filtered MS/MS spectra are searched (without enzyme specificity) against the desired database that includes both regular and reversed protein sequences order to estimate a false positive rate *(21)*. The SEQUEST search algorithm is used to interpret MS/MS spectra as described previously *(15,16,27,36)*. Database search criteria are as follows: mass tolerance of ±3 mass units for the precursor peptide in the SEQUEST database search using its average mass and the monoisotopic masses for the fragment ions (*see* **Note 12**). Cysteine residues are considered to have a static modification of +57 mass units due to their alkylation by iodoacetamide. SEQUEST results are then filtered with DTASelect v2.0 using a minimum number of two peptides to identify proteins and depending on the type of digestion, the tryptic status as to half or fully tryptic for Lys-C/trypsin digestions and non-tryptic for Proteinase K digestions. This filtering is done in combination with a false positive rate set at 5% [Cociorva, D., and Yates, J. R., 3rd. (2006) Manuscript in preparation].

4. Notes

1. If the sample contains a lot of insoluble material, CNBr digestion might be considered: Add 100 µL of 90% formic acid and incubate for 5 min at room temperature. Next, add 100 µg of CNBr and incubate overnight in the dark at room temperature. Finally, neutralize with solid ammonium bicarbonate to pH 8.5. Be careful during this last step, because it generates gas. Proceed with **step 3** from the sample preparation. CNBr is extremely toxic, handle with appropriate care! All operations are to be performed in a hood.

2. When post-translational modifications are of special interest, a triple-digestion procedure may be considered *(22)*. The idea behind this strategy is that multiple overlapping modified peptides give more confidence to the identified modification than just a single modified peptide and increases the level of sequence coverage. Proteases to be used are trypsin, subtilisin, and elastase. Sample preparation is done as described in the **Heading 3**. For trypsin, samples are diluted to 2 M urea with 100 mM Tris (pH 8.5), and $CaCl_2$ is added to a final concentration of 2 mM. Finally trypsin is added in a 1:100 (enzyme : sample) ratio. For elastase, samples are diluted to 2 M urea with 100 mM Tris (pH 8.5), and elastase is added in a 1:100 (enzyme : sample) ratio. Elastase and trypsin samples are incubated overnight in the dark with shaking at 37°C. For subtilisin, samples are diluted to 4 M urea with 100 mM Tris (pH 8.5), and elastase is added in a 1:100 (enzyme : sample) ratio and incubated for 2 h in the dark with shaking at 37°C.

3. Make sure to use undeactivated fused silica for making the fritted loading column, because deactivated fused silica will not hold the potassium silicate solution.

4. The pulling of the tips may leave the two resulting tips closed. The tip end can be gently and carefully scribed with a ceramic capillary cutter to open it. Tip openings may need to be checked under the microscope.

5. Make sure to slowly pressurize the bomb by opening the valve very slowly. Not doing so may result in a ruptured microcentrifuge tube.

6. While slurry packing fused silica capillary columns, the packing material may stop filling the column. Hold the column while very carefully loosening the ferrule holding the capillary. The pressure is still on! Gently tap the column on the bottom of the microcentrifuge tube and watch the packing continue. Tighten the ferrule holding the capillary and continue packing until desired column length is acquired.

7. While sample loading, the column might get clogged. One possible solution is to try to get flow through the column with buffer A, and if that does not solve the problem, gently heat the column with a blower/heat gun while trying to get flow through the column with buffer A.

8. While measuring the flow through the column may not always be necessary, visually checking for flow through the column should always be done. The optimal flow through the column is very critical. The optimal linear velocity in HPLC is ~1 mm/s. Separation efficiency decreases very fast at lower than optimal velocities. Low flow rates through the column may be due to one of the following events: (1) too high a pressure drop in the connecting tubing resulting in a lower than required net pressure at the top of the column; (2) clogging of the loading column; (3) a clogged analytical column; (4) or a clogged spray tip. Start to locate the clogging, beginning up-stream at the pump. Possible solutions are (1) replacing the connecting tubing; (2) gently heating of the columns with a blower; (3) gently and carefully scribing the tip with a ceramic capillary cutter to open it.

9. The backpressure on the HPLC pump should also be recorded and regularly checked. Pressure may increase in time and is almost always due to a (partly) clogged flow splitter.

10. While running the gradient, the programmed HPLC pump flow may also be increased slightly. Because of the changing solvent composition while running the gradient, the split ratio changes. This results in a lower flow through the column. Increasing the HPLC pump flow from 300 to 350 µL/min should keep the flow through the column approximately constant throughout the whole run.

11. Choosing the gradient and number of salt steps not only depends on sample complexity but also on type of mass spectrometer used. A LTQ linear ion trap mass spectrometer (or LTQ-FT hybrid mass spectrometer or LTQ-Orbitrap hybrid mass spectrometer) is more sensitive and much faster than the LCQ-Deca 3D-ion trap tandem mass spectrometer. For this reason, a LTQ linear ion trap mass spectrometer can handle co-eluting peptides much better and fewer salt steps or shorter gradients may be used.

12. Database searching depends heavily on the type of mass spectrometer used and the type of sample (post-translational modifications). Different mass spectrometers have different resolutions and different mass accuracies, and this has to be taken into account while performing database searches.

Acknowledgments

The authors thank James Wohlschlegel and Cristian Ruse for critically reading the manuscript. J.R.Y. acknowledges support from National Institutes of Health grant P41 RR11823-09 (University of Washington Yeast Resource Center).

References

1. McPherson, J. D., Marra, M., Hillier, L., Waterston, R. H., Chinwalla, A., Wallis, J., et al. (2001) A physical map of the human genome. *Nature* **409**, 934–941.

2. Venter, J. C., Adams, M. D., Myers, E. W., Li, P. W., Mural, R. J., Sutton, G. G., et al. (2001) The sequence of the human genome. *Science* **291**, 1304–1351.

3. Link, A. J., Eng, J., Schieltz, D. M., Carmack, E., Mize, G. J., Morris, D. R., et al. (1999) Direct analysis of protein complexes using mass spectrometry. *Nat. Biotechnol.* **17**, 676–682.

4. Yates, J. R., 3rd, McCormack, A. L., Link, A. J., Schieltz, D., Eng, J., and Hays, L. (1996) Future prospects for the analysis of complex biological systems using microcolumn liquid chromatography-electrospray tandem mass spectrometry. *Analyst* **121**, 65R–76R.

5. Yates, J. R., 3rd, Carmack, E., Hays, L., Link, A. J., and Eng, J. K. (1999) Automated protein identification using microcolumn liquid chromatography-tandem mass spectrometry. *Methods Mol. Biol.* **112**, 553–569.

6. Meiring, H. D., van der Heeft, E., Ten Hove, G. J., De Jong, A. P. J. M. (2002) Nanoscale LC-MS(n): technical design and applications to peptide and protein analysis. *J. Sep. Sci.* **25**, 557–568.

7. McDonald, W. H., Ohi, R., Miyamoto, D. T., Mitchison, T. J., and Yates, J. R. (2002) Comparison of three directly coupled HPLC MS/MS strategies for identification of proteins from complex mixtures: single-dimension LC-MS/MS, 2-phase MudPIT, and 3-phase MudPIT. *Int. J. Mass Spectrom.* **219**, 245–251.

8. Mann, M. and Jensen, O. N. (2003) Proteomic analysis of post-translational modifications. *Nat. Biotechnol.* **21**, 255–261.

9. Romijn, E. P., Christis, C., Wieffer, M., Gouw, J. W., Fullaondo, A., van der Sluijs, P., et al. (2005) Expression clustering reveals detailed co-expression patterns of functionally related proteins during B cell differentiation: a proteomic study using a combination of one-dimensional gel electrophoresis, LC-MS/MS, and stable isotope labeling by amino acids in cell culture (SILAC). *Mol. Cell. Proteomics* **4**, 1297–1310.

10. De Boer, E., Rodriguez, P., Bonte, E., Krijgsveld, J., Katsantoni, E., Heck, A., et al. (2003) Efficient biotinylation and single-step purification of tagged transcription factors in mammalian cells and transgenic mice. *Proc. Natl. Acad. Sci. U.S.A.* **100**, 7480–7485.

11. Gavin, A. C., Bosche, M., Krause, R., Grandi, P., Marzioch, M., Bauer, A., et al. (2002) Functional organization of the yeast proteome by systematic analysis of protein complexes. *Nature* **415**, 141–147.

12. Ho, Y., Gruhler, A., Heilbut, A., Bader, G. D., Moore, L., Adams, S. L., et al. (2002) Systematic identification of protein complexes in Saccharomyces cerevisiae by mass spectrometry. *Nature* **415**, 180–183.

13. Wu, S. L., Amato, H., Biringer, R., Choudhary, G., Shieh, P., and Hancock, W. S. (2002) Targeted proteomics of low-level proteins in human plasma by LC/MSn: using human growth hormone as a model system. *J. Proteome Res.* **1**, 459–465.

14. Taylor, S. W., Fahy, E., Zhang, B., Glenn, G. M., Warnock, D. E., Wiley, S., et al. (2003) Characterization of the human heart mitochondrial proteome. *Nat. Biotechnol.* **21**, 281–286.

15. Washburn, M. P., Wolters, D., and Yates, J. R. (2001) Large-scale analysis of the yeast proteome by multidimensional protein identification technology. *Nat. Biotechnol.* **19**, 242–247.

16. Wolters, D. A., Washburn, M. P., and Yates, J. R. (2001) An automated multidimensional protein identification technology for shotgun proteomics. *Anal. Chem.* **73**, 5683–5690.

17. Link, A. J. (2002) Multidimensional peptide separations in proteomics. *Trends Biotechnol.* **20**, S8–S13.

18. Washburn, M. P. (2004) Utilisation of proteomics datasets generated via multidimensional protein identification technology (MudPIT). *Brief Funct. Genomic Proteomic.* **3**, 280–286.

19. Washburn, M. P., Ulaszek, R., Deciu, C., Schieltz, D. M., and Yates, J. R. (2002) Analysis of quantitative proteomic data generated via multidimensional protein identification technology. *Anal. Chem.* **74**, 1650–1657.

20. Washburn, M. P., Ulaszek, R. R., and Yates, J. R. (2003) Reproducibility of quantitative proteomic analyses of complex biological mixtures by multidimensional protein identification technology. *Anal. Chem.* **75**, 5054–5061.

21. Peng, J., Elias, J. E., Thoreen, C. C., Licklider, L. J., and Gygi, S. P. (2003) Evaluation of multidimensional chromatography coupled with tandem mass spectrometry (LC/LC-MS/MS) for large-scale protein analysis: the yeast proteome. *J. Proteome Res.* **2**, 43–50.

22. MacCoss, M. J., McDonald, W. H., Saraf, A., Sadygov, R., Clark, J. M., Tasto, J. J., et al. (2002) Shotgun identification of protein modifications from protein complexes and lens tissue. *Proc. Natl. Acad. Sci. U.S.A.* **99**, 7900–7905.

23. Choudhary, G., Wu, S. L., Shieh, P., and Hancock, W. S. (2003) Multiple enzymatic digestion for enhanced sequence coverage of proteins in complex proteomic mixtures using capillary LC with ion trap MS/MS. *J. Proteome Res.* **2**, 59–67.

24. Hu, Q., Noll, R. J., Li, H., Makarov, A., Hardman, M., and Graham Cooks, R. (2005) The Orbitrap: a new mass spectrometer. *J. Mass Spectrom.* **40**, 430–443.

25. Schirmer, E. C., Florens, L., Guan, T., Yates, J. R., 3rd, and Gerace, L. (2003) Nuclear membrane proteins with potential disease links found by subtractive proteomics. *Science* **301**, 1380–1382.

26. Wu, C. C., MacCoss, M. J., Howell, K. E., and Yates, J. R. (2003) A method for the comprehensive proteomic analysis of membrane proteins. *Nat. Biotechnol.* **21**, 532–538.

27. Eng, J., McCormack, A., and Yates, J. R. (1994) An approach to correlate tandem mass-spectral data of peptides with amino-acid-sequences in a protein database. *J. Am. Soc. Mass Spectrom.* **5**, 976–989.

28. Keller, L. C., Romijn, E. P., Zamora, I., Yates, J. R., 3rd, and Marshall, W. F. (2005) Proteomic analysis of isolated chlamydomonas centrioles reveals orthologs of ciliary-disease genes. *Curr. Biol.* **15**, 1090–1098.

29. Skop, A. R., Liu, H., Yates, J., 3rd, Meyer, B. J., and Heald, R. (2004) Dissection of the mammalian midbody proteome reveals conserved cytokinesis mechanisms. *Science* **305**, 61–66.

30. Wu, C. C., MacCoss, M. J., Mardones, G., Finnigan, C., Mogelsvang, S., Yates, J. R., 3rd, et al. (2004) Organellar proteomics reveals Golgi arginine dimethylation. *Mol. Biol. Cell* **15**, 2907–2919.

31. Yates, J. R., 3rd, Gilchrist, A., Howell, K. E., and Bergeron, J. J. (2005) Proteomics of organelles and large cellular structures. *Nat. Rev. Mol. Cell Biol.* **6**, 702–714.

32. Wu, C. C., Yates, J. R., 3rd, Neville, M. C., and Howell, K. E. (2000) Proteomic analysis of two functional states of the Golgi complex in mammary epithelial cells. *Traffic* **1**, 769–782.

33. Roepstorff, P. and Fohlman, J. (1984) Proposal for a common nomenclature for sequence ions in mass spectra of peptides. *Biomed. Mass Spectrom.* **11**, 601.

34. McDonald, W. H., Tabb, D. L., Sadygov, R. G., MacCoss, M. J., Venable, J., Graumann, J., et al. (2004) MS1, MS2, and SQT-three unified, compact, and easily parsed file formats for the storage of shotgun proteomic spectra and identifications. *Rapid Commun. Mass Spectrom.* **18**, 2162–2168.

35. Bern, M., Goldberg, D., McDonald, W. H., and Yates, J. R., 3rd. (2004) Automatic quality assessment of Peptide tandem mass spectra. *Bioinformatics* **20**(Suppl 1), I49–I54.

36. Yates, J. R., 3rd, Morgan, S. F., Gatlin, C. L., Griffin, P. R., and Eng, J. K. (1998) Method to compare collision-induced dissociation spectra of peptides: potential for library searching and subtractive analysis. *Anal. Chem.* **70**, 3557–3565.

I

ANALYSIS OF ORGANELLES, VESICLES, AND PROTEIN MACHINERIES

2

Purification and Proteomic Analysis of Chloroplasts and their Sub-Organellar Compartments

Daniel Salvi, Norbert Rolland, Jacques Joyard, and Myriam Ferro

Summary

Sub-cellular proteomics has proven to be a powerful approach to link the information contained in sequenced genomes from eukaryotic cells to the functional knowledge provided by studies of cell compartments. Chloroplasts are plant-specific organelles and are the site of photosynthesis and also of many other essential metabolic pathways, like syntheses of amino acids, vitamins, and pigments. They contain several sub-organellar compartments: the envelope (the two-membrane system surrounding the organelle), the stroma (the internal soluble phase), and the thylakoid membranes (the internal membrane system). There is a link between these compartments and the functions of their constitutive proteins. One way to bring into view the sub-proteomes of the chloroplast is to develop proteomic analyses based (1) on the use of highly purified sub-fractions of the chloroplast and (2) on mass spectrometry (MS)-based analyses for protein identification. To illustrate such strategies, this chapter describes the methods for purification of chloroplasts from *Arabidopsis* leaves and for the specific recovery of highly pure sub-organellar fractions of envelope, stroma, and thylakoids. Subsequently, methods are described to analyze by MS the proteins recovered from these fractions.

Key Words: Chloroplast; chloroplast envelope; stroma; thylakoids; mass spectrometry; proteome.

1. Introduction

Plastids are semiautonomous organelles found in plants and protists. They are generally considered to have originated from endosymbiotic cyanobacteria. In plant leaves, plastids differentiate into chloroplasts and become photosynthetically active. These chloroplasts have a specific sub-organellar organization

From: *Methods in Molecular Biology, vol. 432: Organelle Proteomics*
Edited by: D. Pflieger and J. Rossier © Humana Press, Totowa, NJ

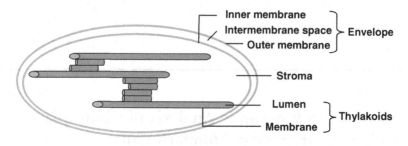

Fig. 1. Schema of a chloroplast.

(*see* **Fig. 1**). They are surrounded by a two-membrane system or envelope. This envelope system is composed of the inner and outer membranes and of an intermembrane space located between these two layers. The chloroplasts contain a soluble phase, called the stroma, and an internal membrane system, called the thylakoids.

While containing their own genome, plastids only synthesize very few proteins (less than a hundred) and import two to three thousands of nucleus-encoded proteins synthesized outside the organelle. Translocation of most of these nucleus-encoded proteins across the envelope is achieved by the joint action of Toc and Tic translocons located, respectively, in the outer and inner envelope membranes of the chloroplast envelope. Moreover, the envelope is also involved in the controlled exchange of a variety of ions and metabolites between the cytosol and the chloroplast. For instance, being the sole site of biosynthesis of most amino acids (with the exception of sulfur-containing amino acids), chloroplasts must export these compounds for protein synthesis in the cytosolic and mitochondrial compartments. The chloroplast envelope is also the site of specific biosynthetic functions such as synthesis of plastid membrane components (glycerolipids, pigments, and prenylquinones) or chlorophyll breakdown. The stromal phase contains enzymes required for the photosynthesis reactions (the Calvin cycle), the synthesis of amino acids or vitamins, the plastid transcription and translation machineries. Within the stroma are also found stacks of thylakoids, the sub-organelle membrane system where the light phase of photosynthesis takes place. Within these thylakoid vesicles is found another sub-organellar fraction called the thylakoid lumen.

According to their various sub-plastidial localization (inner or outer envelope membranes, intermembrane space, stroma, thylakoid membrane, or lumen), chloroplastic proteins will have specific functions in the organelle or in the plant cell. To get access to the protein content of these different chloroplastic sub-compartments, protocols exist that allow the collection of highly pure fractions of envelope, stroma, or thylakoids. The purpose of this article is to describe the procedures developed to obtain highly pure sub-organellar chloroplastic compartments and the methods used for further analysis of these fractions using mass spectrometry (MS)-based proteomic investigations *(1–4)*.

2. Materials

2.1. Preparation of Arabidopsis Leaves

Arabidopsis rosette leaves are obtained from 3- to 4-week-old *Arabidopsis thaliana* plantlets (*see* **Note 1**).

1. Fill large (30 cm × 45 cm) plastic cases with compost and water.
2. Sow seeds onto the surface of the compost by scattering them carefully at a high density (around 30 mg of seeds for a whole box).
3. Grow *A. thaliana* plantlets in growth rooms (12-h light cycle) at 23°C (day)/18°C (night) with a light intensity of 150 μmol/m²/s.
4. Four to six boxes containing 3- to 4-week-old *Arabidopsis* plantlets are expected to provide 400–500 g of rosette material.

2.2. Purification of Arabidopsis Chloroplasts

1. Muslin or cheesecloth, 80 cm large.
2. Nylon blutex (50 μm aperture) (Tripette et Renaud, Sailly Saillisel, France).
3. Beakers (500 mL, 1 L, and 5 L).
4. Ice and ice buckets.
5. Pipettes (1 mL and 10 mL).
6. Percoll (Pharmacia, Uppsala, Sweden).
7. Motor-driven blendor, 3 speeds, 1 gallon (3.785 L) (Waring Blendor).
8. Superspeed refrigerated centrifuge (Sorvall RC5), with the following rotors (and corresponding tubes): fixed angle rotors GS-3 (six 500-mL plastic bottles) and SS34 (eight 50-mL polypropylene tubes) and swinging bucket rotor HB-6 (six 50-mL polycarbonate tubes) or equivalent.
9. Leaf grinding medium: 0.45 M sorbitol, 20 mM Tricine-KOH, pH 8.4, 10 mM ethylenediaminetetraacetic acid (EDTA), 10 mM $NaHCO_3$, and 0.1% (w/v) bovine serum albumin (BSA, defatted).
10. Chloroplast isolation and washing medium: 0.30 M sorbitol, 20 mM Tricine-KOH, pH 7.6, 5 mM $MgCl_2$, and 2.5 mM EDTA.
11. Solution for Percoll gradients: Mix 1 volume of Percoll with 1 volume of medium containing 0.60 M sorbitol, 40 mM Tricine-KOH, pH 7.6, 10 mM $MgCl_2$, and 5 mM EDTA to obtain a 50% Percoll/0.3 M sorbitol solution.

2.3. Fractionation of Chloroplasts

1. Hypotonic medium for chloroplast lysis: 10 mM 3-(N-morpholino) propanesulfonic acid (MOPS)–NaOH, pH 7.8, 4 mM $MgCl_2$, 1 mM phenylmethylsulfonyl fluoride (PMSF), 1 mM benzamidine, and 0.5 mM ε-amino caproic acid.

2. Sucrose gradients for chloroplast fractionation: 10 mM MOPS–NaOH, pH 7.8, and 4 mM MgCl$_2$, with 0.3 M and 0.6 M or 0.93 M sucrose.

3. Chloroplast envelope and thylakoid membranes washing medium: 10 mM MOPS–NaOH, pH 7.8, 1 mM PMSF, 1 mM benzamidine, and 0.5 mM ε-amino caproic acid.

4. Preparative refrigerated ultracentrifuge (Beckman L7), with a SW 41 Ti rotor (six 13.2-mL Ultraclear tubes) or equivalent.

5. Centrifuge (Eppendorf centrifuge 5415D or equivalent) placed in a cold room with 1.5-mL plastic tubes.

6. Branson sonifier model 250 (or equivalent), with 3-mm microtip and ice bucket.

7. Nitrogen (or argon) gas supply (cylinder) with gas pressure regulator connected to a Pasteur pipette through a plastic tube.

2.4. Differential Extraction of Membrane or Soluble Proteins

2.4.1. Chloroform/Methanol Extraction

1. Chloroform/methanol mixtures in the following proportions: 0:9, 1:8, 2:7, 3:6, 4:5, 5:4, 6:3, 7:2, 8:1, and 9:0 (v/v).

2. Cold (–20°C) acetone for a 80% final concentration in water.

2.4.2. Alkaline or Salt Washing of Membranes

1. Na$_2$CO$_3$: 0.1 M final concentration (1 M stock solution).

2. NaOH: 0.1 M or 0.5 M final concentration (2 M stock solution).

3. NaCl: 1 M final concentration (2 M stock solution).

4. Eppendorf centrifuge 5415D or equivalent, with 1.5-mL Eppendorf tubes.

2.4.3. Ammonium Sulfate Fractionation of the Stromal Proteins

1. Sephadex™ G-25M PD-10 columns from GE Healthcare Bio-Sciences, Buckinghamshire, UK Ab or 0.5-mL Zeba™ Desalt Spin Columns from Pierce, West Chester, CA, USA for small volumes.

2.5. Sodium Dodecyl Sulhate–Polyacrylamide Gel Electrophoresis

1. Gel electrophoresis apparatus (Bio-Rad Protean 3 or equivalent), with the different sets of accessories for protein separation by electrophoresis (combs, plates, and casting accessories).

2. Acrylamide stocks: 30% (w/v) acrylamide: 0.8% bisacrylamide—300 g of acrylamide, 8 g of bisacrylamide, and H$_2$O to 1 L. 60% (w/v) acrylamide: 0.8% bisacrylamide—600 g of acrylamide, 8 g of bisacrylamide, and H$_2$O to 1 L and store in amber bottles at 4°C.

3. Sodium dodecyl sulphate (SDS) stock solution: 10% (w/v) SDS—10 g of SDS, H$_2$O to 1 L, and store at room temperature.

4. Gel buffers: 4× Laemmli stacking gel buffer (0.5 M Tris–HCl, pH 6.8)—363 g of Tris, H_2O to 900 mL, adjust to pH 8.8 at 25°C with concentrated HCl, make up volume to 1 L, and store at room temperature. 8× Laemmli resolving gel buffer (3 M Tris–HCl, pH 8.8): 60.6 g of Tris, H_2O to 900 mL, adjust to pH 6.8 at 25°C with concentrated HCl, make up volume to 1 L, and store at room temperature.

5. Stacking gel (5% acrylamide): 5 mL of 30% acrylamide—0.8 % bisacrylamide stock solution, 7.5 mL of 4× Laemmli stacking gel buffer, 17.1 mL of H_2O, 40 μL TEMED, 4 mL of 10% ammonium persulfate solution (10 g ammonium persulfate, H_2O to 100 mL, stored at 4°C, prepare fresh every month), total volume: 30 mL.

6. Single acrylamide concentration gels (10, 12, or 15% acrylamide): (1) For 10% acrylamide gel: 33.3 mL of 30% acrylamide—0.8% bisacrylamide stock solution, 12.5 mL of 8× Laemmli resolving gel buffer, 54 mL of H_2O, 20 μL of TEMED, 0.2 mL of 10% ammonium persulfate, total volume: 100 mL; (2) for 12% acrylamide gel: 40 mL of 30% acrylamide—0.8% bisacrylamide stock solution, 12.5 mL of 8× Laemmli resolving gel buffer, 47.3 mL of H_2O, 20 μL TEMED, 0.2 mL of 10% ammonium persulfate, total volume: 100 mL; and (3) for 15% acrylamide gel: 50 mL of 30% acrylamide—0.8% bisacrylamide stock solution, 12.5 mL of 8× Laemmli resolving gel buffer, 37.3 mL of H_2O, 20 μL of TEMED, 0.2 mL of 10% ammonium persulfate, total volume: 100 mL.

7. Protein solubilization: 4× stock solution—200 mM Tris–HCl, pH 6.8, 40% (v/v) glycerol, 4% (w/v) SDS, 0.4% (w/v) bromophenol blue, and 100 mM dithiothreitol.

8. Gel reservoir buffer: 38 mM glycine, 50 mM Tris, and 0.1% (w/v) SDS (about 400 mL in each reservoir).

9. Gel staining solution: Acetic acid/isopropanol/water, 10/25/65 (v/v/v), containing 2.5 g/L of Coomassie brilliant blue R250.

10. Gel destaining solution: Acetic acid/ethanol/water, 7/40/53 (v/v/v).

11. Pre-stained sodium dodecyl sulphate–polyacrylamide gel electrophoresis (SDS–PAGE) markers low range from Bio-Rad, Hercules, CA, USA or equivalent.

2.6. In-Gel Digestion

All chemicals must be of the highest degree of purity [e.g., high-performance liquid chromatography (HPLC) grade].

1. Solution A: 25 mM ammonium bicarbonate in water.
2. Solution B: 25 mM ammonium bicarbonate in acetonitrile/water, 50/50, v/v.
3. Acetonitrile.
4. Trypsin (Promega, Madison, WI, USA).
5. 5% (v/v) formic acid in water.
6. Thermostatic oven at 37°C.
7. Laminar flow hood (optional).
8. Ice bucket.

2.7. Mass Spectrometry Analysis

All chemicals must be of the highest degree of purity.

1. A QTOF Ultima (Waters, Milford, CA) is used in our laboratory. Other tandem mass spectrometers capable of automated acquisition of tandem mass spectra should also be suitable for this purpose.
2. Reverse-phase HPLC is performed with a CapLC system (Waters). Other HPLC systems capable of delivering ~200 nL/min flow rate should be suitable.
3. CapLC Autosampler (Waters).
4. Columns: C18 PepMap, 75 µm ID, 15 cm length, and 3-µm silica beads with 300-Å pores (Dionex, LCPackings, Amsterdam, the Netherlands).
5. Guard columns: C18 PepMap, 300 µm ID, 5 cm length, and 5-µm silica beads with 100-Å pores (Dionex).
6. Buffer A: Acetonitrile/formic acid/water, 2/0.1/98 (v/v/v).
7. Buffer B: Acetonitrile/formic acid/water, 80/0.08/20 (v/v/v).
8. Buffer C: Acetonitrile/formic acid/water, 5/0.2/95 (v/v/v).

2.8. Software

1. MassLynx 4.0 (Waters) for QTOF data processing. For other instruments, the corresponding data processing software must be used.
2. Mascot 2.0 available on the intranet or Internet(Matrix Science, London, UK) for database searching. Other database searching tools, such as Sequest, may also be used.

3. Methods

3.1. Purification of Arabidopsis Chloroplasts

All operations are carried out at 0–5°C.

1. Prior to the experiment, prepare six tubes containing 30 mL of a 50% Percoll/0.3 M sorbitol solution. Preform Percoll gradients for chloroplast purification by centrifugation at 38,700 g for 55 min (Sorvall SS-34 rotor) (*see* **Note 2**). Store the tubes containing preformed Percoll gradients in the cold room until use.
2. Harvest (*see* **Note 3**) 400–500 g of rosette leaves. Wash them with deionized water. Blot the washed leaves on paper tissue and transfer them in a cold room for the next step.
3. Homogenize the leaf material (400–500 g of leaves and 2 L of grinding medium) two times for 2 s in a Waring Blendor at low speed (*see* **Note 4**). Filter the homogenate rapidly through four to five layers of muslin and one layer of nylon blutex.
4. Distribute equally the filtered suspension into six bottles for centrifugation (500 mL each), and centrifuge them at 2070 g for 2 min (Sorvall GS 3 rotor) (*see* **Note 5**).

5. Suck up the supernatant with a water trump and carefully resuspend each pellet, containing a crude chloroplast fraction, by addition of a minimal volume (36 mL final volume) of washing medium (use a spatula).

6. Load the chloroplasts suspension (6 mL per tube) on the top of the preformed Percoll gradients. Centrifuge the gradients at 13,300 g for 10 min (Sorvall swinging HB-6 rotor) (*see* **Note 6**). At the conclusion of this step, aspire the upper part of the gradient (*see* **Note 7**) and then recover intact chloroplasts (a broad dark green band in the lower part of the gradient) (*see* **Note 8**) with a pipette.

7. Dilute three to four times the intact chloroplasts suspension with 200–300 mL of washing medium. Centrifuge the suspension at 2070 g for 2 min (Sorvall SS-34 rotor).

8. Recover each pellet, containing washed purified intact chloroplasts, for chloroplast envelope preparation. At this stage, the yield of intact chloroplasts is 50–60 mg of protein (*see* **Note 9**) (*see* **Fig. 2**).

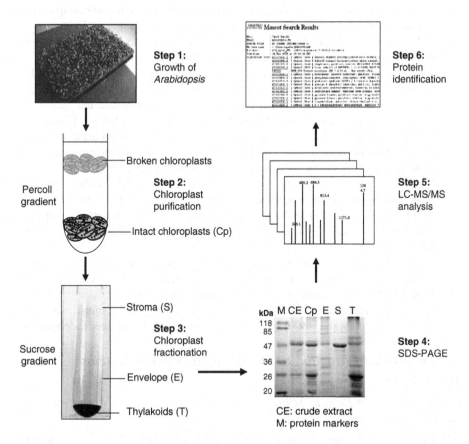

Fig. 2. Outline of the strategy.

3.2. Fractionation of Arabidopsis Chloroplasts

All operations re carried out at 0–5°C.

1. Prior to the experiment, prepare six tubes (13.2 mL, Ultraclear; Beckman, Fullerton, CA, USA) for sucrose gradients consisting of three layers: 3 mL of 0.93 M, 2.5 mL of 0.6 M, and 2 mL of 0.3 M sucrose. Each layer is carefully overlaid with a pipette (*see* **Note 10**) on top of the other, starting with the densest one (0.93 M, at the bottom) and finishing with the lightest one. Store the tubes in the cold room until use.

2. Lyse purified and washed intact chloroplasts (obtained as described in section 3.1) by adding to the pellets hypotonic medium (adjust for a final volume of 21 mL) containing protease inhibitors (10 mM MOPS–NaOH, pH 7.8, 4 mM $MgCl_2$, 1 mM PMSF, 1 mM benzamidine, and 0.5 mM ε-amino caproic acid).

3. Load lysed chloroplasts (3.5 mL per tube) on top of the sucrose gradients. Centrifuge the tubes at 70,000 g for 1 h (Beckman SW41-Ti rotor) (*see* **Note 11**). After centrifugation, the envelope membrane and the thylakoids are present as a yellow band at the 0.93/0.6 M interface and as a dark green band at the bottom of the tube, respectively. The soluble fraction containing the stroma remains on top of this gradient.

4. To recover the soluble stromal proteins, carefully aspirate the upper part of the gradient with a pipette (3 mL per tube) (*see* **Note 12**).

5. To purify envelope membranes, recover the yellow band containing the envelope with a pipette, dilute the suspension three to four times in 10 mM MOPS–NaOH pH 7.8 buffer (containing protease inhibitors), and concentrate the membranes as a pellet by centrifugation at 110,000 g for 1 h (Beckman SW 41 Ti rotor).

6. Add a minimum volume of washing medium (containing protease inhibitors) to the envelope pellet. Take an aliquot for protein amount determination (*see* **Note 13**). Store envelope membrane preparations in liquid nitrogen.

7. For further thylakoid purification (*see* **Note 14**), carefully collect the dark green bottom of the sucrose gradient with a pipette and dilute the suspension in 10 volumes of 10 mM MOPS–NaOH, pH 7.8, buffer (containing protease inhibitors). Concentrate the membranes as a pellet by centrifugation at 110,000 g for 1 h (Beckman SW 41 Ti rotor).

8. Add a minimum volume of washing medium (containing protease inhibitors) to the thylakoid pellet. Take an aliquot for protein amount determination (*see* **Note 13**). Store thylakoid preparations in liquid nitrogen.

9. From such preparations, an average of ~30 mg of stroma proteins, ~20 mg of thylakoid proteins, and ~300 µg of envelope proteins can be obtained (*see* **Note 15**).

3.3. Differential Extraction of Membrane or Soluble Proteins (see Note 16)

3.3.1. Membrane Protein Solubilization with Chloroform/Methanol Mixtures (see Note 17)

1. Dilute slowly 1 volume of the membrane preparation (0.5 to 1 mg of protein in 0.1 mL of original buffer) (*see* **Note 18**) in 9 volumes of cold chloroform/methanol (5/4, v/v) mixtures in Eppendorf tubes (1.5 mL) (*see* **Note 19**).

2. Store the resulting mixtures for 15 min on ice before centrifugation (4°C) for 15 min at 15,000 *g*.
3. Recover the organic phase (discard the white pellet containing less hydrophobic proteins). The pellet contains the chloroform/methanol-insoluble proteins (or organic solvent insoluble fraction). The supernatant contains the chloroform/methanol-soluble proteins (or organic solvent soluble fraction).
4. Then evaporate (*see* **Note 20**) the organic phase under nitrogen (down to 200 μL for large amounts of proteins or 100 μL when original protein concentration is limited). Directly precipitate the proteins by adding 4 volumes (800–400 μL) of cold (–20°C) acetone (80% final acetone concentration) directly to the remaining volume of chloroform/methanol.
5. Store the resulting mixtures for 15 min on ice before centrifugation (4°C) for 15 min at 15,000 *g*.
6. Eliminate the organic supernatant and dry the protein pellet (*see* **Notes 21** and **22**). Resuspend (*see* **Note 23**) the protein pellets in 20 μL of concentrated SDS–PAGE buffer (4×) and store the protein mixtures in liquid nitrogen.
7. Analyze the proteins by SDS–PAGE (various volumes on separate lanes).

3.3.2. Alkaline or Salt Washing of the Membrane Fractions

1. Dilute slowly 1 volume of the membrane preparation (0.5 to 1 mg of protein in 0.1 mL) to 0.5 mL with Na_2CO_3, NaOH, or NaCl stock solutions to obtain 0.1 M, 0.5 M, or 1 M final concentrations, respectively (*see* **Note 24**).
2. Sonicate the resulting mixtures two to five times for 10 s, the power set at 40% duty cycle, output control at 5 on ice.
3. Store the mixtures for 15 min on ice before centrifugation (4°C) for 20 min at 15,000 *g*.
4. Recover insoluble proteins as pellets (*see* **Note 25**) and resuspend them in 20 μL of SDS–PAGE buffer (4×). Store the protein extracts in liquid nitrogen.
5. Analyze the proteins by SDS–PAGE (see **Subheading 3.4**).

3.3.3. Ammonium Sulfate Fractionation of the Stromal Proteins (see Note 26)

1. Slowly add 164 mg (for 30% ammonium sulfate saturation) of solid ammonium sulfate to the purified stromal proteins (2.5–3 mg of proteins in 1 mL) (*see* **Note 27**).
2. Place the tube at 4°C and mix for 15 min before centrifugation (4°C) for 5 min at 10,000 *g*.
3. Remove the supernatant and store on ice for further fractionation. Recover insoluble proteins as a pellet; resuspend them in 100 μL of MOPS–NaOH buffer pH 7.8; and store this sample in liquid nitrogen until use at **step 7**.
4. Slowly add 181 mg (for 60% ammonium sulfate saturation) of solid ammonium sulfate to the supernatant resulting from the 30% ammonium sulfate fractionation.

5. Place the tube at 4°C and mix for 15 min before centrifugation (4°C) for 5 min at 10,000 *g*.
6. Remove the supernatant and store it in liquid nitrogen until use at **step 7** for further analyses. Recover insoluble proteins as a pellet; resuspend them in 100 µL of MOPS–NaOH buffer, pH 7.8; and store this sample in liquid nitrogen until use at **step 7**.
7. Desalt the protein samples using pre-packed desalting columns (e.g., Sephadex™ G-25M PD-10 Columns or 0.5-mL Zeba™ Desalt Spin Columns for small volumes) according to manufacturer's instructions.
8. Take aliquots of the respective fractions for protein amount determination (*see* **Note 13**), add SDS–PAGE buffer (4×), and analyze the proteins by SDS–PAGE (see below).

3.4. Separation of membrane proteins by 1D SDS–PAGE (see Note 28)

1. Prior to the experiment, prepare slab gels for protein electrophoresis (*see* **Note 29**):

   a. Prepare the gel apparatus according to the manufacturer's specifications (*see* **Note 30**).
   b. Prepare the different gel solutions (stacking gel, 10, 12, or 15% acrylamide separation gel). The volumes to be used are determined by gel dimensions and therefore by the specifications of the apparatus.

2. Heat the protein samples at 95°C for 5 min to solubilize the proteins. Add bromophenol blue dye in the samples. Place protein samples (20 µL) into gel slots by means of a pipette. Load the molecular weight markers in another slot.
3. Set the conditions for the electrophoresis at 150 V. Run gels for 1 h at room temperature (until the bromophenol blue dye reaches the lower part of the gel) (*see* **Note 31**).
4. After electrophoresis, remove the gels; place them in plastic boxes in the presence of staining solution. Shake the box gently for 30 min. Pour off the staining solution and replace it by the destaining solution. Shake the box gently for 15 min. Repeat the washing step once or twice (*see* **Notes 32** and **33**).

3.5. In-Gel Digestion (see Note 33)

3.5.1. Washing and Destaining of Gel Pieces (see **Note 34**)

1. Excise gel bands with a clean scalpel and store them individually in 0.5-mL Eppendorf tubes (*see* **Note 35**).
2. Wash the gel pieces with 200 µL of solution A for 15 min with vortexing. Remove excess liquid.
3. Wash the gel pieces with 200 µL of solution B for 15 min with vortexing. Remove excess liquid.

4. Wash the gel pieces with 200 µL of Ultrapure HPLC grade water for 15 min with vortexing. Remove excess liquid.
5. Add 200 µL of Ultrapure HPLC grade acetonitrile to shrink the gel pieces for 15 min with vortexing. Remove excess liquid and dry the gel pieces in a vacuum centrifuge (*see* **Note 36**).
6. Add 200 µL of 7% (v/v) H_2O_2 for 15 min (*see* **Note 37**).
7. Discard excess liquid.
8. Add 200 µL of Ultrapure HPLC grade water and vortex.
9. Discard excess liquid.
10. Shrink the gel pieces with acetonitrile for 15 min.
11. Remove excess liquid and dry the gel pieces in a vacuum centrifuge (*see* **Note 38**).

3.5.2. Application of Trypsin

1. Prepare stock solutions of trypsin from 20 µg of trypsin solubilized in 50 µL of 50 mM acetic acid (*see* **Note 39**). Freeze the aliquots and store them at –20°C before use (*see* **Note 40**).
2. Rehydrate gel pieces with 25 µL of a 6 ng/µL trypsin solution prepared in a 25 mM NH_4HCO_3 buffer at 4°C (use ice bucket) for 15 min (*see* **Note 41**).
3. Add 30 µL of 25 mM NH_4HCO_3 buffer for a complete rehydration.
4. Leave the samples at 37°C with shaking during 30 min.
5. If necessary, cover gel pieces with additional 25 mM NH_4HCO_3 buffer to keep them wet during enzymatic digestion.
6. Leave samples in a thermostatic oven at 37°C for 3–5 h.

3.5.3. Extraction of Peptides

1. Withdraw the supernatant and keep it in a separate Eppendorf tube.
2. Add 30 µL of acetonitrile/water, 50/50 (v/v), in the gel-containing tube for 15 min with shaking.
3. Spin gel pieces down and collect the supernatant to pool it with the first supernatant (*see* **step 1**).
4. Evaporate the pooled supernatants using a vacuum centrifuge.
5. Add 30 µL of 5% (v/v) formic acid in water to the gel pieces for 15 min with shaking.
6. Spin gel pieces down and collect the supernatant to pool it with the two previous supernatants (*see* steps 1 and 3).
7. Evaporate the pooled supernatants using a vacuum centrifuge.
8. Add 30 µL of acetonitrile on the gel pieces for 15 min with shaking.
9. Spin gel pieces down and collect the supernatant to pool it with the two previous supernatants (*see* steps 1, 3, and 6).
10. Evaporate the pooled supernatants using a vacuum centrifuge.

3.6. Protein Identification

3.6.1. LC-MS/MS Analysis

1. Resuspend the dried sample with 5 μL of 5% (v/v) TFA in water. Sonicate and add 5 μL of buffer C (*see* **Note 42**).
2. Add 10 μL of the sample to a vial that is adapted to the autosampler.
3. Set up the injection system for an injection of 6.4 μL in the microlitre pickup mode.
4. Set up the LC method in the MassLynx 4.0 CapLC diagnostic window (*see* **Note 43**).
5. Set up the MS/MS method for the QTOF Ultima instrument (*see* **Note 44**).
6. Launch the LC-MS/MS analysis.

3.6.2. Database Searching for Protein Identification

1. Process the raw data files from the ProteinLynx set up window of MassLynx to generate a peak list file (*see* **Note 45**).
2. Select the parameters for the Mascot search engine: Acetyl (N-term), *N*-acetyl (protein), oxidation of methionine, sulphone, cysteic acid as variable modifications, trypsin/P, two missed cleavages allowed, 2+ and 3+ charge states, and monoisotopic masses.
3. Perform a database search using a contaminant database (trypsin and keratin) (*see* **Note 46**).
4. Remove the queries that are assigned to contaminants and use the "Search Selected" option to search again using an *A. thaliana* database.
5. Protein identifications are validated using the following criteria: proteins that are identified with at least two peptides both showing a score higher than the identity significant threshold given by Mascot are accepted without manual validation. For proteins identified by only one peptide having a score higher than this significant threshold, the peptide sequence is checked using human expertise. Peptides with scores higher than 20 and lower than the significant threshold are systematically checked and/or interpreted manually to confirm or reject the MASCOT suggestion (*see* **Note 47**).

4. Notes

1. *Arabidopsis* Columbia (Col) or Wassilewskija (WS) ecotypes have been generally used for ESTs or genomic sequencing projects but also to generate T-DNA insertion mutants using *Agrobacterium tumefaciens*. For these reasons, they might also be considered as reference ecotypes for *Arabidopsis* proteomic analyses. The procedures described in this chapter were applied efficiently with both Col and WS ecotypes.
2. Vertical rotors can easily be used to obtain preformed Percoll gradients and subsequently purify chloroplasts *(5)*.

3. The number of starch granules present in chloroplasts is critical for the preparation of intact chloroplasts: Chloroplasts containing large starch grain will generally be broken during centrifugation *(5)*. Therefore, prior to the experiment, the plants should be kept in a dark and cold room (4°C) to reduce the amount of starch. A good way to proceed is to place the plants under such conditions the day before the extraction (we usually do this at the beginning of the afternoon prior to the day of the experiment).

4. It is critical the grinding process be as short as possible. Longer blending improves the yield of recovered chlorophyll but increases the proportion of broken chloroplasts.

5. It is essential to equilibrate 2 by 2 on a balance the different tubes prior to centrifugation.

6. It is recommended to disconnect the brake or to use the automatic rate controller (if available) to prevent mixing of the gradients at the critical stage of deceleration.

7. It is important to remove carefully the top of the tube by aspiration with a water trump, then to recover the intact chloroplasts with a pipette.

8. Broken chloroplasts are present in the upper part as a broad band. A small pellet containing cell pieces and large debris is found at the bottom of the tube.

9. The excellent purity of the *Arabidopsis* chloroplasts prepared by Percoll purification step was confirmed through proteomic analysis: only 5% (6 out of 112) of the *Arabidopsis* proteins identified by Ferro et al. *(2)* may correspond to non-plastid proteins. Considering the high sensitivity of the present mass spectrometers, it is not surprising to detect minute amounts of these few extra-plastidial contaminants, which are major proteins in their respective sub-cellular compartment. It is also important to notice that none of the proteins identified in *Arabidopsis* envelope membranes by Ferro et al. *(2)* appears to derive from mitochondria, a classical contaminant of plastid preparations.

10. The use of a peristaltic pump to prepare the gradients is recommended, as some expertise is needed to load the different layers by hand.

11. It is recommended to disconnect the brake to prevent mixing of the gradients at the critical stage of deceleration.

12. Note that the protein from the stroma will be recovered in the hypotonic medium used to lyse the chloroplasts and will contain protease inhibitors (10 mM MOPS–NaOH, pH 7.8, 4 mM $MgCl_2$, 1 mM PMSF, 1 mM benzamidine, and 0.5 mM ε-amino caproic acid). Further desalting of these soluble proteins may be performed using G-25 columns (e.g., PD-10; Amersham Buckinghamshire, UK) if required.

13. Protein contents of fractions are estimated using the Bio-Rad protein assay reagent *(6)*.

14. The major possible contaminations of crude thylakoid membrane preparations derive from soluble stroma proteins and envelope membrane proteins. It is estimated that 50% of the envelope membrane vesicles are recovered in the crude thylakoid fraction. However, since the ratio of envelope to thylakoid membranes

is 1/50 in the chloroplast, these contaminations will be limited to less than 1% of the crude thylakoid fraction.

15. At this stage, the major possible contaminants of envelope preparations are soluble stroma proteins and small pieces of thylakoid membranes. Ferro et al. *(2)* have extensively analyzed such cross-contaminations. Being the most likely source of membrane contamination of the purified envelope fraction, thylakoid cross-contamination needs to be precisely assessed. The yellow color of purified envelope vesicles first indicates that this membrane system contains almost no chlorophyll and therefore very few contaminating thylakoids. Indeed, by western blot analyses using antibodies raised against Light Harvesting Complex Protein (LHCP), Ferro et al. *(2)* demonstrated that several independent *Arabidopsis* envelope preparations contained between 1 and 3% thylakoid proteins.

16. The protocols of **Subheadings 3.3.1** and **3.3.2** can be either applied to envelope or thylakoid fractions. We analyze here fractions containing the most hydrophobic proteins: The chloroform/methanol-soluble proteins or the proteins remaining in the membrane after its treatment by NaOH. The discarded fractions contain a large variety of rather hydrophilic proteins, some being of high interest, for instance, those of the stroma or those of the thylakoid lumen *(7–9)*. Percoll-purified chloroplasts (and chloroplast sub-fractions) are largely devoid of contamination deriving from other cell compartments *(5,10)*. However, since some minor contaminant can be present in the purified membranes, caution must be taken when assigning a protein identified by proteomics to a sub-cellular location although the cell compartment of interest can be highly purified. Other complementary approaches are necessary to assert a protein location (WB, GFP fusion, etc.). In particular, here we are talking about the soluble proteins discarded from a membrane fraction. Since a membrane is at the interface of two soluble media and since soluble proteins from both sides of the membrane may be trapped within membrane vesicles, it may be difficult to conclude about their precise location.

17. Alternatively, a wide variety of detergents can be used: Triton X-100, CHAPS, Triton X-114, sulfobetains, and so on. The reader is referred to articles by Santoni et al. *(11,12)* for the detailed analysis of membrane treatment by detergents.

18. Most of the time, we use MOPS 10 mM, pH 7.0, as a buffer.

19. The volume ratio between chloroform and methanol for an optimal extraction can be determined by comparing the SDS–PAGE polypeptide profile of the organic phase soluble proteins prepared as follows: membranes (5 mg of proteins in 1 mL storage buffer) are divided in 10 fractions of 0.1 mL (in 1.5-mL Eppendorf tubes). The membrane fraction is then slowly diluted by addition of 0.9 mL of cold chloroform/methanol solutions (0:9, 1:8, 2:7, 3:6, 4:5, 5:4, 6:3, 7:2, 8:1, and 9:0, v/v). In general, the total volume of the mixture is 1 mL. If necessary, this can be increased to a much higher value when more membrane material is available.

20. Do not completely dry the sample!

21. Due to acetone precipitation (and removal of pigments), the pellet turns white, be careful not to lose it.

22. Trace amounts of solvent strongly limit protein solubilization.

23. Be patient, wash tube walls and avoid bubbles!

24. Treatment of membranes with these various compounds does not result in the extraction of the same proteins *(2,13,14)*. Na_2CO_3 or NaCl extract proteins that are rather weakly associated with the membrane, whereas NaOH removes proteins that are more tightly associated. It is therefore recommended trying several of these compounds to achieve analyses that are more comprehensive.

25. The supernatant contains the proteins removed from the membrane by alkaline or salt treatment, that is, the less hydrophobic membrane proteins.

26. As stromal proteins are soluble proteins, 2D gel electrophoresis may be performed to further fractionate this protein sample. Alternatively and since further proteomic analyses may be limited due to the large representation (50%) of the Rubisco protein in the chloroplast stroma, fractionation with ammonium sulfate may be applied. Sequential precipitations with increasing amounts of ammonium sulfate can be performed to sub-fractionate the stromal proteins. Note that the Rubisco protein is only recovered in the 30–60% ammonium sulfate saturation step (and thus totally removed from the other fractions).

27. Addition of ammonium sulfate to the protein solution should be performed step by step to limit protein precipitation due to local acidity.

28. Classical proteomic methodologies based on the use of 2D gel electrophoresis proved to be rather inefficient on membrane proteins. In general, almost no high hydrophobic protein, as defined by average hydrophobicity values, is found on 2D gel separations of membrane proteins. Adessi et al. *(15)* observed that loading 2D gels with high amounts of membrane proteins resulted in the severe loss of hydrophobic proteins and therefore in the artifactual enrichment of the less hydrophobic components (and hydrophilic contaminants of the purified membrane fraction). In this case, hydrophobic proteins probably precipitated at their isoelectric point in the first dimension, thus preventing any further migration and separation in the second dimension *(15)*. In contrast, 2D-gel electrophoresis is very efficient to analyze peripheral membrane proteomes *(7–9,16)*. Strategies for membrane proteomics based on 2D-gel electrophoresis combined with a wide diversity of detergents have been extensively analyzed by Santoni et al. *(11,12)*. Therefore, thylakoid and envelope fractions are preferably analyzed by SDS–PAGE. The stroma and the chloroplast fraction can be either analyzed by SDS–PAGE or 2D-gel electrophoresis.

29. We routinely use the procedure described by Chua *(17)* to separate membrane proteins by SDS–PAGE. This chapter describes in detail all stock solutions, medium for stacking, and separation gels.

30. We use a Bio-Rad apparatus, with 7 cm long gels.

31. For some analyses, protein migration can be stopped just between the stacking and the separating gels so that proteins are concentrated in a very thin band for further nano-LC-MS/MS analyses.

32. Gel pieces can be stored in the cold room until proteomic analyses.
33. To avoid contaminations (keratins and polymeric detergents), gloves should be worn during operations with gels (staining, scanning, and excision of bands) and sample preparation for MS analysis. It is advisable to perform all operations under a laminar flow hood to prevent dust contamination.
34. The procedure described for destaining is applicable to bands excised from gels stained with Coomassie blue or with silver. Nevertheless, for silver-stained gels, caution must be taken not to use the crosslinking reagent glyceraldehyde.
35. To obtain protein identification for the whole sample, cut out bands of ~1–5 mm from the top to the bottom of the gel. Otherwise, just cut out discrete bands of interest.
36. Repeat steps 2–4 if the staining is intense, especially for Coomassie blue-stained gels.
37. Treatment with H_2O_2 allows both gel destaining and oxidation of cystein residues. During this treatment, methionine can also be oxidized. Consequently, when running database searching, cysteic acid, oxidized methionine, and sulphone must be specified as variable modifications.
38. At this stage, if trypsin digestion is not performed right away, gel pieces must be stored at –20°C.
39. If trypsin is not purchased from Sigma, refer to the manufacturer's instructions.
40. The aliquot can be stored for several weeks at –20°C.
41. The ratio of enzyme/proteins must be between 1/50 and 1/20. When adding 25 μL of a 6 ng/μL trypsin solution, about 3–5 μg of proteins should be present in the gel band. The concentration of trypsin must be increased or decreased according to the amount of proteins loaded onto the gel and to the gel volume.
42. It is sometimes necessary to dilute the sample to avoid saturating the LC-MS system. Refer to your instrument specifications to check saturation issues.
43. Before sample loading, the HPLC system must be adjusted to a 200–300 nL/min flow rate. The CapLC system delivers a 4–5 μL/min flow rate. A split is set upstream the column to get a final 200–300 nL/min flow rate. Optionally, a guard column is set close to the injection system to concentrate, filter, and desalt the sample before loading it onto the separation capillary column. A 20 μL/min flow rate of buffer C is run on the guard column to load the sample. A typical chromatographic separation uses the following gradients: (1) from 10–40%, buffer B in 40 min; (2) from 40–90%, buffer B in 5 min; (iii) from 90–10%, buffer B in 10 min; and (4) 10%, buffer B for 5 min for further column equilibration. The remaining percentage of the elution solvent is made of buffer A. For maintenance purposes and for checking the LC-system (flow-rate, pressure, and leakages), refer to the manufacturer's instructions.
44. Prior to MS analysis, the calibration and the sensitivity of the instrument used for MS/MS analysis must be checked according to the manufacturer's instructions. For automatic LC-MS/MS analysis, the QTOF Ultima instrument is run in survey mode with the following parameters: 1 s scan time and 0.1 s interscan delay for both MS survey and MS/MS scans; 400–1600 Da and 50–2000 Da mass

ranges for survey and MS/MS scans, respectively; five components selected for fragmentation per cycle, MS/MS to MS switch after 5 s; switchback threshold: 10 counts/s, include charge states 2, 3, and 4 with the corresponding optimized collision energy profiles. Optionally, a list of the *m/z* values corresponding to the most intense peptides of autodigested trypsin can be set as an exclusion list.

45. To generate a peak list file with ProteinLynx, we use the following parameters: Electrospray instrument type; QA threshold: 10; Mass Measure option with no background substract; Savitzky Golay Smooth window: 3; number of smooths: 2, 4 channels; and 80% centroid top. Alternatively, other processing tools, such as Mascot Distiller, can be used with equivalent parameters.

46. We have compiled a database of contaminants containing trypsin and common human keratins. We run a first search using this database to avoid false protein identifications when searching in an *A. thaliana* database.

47. Depending on data quality, the instrument used, the processing parameters, the processing tools, and the database searching tool, the criteria for validating protein identifications can be different and rely on the expertise of the people checking the results.

References

1. Ferro, M., Salvi, D., Rivière-Rolland, H., Vermat, T., Seigneurin-Berny, D., Grunwald, D., et al. (2002) Integral membrane proteins of the chloroplast envelope: identification and subcellular localization of new transporters. *Proc. Natl. Acad. Sci. U.S.A.* **99**, 11487–11492.

2. Ferro, M., Salvi, D., Brugière, S., Miras, S., Kowalski, S., Louwagie, M., et al. (2003) Proteomics of the chloroplast envelope membranes from *Arabidopsis thaliana*. *Mol. Cell. Proteomics* **2**, 325–345.

3. Ferro, M., Seigneurin-Berny, D., Rolland, N., Chapel, A., Salvi, D., Garin, J., et al. (2002) Organic solvent extraction as a versatile procedure to identify hydrophobic chloroplast membrane proteins. *Electrophoresis* **21**, 3517–3526.

4. Seigneurin-Berny, D., Rolland, N., Garin, J., and Joyard, J. (1999) Differential extraction of hydrophobic proteins from chloroplast envelope membranes: a subcellular-specific proteomic approach to identify rare intrinsic membrane proteins. *Plant J.* **19**, 217–228.

5. Douce, R. and Joyard, J. (1982) Purification of the chloroplast envelope. In *Methods in Chloroplast Molecular Biology* (Edelman, M., Hallick, R., and Chua, N. H., eds.), Elsevier/North-Holland, Amsterdam, pp. 139–256.

6. Bradford, M. M. (1976) A rapid and sensitive method for the quantitation of microgram quantities of protein utilizing the principle of protein-dye binding. *Anal. Biochem.* **72**, 248–254.

7. Peltier, J. B, Emanuelsson, O., Kalume, D. E., Ytterberg, J., Friso, G., Rudella, A., et al. (2002) Central functions of the lumenal and peripheral thylakoid proteome of Arabidopsis determined by experimentation and genome-wide prediction. *Plant Cell* **14**, 211–236.

8. Peltier, J. B., Friso, G., Kalume, D. E., Roepstorff, P., Nilsson, F., Adamska, I., et al. (2000) Proteomics of the chloroplast: systematic identification and targeting analysis of lumenal and peripheral thylakoid proteins. *Plant Cell* **12**, 319–341.

9. Schubert, M., Petersson, U. A., Haas, B. J., Funk, C., Schroder, W. P., and Kieselbach, T. (2002) Proteome map of the chloroplast lumen of *Arabidopsis thaliana*. *J. Biol. Chem.* **277**, 8354–8365.

10. Seigneurin-Berny, D., Gravot, A., Auroy, P., Mazard, C., Kraut, A., Finazzi, G., et al. (2006) HMA1, a new Cu-ATPase of the chloroplast envelope, is essential for growth under adverse light conditions. *J. Biol. Chem.* **281**, 2882–2892.

11. Santoni, V., Molloy, M., and Rabilloud, T. (2000a) Membrane proteins and proteomics: un amour impossible? *Electrophoresis* **21**, 1054–1070.

12. Santoni, V., Kieffer, S., Desclaux, D., Masson, F., and Rabilloud, T. (2000b) Membrane proteomics: use of additive main effects with multiplicative interaction model to classify plasma membrane proteins according to their solubility and electrophoretic properties. *Electrophoresis* **21**, 3329–3344.

13. Brugière, S., Kowalski, S., Ferro, M., Seigneurin-Berny, D., Miras, S., Salvi, D., et al. (2004) The hydrophobic proteome of mitochondrial membranes from *Arabidopsis* cell suspensions. *Phytochemistry* **23**, 1693–1707.

14. Marmagne, A., Rouet, M. A., Ferro, M., Rolland, N., Alcon, C., Joyard, J., et al. (2004) Identification of new intrinsic proteins in *Arabidopsis* plasma membrane proteome. *Mol. Cell. Proteomics* **3**, 675–691.

15. Adessi, C., Miège, C., Albrieux, C., and Rabilloud, T. (1997) Two-dimensional electrophoresis of membrane proteins: a current challenge for immobilized pH gradients. *Electrophoresis* **18**, 127–135.

16. Friso, G., Giacomelli, L., Ytterberg, A. J., Peltier, J. B., Rudella, A., Sun, Q., et al. (2004) In-depth analysis of the thylakoid membrane proteome of *Arabidopsis thaliana* chloroplasts: new proteins, new functions, and a plastid proteome database. *Plant Cell* **16**, 478–499.

17. Chua, N. H. (1980) Electrophoretic analysis of chloroplast proteins. *Methods Enzymol.* **69**, 434–436.

3

Proteomic Analysis of the Stacked Golgi Complex

Mark S. Lowenthal, Kathryn E. Howell, and Christine C. Wu

Summary

Although a great deal is known about the structure and function of most mammalian organelles, comprehensive proteomes are necessary to provide a molecular framework to integrate this information. The Golgi complex is the central organelle of the secretory pathway and functions to posttranslationally modify newly synthesized proteins and lipids and to sort them to their sites of function. The methods described in this chapter facilitate the isolation of an enriched stacked Golgi fraction (GF) from rat liver *(1,2)* and the shotgun proteomic analysis of the fraction using multidimensional protein identification technology (MudPIT) *(3,4)*.

Key Words: Golgi complex; cell fractionation; MudPIT; proteomics.

1. Introduction

The Golgi complex is the central organelle of the secretory pathway. In mammalian cells, the Golgi is organized into stacked cisternae and responsible for the posttranslational modification of newly synthesized proteins and lipids and the sorting and trafficking of these molecules to their sites of function. Though much is known about the Golgi, there remain heated debates concerning the mechanisms by which molecules are trafficked within the organelle *(5,6)*. Because current proteomics techniques are capable of resolving and identifying complex proteomes with greater sensitivity than ever before, we undertook a comprehensive proteomic analysis of an enriched stacked GF to obtain a list of the proteins that are resident and/or associated with and drive the function of this organelle *(7,8)*.

From: *Methods in Molecular Biology, vol. 432: Organelle Proteomics*
Edited by: D. Pflieger and J. Rossier © Humana Press, Totowa, NJ

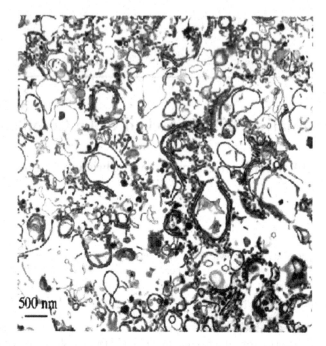

Fig. 1. Electron micrograph of an enriched, stacked Golgi fraction obtained from organellar fractionation of rat liver.

In this chapter, we describe a Golgi isolation approach that results in a highly enriched GF that maintains its stacked structure (*see* **Fig. 1**) *(1,2)*. Because all membrane proteins that are destined for organelles of the secretory and endocytic pathways traffic through the Golgi, we minimized their presence by pretreating the animals with cycloheximide *(1,7,8)*. Cycloheximide blocks protein synthesis. Therefore, newly synthesized molecules moving through the Golgi continue to traffic to their site of function, thus reducing the overall protein content by ~50% and enriching for bona fide Golgi-resident proteins *(1)*. The protein complement of this fraction is then analyzed using multidimensional protein identification technology (MudPIT) *(3)*.

2. Materials

2.1. Tissue Harvest and Homogenization

1. Male Sprague–Dawley rats (6-week-old, ~250 g) (Harlan, Indianapolis, IN).
2. Cycloheximide (Sigma-Aldrich, St. Louis, MO, USA): 50 mg/kg.
3. 0.22 gauge needles (BD Syringe, Franklin Lakes, NJ).
4. Scalpel, small scissors, dissecting tweezers, and guillotine.

5. No. 22 surgical blades (Feather, Osaka, JP) in dissecting knife handle no. 4 (Fisher, Hampton, NH).
6. Homogenization buffer: 0.5 M sucrose, 5 mM $MgCl_2$, 0.1 M KH_2PO_4/K_2HPO_4 buffer, pH 6.7, and protease inhibitor cocktail in dimethylsulfoxide (DMSO) (250 μL/50 mL buffer) (all from Sigma).
7. Teflon-coated pestle and Pyrex homogenizer (Kontes Glass Co., Vineland, NJ).
8. Polytron PT10/35 with PTA 10TS and 12 mm generator (Brinkmann, Westbury, NY).

2.2. Stacked Golgi Isolation

1. Phosphate buffer: 0.4 M KH_2PO_4/K_2HPO_4 buffer, pH 6.7 (Sigma), stable at 4°C for up to 1 month.
2. 1.0 M $MgCl_2$ (Sigma).
3. 2.0 M sucrose (Sigma): stable at 4°C for up to 1 month.
4. Protease inhibitor cocktail (Sigma).
5. Ultra-Clear 14-mL centrifuge tubes (SW40; Beckman Coulter, Fullerton, CA, USA).
6. Wide-bore standard disposable transfer pipettes (Thermo Fisher Scientific, Waltham, MA, USA).
7. Ultracentrifuge (Beckman Coulter, Fullerton, CA, USA).
8. DC Protein Assay Kit (Bio-Rad, Hercules, CA, USA) and bovine serum albumin (2 M stock solution) (Sigma).

2.3. Biochemical Fractionation of Isolated Stacked Golgi

1. 200 mM Na_2CO_3, pH 11.3 (Sigma).
2. Insulin syringes (Becton Dickinson, Franklin Lakes, NJ).
3. Tris base (Sigma).

2.4. Protein Extraction and Digestion

1. Methanol (MeOH) and chloroform ($CHCl_3$) (Fisher).
2. Urea (ultrapure molecular grade) (Sigma).
3. Dithiothreitol (DTT): prepare 500 mM solution in water and freeze single use (50 μL) aliquots immediately at −20°C.
4. Iodoacetamide (IAA): prepare 500 mM solution in water and freeze single use (50 μL) aliquots immediately at −20°C.

2.4.1. Formic Acid/Cyanogen Bromide Method

1. Cyanogen bromide (CNBr) solution: 500 mg/mL of CNBr in 90% formic acid (both from Sigma).
2. Saturated solution of ammonium bicarbonate.
3. Thermomixer (Eppendorf).

4. Endoproteinase Lys-C (Roche Applied Sciences, Basel, Switzerland): store at 4°C as solid.
5. Trypsin, modified porcine (Promega, Madison, WI, USA): store at –20°C as solid, thaw one time only.
6. Stock 100 mM $CaCl_2$ (Sigma): stable at 25°C.

2.4.2. High pH/Proteinase K Method

1. Insulin syringes from BD Syringe with 28-gauge needles.
2. Proteinase K (Roche Applied Sciences): aliquot in water (1 mg/mL) and freeze single use (50 µL) aliquots immediately at –20°C.
3. Formic acid, 90% (Mallinckrodt Baker, Phillipsburg, NJ, USA).

2.4.3. RapiGest/Trypsin Method

1. Ammonium bicarbonate: 50 mM NH_4HCO_3 (JT Baker) in H_2O.
2. RapiGest SF (Waters, Milford, MA, USA).
3. Trypsin, modified porcine (Promega): store at –20°C as solid, thaw one time only.
4. Stock 100 mM $CaCl_2$ (Sigma): stable at 25°C.
5. Concentrated HCl solution.

2.5. Multidimensional Protein Identification Technology

1. Agilent 1100 series HPLC pump with degasser, autosampler.
2. ThermoFinnigan LTQ Linear Ion Trap Mass Spectrometer.
3. 250 and 100 µm inner diameter fused silica capillary (Polymicro, Phoenix, AZ).
4. PEEK polymer Inline Microfilter Assembly (Upchurch Scientific, Oak Harbor, WA, USA).
5. Reverse-phase C_{18} packing material, Aqua, 5-µm beads, 100 Å pore size (Phenomenex, Torrance, CA).
6. Strong cation exchange (SCX) packing material, Partisphere (Whatman, Kent, UK).
7. Buffer A: acetonitrile/formic acid/water, 5/0.1/95 (v/v/v) (all from JT Baker). Acetonitrile is later designated MeCN.
8. Buffer B: MeCN/formic acid/water, 95/0.1/5 (v/v/v).
9. Calibrated micropipets 1–10 µL (Drummond Scientific Company, Broomall, PA).
10. Ammonium acetate (NH_4OAc) (JT Baker) at 100, 200, 300, 400, 500, 600, 700, 800, 900, 1000, and 5000 mM salt solutions: stable for up to 1 month at 25°C.
11. High-pressure bomb (in-house).
12. Laser-based micropipette puller Model P-2000 (Sutter Instruments Co., Folsom, CA).

2.6. Data Analysis

1. SEQUEST: see http://www.thermo.org.
2. DTASelect: see http://fields.scripps.edu/DTASelect/.

3. Methods

Rat liver is the preferred tissue for Golgi fractionation because of its tissue homogeneity and minimal amounts of connective tissue *(9,10)*. Cycloheximide may be used several hours prior to killing the animal to clear proteins in transit from the Golgi complex *(1,7,8)*, thereby enriching for Golgi-resident proteins. The Golgi fractionation protocol presented here is similar to that modified from Leelavathi *(2)* by Taylor and colleagues *(1)*.

Subcellular fractionation is highly sensitive to both sample handling and homogenization technique. To optimize the isolation of intact stacked Golgi with reproducibility, it is essential to (1) keep all samples on ice and (2) homogenize gently. A gentle homogenization procedure allows cells to be "popped open" so that stacked Golgi can be collected from the cells intact. Over-homogenization will lead to poor, inconsistent results. (3) Handle gently. The stacked Golgi complex is inherently fragile and prone to shearing after cell lysis. Never pellet or aggressively agitate the GF. GFs should be collected from the gradient with a wide-bore implement. Poor handling will result in vesiculation and poor recovery and yield.

3.1. Tissue Harvest and Homogenization

1. Cycloheximide (50 mg/kg) is administered to rats intraperitoneally 4 h before being killed by cervical dislocation. Livers are quickly removed and placed on ice.
2. One liver is finely minced with surgical blades on a precooled glass Petri dish on ice until pieces are <1 mm^3 and placed into a tared 50-mL conical tube on ice. Determine minced liver wet weight and resuspend at a ratio of 1 g of minced liver per 2 mL of 0.5 M sucrose homogenization buffer.
3. Homogenize the sample in the 50-mL conical tube using a Polytron PT10/35 using a middle setting (~3–4) for <20 s while slowly moving the generator blade from the top to the bottom of the tube with one pass in a circular motion. Stop the homogenization and allow the blade to stop in the sample so as to avoid splashing (*see* **Note 1**).
4. Alternatively, homogenize the sample by hand with a large Teflon-coated pestle/Pyrex tissue grinder. Use 6–12 strokes on small batches of the liver sample (~2 mL) and combine (*see* **Note 2**).
5. The homogenate is combined and subjected to a low-speed centrifugation, at 1500 *g* for 10 min at 4°C, and the resulting postnuclear supernatant (PNS) is collected on ice.

3.2. Stacked Golgi Isolation

1. Load the PNS in the middle of a discontinuous sucrose step gradient as shown in **Fig. 2**. Sucrose buffers are prepared fresh and kept on ice at all times. First, prepare buffers in 50 mL final volumes from stock solutions. Sucrose buffers are prepared as shown in **Table 1**.

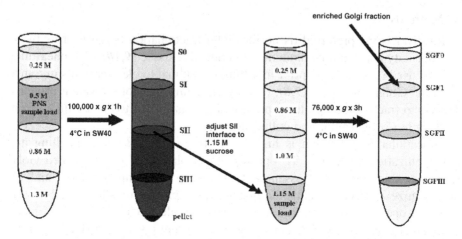

Fig. 2. Representation of sucrose step gradients for preparation of enriched, stacked Golgi fractions from tissue lysates.

Table 1
Sucrose Gradient Solutions

	0.25 M	0.50 M	0.86 M	1.3 M
Phosphate buffer	12.50 mL	12.50 mL	12.50 mL	12.50 mL
1.0 M MgCl$_2$	0.25 mL	0.25 mL	0.25 mL	0.25 mL
2.0 M sucrose	6.25 mL	12.50 mL	21.50 mL	32.50 mL
H$_2$O	31.00 mL	24.75 mL	15.75 mL	4.75 mL
Proteolytic inhibitors	250 µL	250 µL	250 µL	250 µL
Final volume	50 mL	50 mL	50 mL	50 mL

2. Layer the sucrose buffers and sample into ultracentrifuge tubes (SW40 tubes will hold up to ~12.5 mL) with increasing sucrose buffer densities from the bottom to the top (i.e., 1.3 M on the bottom, 0.86 M sample load in 0.5 M sucrose, and 0.25 M on top). Take care not to disturb the interfaces when adding new layers (*see* **Note 3**). Layer ~2.5 mL of 1.3 M sucrose solution, ~3 mL of 0.86 M sucrose solution, PNS (no more than 4–5 mL), and 0.25 M sucrose (2–3 mL) carefully to maintain discrete interfaces. Volumes may vary depending on the cell/tissue PNS volumes. While transporting the gradient to the ultracentrifuge, caution must be taken not to disturb the layers.

3. Precool the rotor and swinging buckets in the ultracentrifuge. Centrifuge the gradient at 100,000 *g* for 1 h at 4°C with the brake off and slow acceleration.

4. Collect fractions of variable volumes from the top of the gradient using a wide-bore transfer pipette. Typically, the SII interface is collected in ~750 µL. **Figure 2** displays the location of the SII (intermediate fraction) which is enriched in stacked

Golgi at the 0.50/0.86 M interface. SI and SIII fractions refer to the interfaces between the 0.25/0.50 M and 0.86/1.3 M fractions, respectively.

5. Adjust the molarity of SII interface fraction to 1.15 M sucrose using 2 M sucrose, and layer this fraction at the bottom of a second sucrose step gradient. Exact densities can be determined using a refractometer. Overlay the SII fraction with equal volumes of 1.0, 0.86, and 0.25 M sucrose and centrifuge the gradient at 76,000 g for 3 h at 4°C. The enriched Golgi fraction (SGF1) is collected at the 0.25M/0.86 M interface. Typically, the SGF1 interface is collected in ~300 μL. Collect and aliquot all fractions (if desired) and store at –80°C.

3.3. Biochemical Fractionation of Isolated Stacked Golgi

1. Collect and dilute the enriched GF with an equal volume of 0.1 M KH_2PO_4/K_2HPO_4 buffer, pH 6.7. The sample is then pelleted for 60 min at 150,000 g in the ultracentrifuge at 4°C. The supernatant is removed and discarded. The subsequent steps (steps 2–6) are used to further fractionate the GF into soluble and membrane GFs. If the distinction between soluble and membrane proteins is not important to your work, proceed directly to **Subheading 3.4** for protein extraction and digestion.

2. Resuspend the GF pellet in 200 mM carbonate wash (Na_2CO_3), pH 11.3, at a concentration no greater than 1 mg/mL (*see* **Note 4**).

3. With the sample on ice, agitate the pellet using an insulin syringe every 15 min. Use 4–5 strokes of the syringe each time. Continue the agitation for at least 1 h. This causes the sealed membranes to open and form sheets instead of closed vesicles *(10,11)*.

4. Pellet the "Golgi membrane fraction" at 150,000 g for 1 h at 4°C.

5. Collect and label the supernatant as the "Golgi-soluble fraction."

6. Re-extract the membrane pellet at 1 mg/mL in 200 mM carbonate wash (Na_2CO_3), pH 11.3, agitate with an insulin syringe every 15 min for 1 hour, and re-pellet at 150,000 g for 1 h at 4°C. The supernatant is discarded and the pellet is labeled as the "Golgi membrane fraction."

3.4. Protein Extraction and Digestion

1. The Golgi membrane fraction can be digested directly using the digestion protocols described below. Alternatively, for all digestion methods below, the membrane fraction can first be extracted (de-lipidated) and the proteins precipitated. The Golgi-soluble fraction can be digested directly using the high pH/proteinase K method or RapiGest method. Alternatively, for all methods below, the proteins in the soluble fraction can first be precipitated.

2. Extract and precipitate the protein from Golgi membrane and soluble fractions by methanol/chloroform precipitation (MeOH/CHCl$_3$ ppt). Add 600 μL of MeOH (4×) and 150 μL of chloroform (1×) to 150 μL of sample in a 1.7-mL centrifuge tube and vortex briefly. Add 450 μL of ddH$_2O$ (3×) to the mixture and vortex. Spin the tube at maximum speed (16,000 g) for 2 min in a desktop microfuge.

3. The proteins will form a thin, opaque interface between the organic and chloroform layers. Remove most of the upper layer and discard.

4. Add an additional 450 μL of ddH$_2$O (3×) to the protein/chloroform mixture and vortex briefly.

5. Pellet the protein in the microfuge at maximum speed for 2 min. Steps 1–4 can be repeated if necessary.

6. Proceed to protein digestion using one of the methods described below. Alternative digestion strategies will facilitate more identifications and more complete coverage of identified proteins using mass spectrometry. Digest separate fractions of your sample and analyze separately for better protein coverage, if initial starting protein levels permit. Typically, ~25–50 μg of undigested protein is adequate for a single MudPIT run.

3.4.1. Formic Acid/CNBr Digestion Method *(3,8)*

1. Resuspend the membrane pellet (or precipitated protein pellet) in 50 μL of 500 mg/mL CNBr in 90% formic acid and incubate in the dark in a fume hood overnight *(1)*. The high concentration of organic acid solubilizes membrane proteins and CNBr cleaves at methionine residues under acidic conditions. Use an insulin syringe to break up the solid pellet if it does not solubilize easily. CNBr is highly toxic and should always be used with care in a chemical fume hood.

2. Adjust the pH of the sample to 8.5 with saturated ammonium bicarbonate. Add solid urea to a final concentration of 8 M. Vortex the sample briefly.

3. Reduce the sample by adding DTT to 25 mM and incubating at 60°C for 30 min on a Thermomixer.

4. Let the sample cool to room temperature. Alkylate the sample by adjusting the solution to 100 mM IAA and incubate the tube in dark for 30 min.

5. Add Endoproteinase Lys-C at a 1:50 (w/w) enzyme : substrate ratio and incubate at 37°C overnight in a Thermomixer.

6. The following morning, adjust the sample to 4 M urea and 1 mM CaCl$_2$ with water. Add modified trypsin at a 1:100 (w/w) enzyme : substrate ratio and incubate at 37°C overnight in a Thermomixer.

7. Adjust sample to 5% (v/v) formic acid. Spin the sample at 16,000 g for 20 min to pellet any insoluble material and save the supernatant for mass spectrometry analysis.

3.4.2. High pH/proteinase K Method *(8,11)*

1. Resuspend the membrane pellet (or precipitated protein pellet) in 200 mM Na$_2$CO$_3$, pH 11.3, at 1 mg/mL concentration. Adjust to 8 M urea, agitate with an insulin syringe if necessary.

2. Reduce the sample by adding DTT to 25 mM and incubating at 60°C for 30 min on a Thermomixer.

3. Let the sample cool to room temperature. Alkylate the sample by adjusting the solution to 100 mM IAA and incubate the tube in dark for 30 min.
4. Add Proteinase K to the solution at a 1:50 (w/w) enzyme : substrate ratio and incubate at 37°C for 3 h with shaking. Proteinase K is a robust non-specific protease that has a attenuated function at pH 11. Use of this enzyme results in peptides with overlapping sequences. Add an additional aliquot of Proteinase K [1:100 (w/w) – enzyme : substrate ratio] and incubate overnight at 37°C with shaking.
5. Quench the reaction by diluting the sample with formic acid to 5% (v/v) final concentration.
6. Microfuge at 14,000 g at 4°C for 30 min to remove any insoluble particulates.

3.4.3. RapiGest/Trypsin Method (12)

1. Make 0.1% (w/v) RapiGest detergent diluted in 50 mM ammonium bicarbonate, pH 7.8 (use 1 mg of RapiGest per 1 mL of buffer).
2. Add 50 μL of 0.1% (w/v) RapiGest to the membrane pellet (or precipitated protein pellet) and vortex. Use an insulin syringe to break up the solid pellet if it does not solubilize easily.
3. Boil the sample for 3 min. Let it cool to room temperature.
4. Reduce the sample by adding DTT to 5 mM and incubating at 60°C for 30 min on a Thermomixer.
5. Let the sample cool to room temperature. Alkylate the sample by adjusting the solution to 15 mM IAA and incubate the tube in the dark for 30 min.
6. Add $CaCl_2$ to a final concentration of 1 mM and vortex.
7. Add trypsin for a final concentration of 1:50 (w/w) enzyme : protein ratio and incubate overnight at 37°C in a Thermomixer.
8. Add concentrated HCl solution to 100 mM. Incubate at 37°C for 45 min to dissociate the acid-cleavable detergent RapiGest. Spin the sample at 16,000 g for 20 min to pellet any insoluble material and save the supernatant for mass spectrometry analysis.

3.5. MudPIT Analysis

1. Prepare loading and analytical columns in-house using a high-pressure bomb and a laser-based micropipette puller. A typical peptide sample load is the equivalent of 25–75 μg of predigested protein sample.
2. Add ~3 mg of reverse-phase C_{18} packing material (Aqua) to 1.0 mL of HPLC-grade methanol in a microfuge tube. Vortex and spin the tube to pellet the packing material.
3. Using a high-pressure bomb, pack ~1.5 cm of the reverse-phase C_{18} beads into a 250 μm inner diameter fused silica capillary (~10 cm total length) using an in-line microfilter to trap the beads. This is the sample loading column. It is important to be consistent with the amount of packing material used in both the loading and analytical columns to reduce experimental error.

4. Equilibrate the loading column with buffers A and B using a typical HPLC gradient (refer below for gradients used in mass spectrometry analysis) with a back pressure of ~25–30 bar to condition the packing material prior to loading the peptide samples.
5. Pressure load protein digests onto the fused silica capillary loading column.
6. Equilibrate the loading column again for 20 min with 100% buffer A.
7. Pull 100 μm inner diameter fused silica capillary columns using a laser-based micropipette puller to create a fine spray tip. Unpacked capillaries should be at least 18–20 cm in length.
8. Prepare a biphasic analytical column by first packing 8 cm of reverse-phase C_{18} packing material followed by 4 cm of SCX packing material into the capillary. It is important to pack the material slowly to create discrete interfaces and avoid mixing of stationary phases. Equilibrate the analytical column similarly to what is done for the loading column.
9. Join the loading column with the biphasic analytical column using the filter assembly and attach to the outlet flow of an Agilent 1100 series high-performance liquid chromatography (HPLC) pump. Operate the pump at a flow rate of 100 μL/min with an active split of ~1:500. Measure the final flow through the column using a calibrated micropipette between 150 and 300 nL/min.
10. Place the analytical column in-line with the heated capillary inlet of the mass spectrometer and monitor a consistent spray in the TunePlus window (*see* **Note 5**).
11. Run an HPLC "first" gradient (*see* **Fig. 3**) while collecting data using the Xcalibur software provided by Thermo. Inject a nominal 10-μL pulse of buffer A onto the column (equivalent to a 0 mM salt pulse) and collect MS^2 spectra in a data-dependent mode. This first gradient serves to move peptides from the loading column to the analytical column for impending MudPIT analysis.

first and last gradient   **mid-gradient**

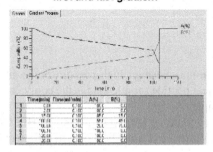
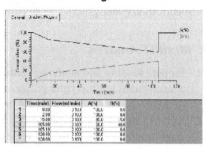

Fig. 3. Reproductions of the Instrument Setup window from Xcalibur 1.8 software package. Gradient programs used during LC separation of peptides from the "first," "mid," and "last" multidimensional protein identification technology (MudPIT) steps are graphically displayed as a change in percentage of each buffer over the course of the 120-min LC run.

12. After the first gradient is completed, remove the loading column from the in-line setup and monitor the spray of the analytical column (with column heater, if possible).

13. Perform MudPIT analysis of the peptide sample by running 11 consecutive HPLC gradients using steps of increasing salt pulses. Each of the 11 gradients is preceded by a 50-μL injection of ammonium acetate (NH_4OAc) ranging from 100 to 5000 mM (i.e., 100, 200, 300, 400, 500, 600, 700, 800, 900, 1000, and 5000 mM) delivered by the autosampler. Salt injections act to move peptides from the SCX packing material to the C_{18} material. The peptides will elute from the C_{18} phase during the acetonitrile gradient.

14. Thoroughly equilibrate the analytical column with buffer A before using it for another MudPIT experiment. Store hydrated columns at room temperature with the tip immersed in buffer A and the back end protected with a closed fitting.

3.6. Data Analysis

1. Convert the RAW files into .ms2 files containing the tandem mass spectra using the Windows-based software program makeMS2. Search the .ms2 files against a rat sequence database concatenated to a randomized database using SEQUEST-Norm *(12–14)*.

2. Assemble the identified peptides into proteins and filter to minimize false positives using the program DTASelect 1.9 *(15)*. General starting parameters for filtering are normalized Xcorr = 0.156, DeltCN = 0.15, % ion = 0.2, and minimum peptide length = 6.

3. Calculate the false discovery rate (FDR) by dividing the total number of hits to the randomized database by the total number of hits to the real, non-randomized database *(12)*. Adjust the DTASelect parameters to maintain an FDR less than 5%.

4. This method results in the identification of hundreds to thousands of proteins and will vary depending on the quality and enrichment of the stacked GF and the sensitivity of the mass spectrometer used. Because no subcellular fraction is ever completely "pure," non-Golgi proteins will also be present in this sample. Typically, however, >25% of the identified proteins are known Golgi-resident proteins and >50% are proteins essential for Golgi function.

4. Notes

1. You can rinse the sample from the blade with buffer and save.

2. Be careful not to introduce air bubbles into the samples. It is essential that the homogenization does not lead to the breakage of nuclei indicated by a significant increase in the viscosity of the sample. Over-homogenization will result in poor separation of membranes.

3. Pipet buffers slowly with the tip angled against the side of the centrifuge tube – on ice.

4. Make sure to remove all of the residual liquid from the top of the pellet before adding Na_2CO_3 buffer, as this will lower the overall pH of the carbonate wash. If the GF is not in pellet form, be sure that the final pH of the GF is adjusted to 11.3.

5. Optional – use of a column heater set at $\sim$45°C will improve the recovery of hydrophobic peptides *(16)*.

References

1. Taylor, R. S., Jones, S. M., Dahl, R. H., Nordeen, M. H., Howell, K. E. (1997) Characterization of the Golgi complex cleared of proteins in transit and examination of calcium uptake activities. *Mol. Biol. Cell* **8**, 1911–1931.
2. Leelavathi, D. E., Estes, L. W., Feingold, D. S., Lombardi, B. (1970) Isolation of a Golgi-rich fraction from rat liver. *Biochim. Biophys. Acta* **211**, 124–138.
3. Washburn, M. P., Wolters, D., Yates, J. R. III. (2001) Large-scale analysis of the yeast proteome by multidimensional protein identification technology. *Nat. Biotechnol.* **19**, 242–247.
4. Wolters, D. A., Washburn, M. P., Yates, J. R. III. (2001) An automated multidimensional protein identification technology for shotgun proteomics. *Anal. Chem.* **73**, 5683–5690.
5. Marsh, B. J., Mastronarde, D. N., McIntosh, J. R., Howell, K. E. (2001) Structural evidence for multiple transport mechanisms through the Golgi in the pancreatic beta-cell line, HIT-T15. *Biochem. Soc. Trans.* **29**, 461–467.
6. Ladinsky, M. S., Mastronarde, D. N., McIntosh, J. R., Howell, K. E., Staehelin, L. A. (1999) Golgi structure in three dimensions: functional insights from the normal rat kidney cell. *J. Cell Biol.* **144**, 1135–1149.
7. Taylor, R. S., Fialka, I., Jones, S. M., Huber, L. A., Howell, K. E. (1997) Two-dimensional mapping of the endogenous proteins of the rat hepatocyte Golgi complex cleared of proteins in transit. *Electrophoresis* **18**, 2601–2612.
8. Wu, C. C., MacCoss, M. J., Mardones, G., Finnigan, C., Mgelsvang, S., Yates, J. R. III, et al. (2004) Organellar proteomics reveals Golgi arginine dimethylation. *Mol. Biol. Cell* **15**, 2907–2919.
9. Ehrenreich, J. H., Bergeron, J. J. M., Siekevitz, P., Palade, G. E. (1973) Golgi fractions prepared from rat liver homogenates. I. Isolation procedure and morphological characterization. *J. Cell Biol.* **59**, 45–72.
10. Howell, K. E., Palade, G. E. (1982) Hepatic Golgi fractions resolved into membrane and content subfractions. *J. Cell Biol* . **92**, 822–832.
11. Wu, C. C., MacCoss, M. J., Howell, K. E., and Yates, J. R. III (2003) A method for the comprehensive proteomic analysis of membrane proteins. *Nat. Biotechnol.* **21**, 532–538.
12. Blackler, A. R., Klammer, A. A., MacCoss, M. J., Wu, C. C. (2006) Quantitative comparison of proteomic data quality between a 2D and 3D quadrupole ion trap. *Anal. Chem.* **78**, 1337–1344.

13. Eng, J. K., McCormack, A. L., Yates III, J. R. (1994) An approach to correlate tandem mass spectra of modified peptides to amino acid sequences in a protein database. *J. Am. Soc. Mass Spectrom.* **5**, 976–989.

14. MacCoss, M. J., Wu, C. C., Yates, J. R. III (2002) Probability-based validation of protein identifications using a modified SEQUEST algorithm. *Anal. Chem.* **74**, 5593–5599.

15. Tabb, D. L., McDonald, W. H., Yates, J. R. III (2002) DTASelect and Contrast: tools for assembling and comparing protein identifications from shotgun proteomics. *J. Proteome Res.* **1**, 21–26.

16. Speers, A. E., Blackler, A. R., Wu, C. C. (2007). Elevated temperature facilitates shotgun proteomic analysis of membrane proteins. *Anal. Chem.* **79**(12), 4613–4620.

4

Purification of *Saccharomyces cerevisiae* Mitochondria by Zone Electrophoresis in a Free Flow Device

Hans Zischka*, Norbert Kinkl*, Ralf J. Braun, and Marius Ueffing

Summary

This chapter describes the isolation of yeast mitochondria by differential centrifugation followed by mitochondrial purification through zone electrophoresis (ZE) using a free flow device (FFE). Starting from a yeast colony, cultures are grown under respiratory conditions to logarithmic phase. Cells are collected, their cell walls enzymatically disintegrated and the resulting spheroplasts are homogenized. Mitochondria are pre-fractionated from this homogenate by differential centrifugation. With the focus on further purification, pre-fractionated mitochondria are subjected to ZE-FFE. In ZE-FFE, mitochondria are transported with the buffer flow through the separation chamber and purified from contaminants by specific deflection through a perpendicularly oriented electric field. The purified mitochondria can be collected by centrifugation and used for further experiments and analysis such as sodium dodecyl sulphate–polyacrylamide gel electrophoresis (SDS–PAGE), immunoblotting, 2-DE or electron microscopy.

Key Words: Mitochondria; free flow electrophoresis; zone electrophoresis; *Saccharomyces cerevisiae*.

1. Introduction

Mitochondria are of fundamental importance in cellular metabolism and comprise a multitude of meta- and anabolic functions (*1*). Many of these enzymatic reaction pathways are studied in mitochondria isolated from baker's yeast (*2*). In total, about 700–800 different proteins are thought to make up the

---

*Both authors share equal first authorship.

From: *Methods in Molecular Biology, vol. 432: Organelle Proteomics*
Edited by: D. Pflieger and J. Rossier © Humana Press, Totowa, NJ

Saccharomyces cerevisiae mitochondrial proteome *(3,4)*. An essential prerequisite for its analysis is an organelle preparation of high purity. Yeast mitochondria are classically pre-fractionated by differential centrifugation after an enzymatic digestion of the yeast cell wall and homogenization *(5)*. Mitochondrial purity can be further increased by subsequent density gradient centrifugation *(6)*. As an alternative approach, mitochondria can be purified by an electrophoretic method *(7)*. Advantages of the latter approach are a possible continuous sample application combined with a gentle highly efficient purification of yeast mitochondria.

This chapter describes the isolation of yeast mitochondria by differential centrifugation followed by mitochondrial purification through zone electrophoresis (ZE) using a free flow device (FFE). ZE-FFE was introduced by Kurt Hannig *(8)* about 40 years ago. Mitochondrial pre-fractionation by differential centrifugation is carried out following the protocol described by Herrmann et al. *(5)*, with the modifications given below. With the focus on further purification, this mitochondrial pre-fraction is subsequently subjected to ZE-FFE. In ZE-FFE, mitochondria are purified according to their surface charge and hydrodynamic properties. Mitochondria are transported with the buffer flow through the separation chamber and purified from contaminants by specific deflection through a perpendicularly oriented electric field *(7,9)*. This purification method allows a marked reduction of non-mitochondrial contaminants originating from cytosol, endoplasmic reticulum, Golgi and peroxisomes *(7)* (*see* **Fig. 1**).

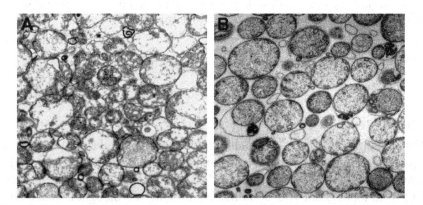

Fig. 1. Electron micrographs of mitochondrial preparations (magnification 15750). (**A**) Typical micrograph of pre-fractionated yeast mitochondria isolated by differential centrifugation. The appearance of the preparation is rather heterogeneous, showing several damaged organelles and other vesicular structures. (**B**) Typical micrograph of mitochondria purified by zone electrophoresis in a free flow device (ZE-FFE) demonstrating a homogeneous organelle population.

A schematic overview of the FFE chamber is given in **Fig. 2**. The chamber consists of a front and back plate separated by a 0.5 mm spacer. Within this intermediate space, a laminar buffer flow is created by a constant injection of separation media through the inlets M2–M6 at the starting side of the chamber. The separation buffer is blocked by the counterflow (*see* **Fig. 2**, no. 8) injected at the end side of the chamber and deviated through the fraction collector into a 96-well plate (*see* **Fig. 2**, no. 7). Cathode and anode (*see* **Fig. 2**, no. 3 and 4, respectively) are oriented parallel to this buffer stream. The electrodes are flushed to avoid electrode products gathering and separated from the separation chamber by electrophoresis membranes (*see* **Fig. 2**, no. 6). To protect these membranes, stabilization media is flushed around the electrode chambers through media inlets M1 and M7.

The pre-fractionated mitochondria are injected though the sample inlet (*see* **Fig. 2**, no. 1) into the electrophoresis chamber, transported with the laminar buffer flow and deflected towards the anode as they are negatively charged

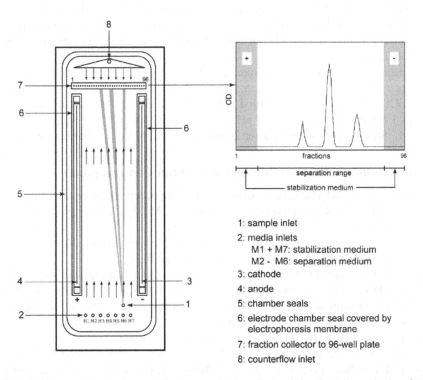

1: sample inlet
2: media inlets
   M1 + M7: stabilization medium
   M2 - M6: separation medium
3: cathode
4: anode
5: chamber seals
6: electrode chamber seal covered by
   electrophoresis membrane
7: fraction collector to 96-well plate
8: counterflow inlet

Fig. 2. Schematic drawing of the separation section of the free flow device (FFE) chamber. Pre-fractionated mitochondria are injected into the FFE, deflected within the separation range towards the anode and collected through the fraction collector. For details see introduction.

particles at neutral pH *(10)*. With the velocity of the buffer flow applied in our FFE separations, purified mitochondria can be collected after several minutes with the fraction collector (*see* **Fig. 2**, no. 7) in a constant fashion. The separation can be monitored by appropriate detection devices, i.e. 96-well readers. Several parameters of the collected fractions can be measured, such as optical densities, molecular absorptions, protein contents, enzymatic activities and others. The mitochondria-containing fractions can be pooled and organelles concentrated by centrifugation. The purified mitochondria can be stored frozen and/or directly subjected to further analysis.

2. Materials

2.1. Yeast Culture

1. YPGlc agar plates: to prepare 1 L, dissolve 10 g of yeast extract, 20 g of bacto-peptone and 15 g of agar in distilled H_2O and adjust the volume to 840 mL to obtain YP medium. Prepare 160 mL of a 25% glucose solution in distilled H_2O. Sterilize both solutions by autoclaving for 20 min at 120°C. Mix the 840 mL of YP medium and the 160 mL of 25% glucose solution to obtain a final glucose concentration of 4% (YPGlc medium). Pour the plates and store them at 4°C after solidification.

2. Lactate medium: for 5 L medium, dissolve 15 g of yeast extract, 5 g of glucose, 5 g of NH_4Cl, 5 g of KH_2PO_4, 5.5 g of $MgSO_4·7H_2O$, 2.5 g of NaCl and 2.5 g of $CaCl_2·2H_2O$ in 4.5 L of distilled H_2O. Add 110 mL of 90% lactate and 1.5 mL of 1% $FeCl_3$ solution. Adjust pH with 10 M KOH to 5.5 and adjust volume to 5 L. Sterilize by autoclaving for 20 min at 120°C. Store at room temperature.

2.2. Isolation of Mitochondria by Differential Centrifugation

1. Zymolyase 20T (Seikagaku, Corporation, Tokyo, Japan). Store at 4°C.
2. 1.2 M sorbitol: to prepare 1 L, dissolve 218.62 g of sorbitol in 1 L of distilled H_2O.
3. 100 mM phenylmethylsulfonyl fluoride (PMSF): to prepare 10 mL, dissolve 174 mg of PMSF in 10 mL of ethanol. Prepare freshly.
4. 1 M potassium phosphate buffer (KPB) (pH 7.4): prepare solutions of 1 M KH_2PO_4 (68.0 g for 500 mL of water) and of 1 M K_2HPO_4 (87.1 g for 500 mL of water) in distilled H_2O. To 500 mL of KH_2PO_4 solution, slowly add K_2HPO_4 solution until a pH of 7.4 is reached.
5. 1 M Tris–HCl (pH 7.4): to prepare 500 mL, dissolve 60.57 g of Tris in distilled H_2O, adjust pH to 7.4 with 5 M HCl and adjust to a total volume of 500 mL.
6. 1 M Tris–SO_4 (pH 9.4): to prepare 500 mL, dissolve 60.57 g of Tris in distilled H_2O, adjust pH to 9.4 with 5 M H_2SO_4 and adjust to a total volume of 500 mL.
7. 500 mM ethylenediaminetetraacetic acid (EDTA): To prepare 500 mL, dissolve 95 g EDTA in 400 mL of distilled H_2O. Adjust pH to 8.0 with 10 M KOH and fill up to 500 mL with distilled H_2O.

8. 1 M dithiothreitol (DTT): To prepare 2 mL, dissolve 308 mg of DTT in 2 mL of distilled H_2O. Prepare freshly.

9. Zymolyase buffer: 1.2 M sorbitol and 20 mM KPB, pH 7.4. To prepare 1 L, mix 980 mL of 1.2 M sorbitol with 20 mL of 1 M KPB, pH 7.4. Store at 4°C.

10. Homogenization buffer: 0.6 M sorbitol, 10 mM Tris–HCl, pH 7.4, 5 mM EDTA, and 2 mM PMSF. To prepare 500 mL, mix 250 mL of 1.2 M sorbitol with 5 mL of 1 M Tris–HCl, pH 7.4, and adjust to 500 mL with distilled H_2O. Add PMSF and EDTA to a final concentration of 2 mM and 5 mM, respectively, immediately before use. Store at 4°C.

11. Sucrose-EDTA-Tris (SET) buffer: 0.25 M sucrose and 10 mM Tris–HCl, pH 7.4. To prepare 1 L, dissolve 85.6 g of sucrose in distilled H_2O, add 10 mL of 1 M Tris–HCl, pH 7.4, and adjust to a total volume of 1 L. Add PMSF and EDTA to a final concentration of 2 mM and 5 mM, respectively, immediately before use. Store at 4°C.

2.3. Purification of Yeast Mitochondria by ZE-FFE

2.3.1. Free Flow Apparatus

The technique of ZE-FFE was introduced already 40 years ago (8), and several ZE-FFE devices have been commercially available since then. Principle components are a stable high-voltage power supply, high-precision pumps, a well-cooled electrophoresis chamber and an efficient sample collector. In the 1980s, the Elphor VaP by Bender and Hobein (Munich, Germany) was a frequently used instrument, which worked in vertical mode in a downward flow direction. More recently, the OCTOPUS instrument was available from Dr. Weber GmbH (Kirchheim, Germany). This apparatus, in its latest version, was used for the elaboration of the present protocol. The instrument consists of two independent parts, a process unit and a control unit. The process unit consists of an electrophoresis chamber, with dimensions of 500 mm (length), 100 mm (width) and 0.5 mm (thickness). The chamber can be tilted and was run in horizontal mode; collection of the separated fractions was done through a pulse-free splitting device using the counterflow technique. Due to similar work mode the herein method may be adapted to the most recent instrumentation that is the BD™ Free Flow Electrophoresis System from BD (Franklin Lakes, NJ).

2.3.2. Free Flow Buffers

1. 5× electrode solution (5× EL): 500 mM acetic acid (HAc) and 500 mM triethanolamine (TEA), pH 7.4. To prepare 1 L, add 28.6 mL of HAc and 66.65 mL of TEA to 850 mL of ultra-pure H_2O, adjust pH to 7.4 with a solution of 10 M KOH and adjust to a total volume of 1 L. Store at 4°C.

2. Electrode solution: 100 mM HAc and 100 mM TEA. To prepare 1 L, add 200 mL of 5× EL to 800 mL of ultra-pure H_2O. Store at 4°C.
3. Stabilization medium: 100 mM HAc, 100 mM TEA, and 0.28 M sucrose. To prepare 1 L, dissolve 95.8 g of sucrose in 600 mL of ultra-pure H_2O, add 200 mL of 5× EL and adjust to a total volume of 1 L. Store at 4°C.
4. Separation medium: 10 mM HAc, 10 mM TEA, and 0.28 M sucrose. To prepare 1 L, dissolve 95.8 g of sucrose in 700 mL of ultra-pure H_2O, add 100 mL of electrode solution and adjust to a total volume of 1 L. Store at 4°C.
5. Counterflow medium: 0.28 M sucrose. To prepare 1 L, dissolve 95.8 g of sucrose in 800 mL of ultra-pure H_2O and adjust to a total volume of 1 L. Store at 4°C.

3. Methods

3.1. Growth of Yeast Cultures

1. Streak out the yeast strain, e.g. BY4741, onto an YPGlc agar plate and grow for 3 days at 30°C.
2. Inoculate 50 mL lactate medium with a single yeast colony.
3. Grow the culture overnight at 30°C in a baffled flask, shaking at 130 rpm.
4. Use the overnight culture to inoculate fresh lactate medium. Periodically measure the OD_{600} to determine the logarithmic growth phase and expanded cultures (initial OD_{600} of ~0.1).
5. Isolate mitochondria from the cultures in logarithmic growth phase (OD_{600} of 1.0–1.5), for example from five flasks with 2.5 L yeast culture each (total volume 10–12 L).

3.2. Isolation of Mitochondria by Differential Centrifugation

Mitochondria are essentially isolated according to a standard protocol *(5)* with minor modifications.

1. Harvest cells (e.g. 5× 2.5 L) by centrifugation at 4400 *g* for 5 min at 4°C (*see* **Note 1**).
2. Discard supernatant and resuspend cells in 250 mL distilled H_2O.
3. Centrifuge cells at 4400 *g* for 5 min at 4°C and determine the wet weight of the pellet.
4. Resuspend cells in 100 mM Tris–SO_4, pH 9.4, at 2 mL/g wet weight (*see* **Note 2**).
5. Add 10 μL of 1 M DTT per mL of resuspended cells to obtain a final concentration of 10 mM DTT, which will reduce disulfide bonds of the yeast cell walls.
6. Incubate cells for 15 min at 30°C in a shaking water bath. Do not prolong reduction for more than 15 min as this will lead to drastically reduced yield and quality of mitochondria.
7. Centrifuge cells at 4400 *g* for 5 min at 4°C to remove surplus DTT.

8. Resuspend cells in 1.2 M sorbitol using 2 mL/g wet weight.
9. Centrifuge cells at 4400 g for 5 min at 4°C.
10. Resuspend the cell pellet in zymolyase buffer at 7 mL/g wet weight. Add 4 mg of zymolyase per gram of wet weight to digest the yeast cell wall enzymatically and obtain spheroplasts. Remove a small aliquot prior to the addition of the zymolyase for the spheroplast test.
11. Incubate cells at 30°C in a shaking water bath.
12. Check for spheroplast formation by adding 50 μL of cell suspension to 1 mL of distilled H_2O and measuring the OD_{600}. Spheroplasts – in contrast to yeast cells with intact cell walls – will disrupt in hypotonic aqueous solution leading to a decrease in OD_{600}. Incubation is stopped when the OD_{600} is 30–50% of the OD prior to the addition of zymolyase.
13. Centrifuge the spheroplasts at 2000 g for 5 min at 4°C. Spheroplasts are fragile and should therefore be handled with care.
14. Resuspend the spheroplasts gently in (isotonic) 1.2 M sorbitol and centrifuge at 2000 g for 5 min at 4°C to remove surplus zymolyase.
15. Resuspend spheroplasts in homogenization buffer at 7 mL/g wet weight.
16. Transfer the spheroplast suspension to a glass potter and homogenize with a motor-driven Teflon pestle with 10 strokes at 800–1000 rpm. Avoid foaming of the suspension.
17. Centrifuge the suspension at 1700 g for 5 min at 4°C to remove cell debris and nuclei.
18. Transfer the supernatant to a new centrifuge tube and centrifuge again at 1700 g for 5 min at 4°C.
19. Transfer the supernatant into a new centrifuge tube and centrifuge at 17,400 g for 12 min at 4°C to obtain the first crude mitochondrial preparation.
20. Meanwhile, finalize the SET buffer by adding EDTA and PMSF to a final concentration of 5 mM and 2 mM, respectively.
21. Discard the supernatant and resuspend the pellet carefully in SET buffer.
22. Centrifuge at 3000 g for 5 min at 4°C to remove agglomerates.
23. Transfer the supernatant to a new centrifuge tube and centrifuge at 17,400 g for 12 min at 4°C.
24. Resuspend the mitochondrial pellet in a small volume of SET buffer and determine the protein concentration using the Bradford protein assay.
25. Adjust the protein concentration to 10 mg/mL with SET buffer. Samples can be shock frozen with liquid nitrogen and stored at –80°C or used immediately.

3.3. Purification of Yeast Mitochondria by ZE-FFE

3.3.1. FFE Setup

1. Turn on the cooler and set temperature to 4°C (*see* **Note 3**).
2. Set the separation chamber to the vertical position. Place the fractionation plate from the fraction collector housing to the front plate, otherwise the limited length

of the fractionation tubing does hamper the opening of the chamber. Now you can pull open the front (*see* **Note 4**).

3. Carefully inspect the front and back plate for dust particles or protein agglomerates and clean both plates with lint-free paper and by rinsing with water (*see* **Note 5**).

4. Rinse the spacer (0.5 mm) with distilled H_2O and place it on the front plate of the separation chamber. Check for proper alignment and do not cover the separation media inlets.

5. Rinse the electrode membranes and filter paper strips with distilled H_2O and place them on the electrode chambers of the front plate with the smooth side of the membrane facing the electrode seal. Check for proper alignment of the membrane/filter paper strip sandwiches on the electrode seals (*see* **Notes 6** and **7**).

6. Close the separation chamber.

7. Close the middle two clamps simultaneously and then close the remaining clamps pair by pair. Inspect the sample application tubing for proper alignment parallel to the flow direction. Place the fractionation plate on the fraction collector housing located on top of the instrument.

8. Turn on the electrode solution pump and flush the electrode chambers with distilled H_2O. Leakage of the electrode solution into the separation chamber has to be strictly avoided. If necessary, reopen the chamber and adjust the electrode membranes or renew them in case of damage.

9. Tilt the separation chamber to a diagonal position and turn the bubble trap to the filling position (i.e. parallel to the separation chamber). Open the Luer-lock closure on the counterflow tube and place the two parts of the counterflow tube in the fractionation tray.

10. Start the separation media pump and fill the chamber with distilled H_2O. Entrapped air bubbles have to be removed and expelled. Thereto set the media pump on reverse mode and empty the buffer in the chamber until the liquid level is below the entrapped air bubbles. Switch the direction of the media pump to normal mode and refill the chamber (*see* **Note 8**).

11. When excess water drops from both counterflow tubes (i.e. the chamber is completely filled without entrapped air bubbles), reconnect counterflow tubes. Activate the 96-well fraction collector by closure of the outlet opening of the three-way tap on the counterflow tube. Turn the bubble trap to the operating position (i.e. perpendicular to the separation chamber).

12. Check for blocked 96 fractionation tubes. All 96 fractionation tubes should release water droplets freely in a constant manner. Blocked channels can be activated by applying negative pressure with a syringe connected to the tube outlets.

13. Setup is completed if all 96 fractionation tubes are enabled and no entrapped air bubbles are visible in the separation chamber and the sample inlet tubing. Set the separation chamber to a horizontal position.

3.3.2. Quality Tests for Correct Instrument Assembly and Flow Profile

ZE-FFE is a carrier-free electrophoresis. This implies that any disturbance of the laminar buffer stream through the separation device has to be carefully avoided because a direct effect on the separation will occur. Typical problems in the instrument setup are as follows:

- Unevenly tightened clamps.
- Dirt particles, lint or agglomerated protein precipitates in the separation chamber.
- Blocked or defective tubes, either at the fractionation collector or media inlet.
- Entrapped air bubbles.

We therefore recommend carrying out test runs as a quality control for correct instrument setup (*see* **Note 9**).

3.3.2.1. STRIPES TEST

This test reveals disturbances in the uniform laminar buffer stream and should be done after the FFE setup.

1. Prepare a dye solution by adding a few grains of Sulfanilazochromotrop (Sigma-Aldrich, Taufkirchen, Germany) to distilled H_2O.
2. Close the sample inlet with a syringe, stop the media pump.
3. Place the media tubes 2, 4 and 6 in the dye solution and tubes 1, 3, 5, 7 and 8 (corresponding to the counterflow) in distilled H_2O. Do not put the dye solution in tubes 1 or 7 because contamination of the electrode membranes would occur!
4. Restart media pump again. Upon correct setup, alternate colorless and red stripes of identical width and with sharp boundaries will flow in parallel along the separation chamber. Any disturbance of the uniform laminar buffer stream will directly alter this pattern. If the expected stripes are not observed, trouble shooting should be performed after rinsing the separation chamber with distilled water to wash out the dye solution.
5. Stop the media pump, place all media tubes in distilled H_2O and reopen the sample inlet tube.
6. Restart media pump and completely wash out the dye solution of the separation chamber.

3.3.2.2. SAMPLE INLET TEST

In this test, dye solution will enter the separation chamber through the sample inlet and flow in a narrow straight line parallel along the separation chamber. In addition, this test can be used to estimate the time for the sample to be transported from the sample inlet to the fraction collector.

1. Stop the sample pump.
2. Place the sample inlet tube in the dye solution.

3. Restart the sample inlet pump.
4. Upon correct setup, a narrow red stripe flowing as a straight line should reach one to maximum three fractions of the collector depending on the media and sample flow velocity.
5. Stop the sample pump and place the sample inlet tube in distilled H_2O.
6. Completely wash out the dye solution of the separation chamber.

3.3.3. Purification of S. cerevisiae *Mitochondria by ZE-FFE*

1. Sample preparation:

   a. Crude mitochondrial preparations isolated by the differential centrifugation method described above are used for the subsequent ZE-FFE purification. Such preparations can be used immediately or have to be thawed if stored frozen. After thawing or during manipulations, keep mitochondrial suspensions on ice.
   b. Wash the mitochondrial preparations twice in separation medium and pellet again by centrifugation at 4°C, 16000 *g*, for 10 min.
   c. Carefully resuspend the mitochondrial pellet in separation medium at a concentration of 1–3 mg/mL and adjust the sample to the sample inlet pump. The turbidity of the mitochondrial sample at such concentrations allows a visual inspection of the separation; lower concentrations may hamper this. Higher sample concentrations are possible, but agglomeration of the sample should be carefully avoided.

2. Remove the distilled H_2O from the electrode chambers with the electrode solution pump and refill with electrode solution.
3. Turn media pump off. Place the media tubes 1 and 7 in the stabilization medium, tubes 2–6 in separation medium and tube 8 in the counterflow medium. Remove any air bubbles that might remain at the tip of the tubes.
4. Set the rate of media delivery (without counterflow) to 300–400 mL/h.
5. Turn media pump on and rinse the chamber with media for at least 10 min.
6. Set the voltage to 750 V and switch on the high voltage (*see* **Notes 10** and **13**).
7. Wait approximately 10 min for the current to reach a stable minimum. Our typical values lie between 80 and 120 mA.
8. Set the sample pump to a rate of 2 mL/h and turn the sample pump on.
9. Avoid introducing air bubbles through the sample inlet and carefully inspect the entry of the mitochondrial sample solution in the separation chamber. The solution should enter the laminar media stream without clotting or agglomeration and a major line should be visible that deflects towards the anode (*see* **Note 11**).
10. When the sample reaches the end of the separation chamber or a stable sample deflection is observed (usually around 5–10 min), start collecting the sample fractions in a 96-well plate or deep-well plate.
11. From the collected sample fractions, separation profiles can be measured by means of a 96-well plate reader (*see* **Note 12**).

12. Continue to purify the remaining sample. To ensure a stable separation pattern over the whole sample application time, regularly check the sample and the media volumes during the separation run. Inspect the fractionation plate for blocked tubes during the run and avoid introduction of air bubbles (*see* **Note 10**).

13. According to the separation profile corresponding main peaks from the deep well plates can be pooled (*see* **Note 13**). The purified mitochondria can be collected by centrifugation (16,000 *g*, 10 min, 4°C) and used for further experiments (*see* **Subheading 3.3.4.**). An example of a separation profile is shown in **Fig. 3**.

14. At the end of the separation run, switch off the voltage and tilt the separation chamber to a diagonal position.

15. Empty the electrode solution chambers and flush with distilled H_2O. Stop the media and sample pumps, replace the inlet tubes in distilled H_2O and flush the separation chamber for at least 10 min.

16. Place the fractionation plate in a water tray and run the media pump in reverse mode for at least 10 min. Subsequently remove the water tray, open the

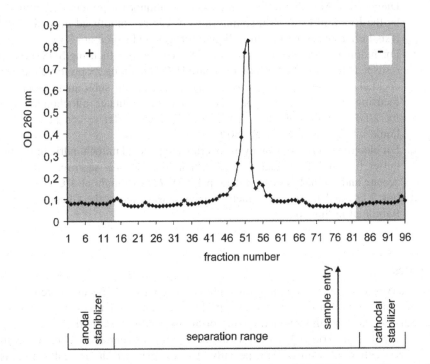

Fig. 3. Zone electrophoresis in a free flow device (ZE-FFE) separation profile of a pre-fractionated mitochondrial preparation. Separation conditions: (1) separation voltage, 750 V; (2) media velocity, 400 mL/h; (3) sample concentration, 3 mg/mL; (4) sample velocity, 2 mL/h and (5) separation temperature, 4°C. Separation was monitored by measuring the OD_{260} of 200 μL aliquots of the collected 96 fractions.

counterflow connection and empty the chamber through the media pump. Empty the electrode solution chambers.

17. Turn all pumps off, tilt the separation chamber to a vertical position and open the separation chamber. Remove spacer, membranes and filter paper strips. Clean the plates with distilled H_2O, isopropanol and distilled H_2O again.

18. Close the chamber with a paper stack between the separation plates. Do not close the clamps. Tilt the separation chamber to a diagonal position and let dry overnight.

3.3.4. Preparation of ZE-FFE-Purified Mitochondria for Further Analysis

1. FFE-purified mitochondrial preparations are sedimented at 16000 g for 10 min.
2. Depending on the subsequent analysis, mitochondrial pellets are treated as follows:

   a. For sodium dodecyl sulphate–polyacrylamide gel electrophoresis (SDS–PAGE) and/or immunoblotting (western) analysis: resuspend the pellet in a small volume of SET buffer containing 1× protease inhibitor cocktail (Roche Diagnostics, Mannheim, Germany). Use an aliquot for protein determination by the Bradford assay. Remaining samples can be solubilized in SDS–PAGE lysis buffer or can be frozen in liquid nitrogen and stored at –80°C.

   b. For bidimensional electrophoresis (2-DE) (isoelectric focusing/SDS–PAGE): resuspend the mitochondria in ultra-pure H_2O containing 2× protease inhibitor cocktail. Sucrose is avoided as it may interfere with subsequent isoelectric focusing. In addition, this hypoosmotic condition facilitates mitochondrial lysis for 2-DE. Samples are solubilized in 2-DE lysis buffer or can be frozen in liquid nitrogen and stored at –80°C.

   c. For electron microscopy analysis, overlay the pelleted mitochondria in fixation buffer (e.g. 4% formaldehyde, 2% glutaraldehyde, 4% sucrose, 2 mM Ca-acetate and 50 mM Na-cacodylate, pH 7.2) *(11)* overnight at 4°C.

   d. For 2-DE and mass spectrometry analysis, we refer to the Chapter 6 by T. Rabilloud in this book.

4. Notes

1. Harvest cells in logarithmic growth phase. Stationary cultures are more resistant towards zymolyase treatment and prolonged incubation times are necessary, thereby decreasing yield and quality of the isolated mitochondria.

2. The pH of the Tris–SO_4 buffer should be 9.4, because enforced reduction of the yeast cell walls with DTT, especially at lower pH, will dramatically decrease the yield and quality of the isolated mitochondria.

3. To avoid strong temperature gradients in the separation chamber, it is of importance to avoid calcinations of the cooling system. Routinely check for appropriate cooling. In addition, we recommend to pre-cool the FFE buffers and to keep them on ice during the run.

4. Although possible in principle, we do not recommend leaving the instrument assembled after the separation but rather to do the setup on a daily basis. This allows a pre-separation check up of the system parts and avoids abrasion of the silicone seals by prolonged pressure.

5. Special care should be taken with the FFE hardware. Inspect the back plate for breakage of glass or ruptures in the backing foil. We recommend regularly checking and calibrating the media pump. In addition, new tubes should be used on a regular basis (i.e. after 6 months or ~200 work hours). This will avoid abrasion of the tubes due to peristaltic pump action and avoid precipitation and bacteria growth in the tubes due to residual buffer.

6. New electrophoresis membranes must be reswollen by overnight incubation in a 1:1 mixture of glycerol/isopropanol. Never let the membranes dry. After use, store them in distilled H_2O with 0.1% formaldehyde.

7. Assemble the wet membranes in correct orientation. The smooth side (i.e. the membrane) should face the electrode chamber seal; the rough side (i.e. the paper backing) should face the filter paper strip. In ZE-FFE, 0.5-mm spacers and 0.8-mm filter paper strips should be used.

8. Avoid the introduction of air bubbles during the media change. Frequently, air bubbles emerge at the tube tips when media are exchanged. A remedy would be to run the media in reverse mode for 2–3 s and then return to normal mode. In case of media change during electrophoretic separation turn off voltage first!

9. We strongly recommend controlling the flow profile after the setup with the stripes test on a daily basis. If this test fails, a consequent unsatisfactory separation is highly likely and remedies should be taken before starting with the separation.

10. Important safety note: the current occurring during separation is high enough to cause severe body harm. It is strongly advised that only visual inspection takes place during an electrophoresis run. In case of problems, switch off power first!

11. ZE-FFE can purify mitochondria significantly. However, if the preceding isolation yielded a high amount of broken and severely damaged organelles, large-scale clotting will occur and the whole procedure will fail. It is therefore important to handle the mitochondria with care: in particular, resuspension should avoid strong shear forces and mitochondria should be kept on ice during the manipulations. If clotting is visible in the starting sample for ZE-FFE, a short spin (e.g. 500 g, 3 min, 4°C) can remove larger aggregates.

12. The progress of the separation should be monitored during the run by collecting 96-well plates and measuring parameters such as optical density, absorption of mitochondrial DNA, protein content, enzymatic activities, etc. by means of an appropriate reader. As the volume in the 96 wells will vary, it is advisable to use a multi-channel pipette and transfer equal volumes (e.g. 200 µL) of the collected fractions to a separate measuring plate.

13. One critical issue that has frequently been discussed is the choice of the two major electrophoresis parameters, i.e. the separation voltage and the media flow velocity. Here, we would like to give the advice to choose these parameters such that the

major peak containing the purified mitochondria would lie in the middle area of the separation profile. Routinely, we use 750 V and a media velocity of 300–400 mL/h. In our experience, an increase of the voltage will only shift the mitochondria more towards the anode but not increase separation efficiency. On the other hand, a decrease in media flow will also cause an anodal shift and will additionally lead to significant peak broadening, thereby hampering organelle purification.

Acknowledgments

The authors thank Dr. E. E. Rojo for critical reading of the manuscript. We also thank Prof. Dr. A. Borst and D. Büringer for their support with electron microscopy. This project was funded by the German Federal Ministry for Education and Research (BMBF) (FKZ: 031U208E, subproject 3; NGFN2 SMP Proteomics, FKZ: 01GR0449, subproject 9, 'Human Brain') and the European Union by EU grant *LSHG-CT-2003-50520* (INTERACTION PROTEOME).

References

1. Voet, D. and Voet, J. G. (1995) *Biochemistry*, J. Wiley & Sons, Inc.
2. Reichert, A. S. and Neupert, W. (2004) Mitochondriomics or what makes us breathe. *Trends Genet.* **20**, 555–562.
3. Prokisch, H., Andreoli, C., Ahting, U., Heiss, K., Ruepp, A., Scharfe, C., et al. (2006) MitoP2: the mitochondrial proteome database–now including mouse data. *Nucleic Acids Res.* **34**, D705–D711.
4. Prokisch, H., Scharfe, C., Camp, D. G., II, Xiao, W., David, L., Andreoli, C., et al. (2004) Integrative analysis of the mitochondrial proteome in yeast. *PLoS Biol.* **2**, e160.
5. Herrmann, J. M., Fölsch, H., Neupert, W., and Stuart, R. A. (1994) Isolation of yeast mitochondria and study of mitochondrial protein translation. In *Cell Biology: A Laboratory Handbook* (Celis, D. E., ed.), Academic Press, San Diego. Vol. 1, pp. 538–544.
6. Meisinger, C., Sommer, T., and Pfanner, N. (2000) Purification of Saccharomcyes cerevisiae mitochondria devoid of microsomal and cytosolic contaminations. *Anal. Biochem.* **287**, 339–342.
7. Zischka, H., Weber, G., Weber, P. J., Posch, A., Braun, R. J., Buhringer, D., et al. (2003) Improved proteome analysis of Saccharomyces cerevisiae mitochondria by free-flow electrophoresis. *Proteomics* **3**, 906–916.
8. Hannig, K. and Wrba, H. (1964) Isolation of vital tumor cells by carrier-free electrophoresis. *Z. Naturforsch.* **19**, 860.
9. Hannig, K. and Heidrich, H. G. (1990) *Free-Flow Electrophoresis*. GIT Verlag, Darmstadt.
10. Ericson, I. (1974) Determination of the isoelectric point of rat liver mitochondria by cross-partition. *Biochim. Biophys. Acta* **356**, 100–107.
11. Lewis, P. R. and Shute, C. C. (1969) An electron-microscopic study of cholinesterase distribution in the rat adrenal medulla. *J. Microsc.* **89**, 181–193.

5

Preparation of Respiratory Chain Complexes from *Saccharomyces cerevisiae* Wild-Type and Mutant Mitochondria

Activity Measurement and Subunit Composition Analysis

Claire Lemaire and Geneviève Dujardin

Summary

The mitochondrial oxidative phosphorylation involves five multimeric complexes imbedded in the inner membrane: complex I (Nicotinamide Adenine Dinucleotide (NADH) quinone oxidoreductase), II (succinate dehydrogenase), III (ubiquinol cytochrome *c* oxido reductase or *bc1* complex), IV (cytochrome *c* oxidase), and V (ATP synthase). These respiratory complexes are conserved from the yeast *Saccharomyces cerevisiae* to human with the exception of complex I, which is replaced by three NADH dehydrogenases in *S. cerevisiae*. Here, we provide several protocols allowing an exhaustive characterization of each yeast complex: this chapter describes procedures from mitochondria preparation to measurement of the activity of each complex and analysis of their subunit composition and provides information on the interactions between different complexes.

Key Words: *S. cerevisiae*; mitochondria; respiratory complexes; BN–PAGE; respiratory activities.

1. Introduction

The main function of mitochondria is to generate energy by the oxidative pathway that involves five oligomeric complexes (complexes I–V). These complexes are composed of numerous constitutive subunits, including membrane and extrinsic proteins that are encoded either by the nuclear or the mitochondrial genome. In addition, these complexes contain co-factors like

From: *Methods in Molecular Biology, vol. 432: Organelle Proteomics*
Edited by: D. Pflieger and J. Rossier © Humana Press, Totowa, NJ

hemes, Fe/S clusters, or metallic ions. Mitochondria from all eukaryotes show common features in their biogenesis and in particular the structural organization of the respiratory complexes seems to be conserved through evolution. A defect in the phosphorylation process has been shown in humans to be the primary cause of many neuromuscular or generalized syndrome diseases. The same defect in yeast is not lethal as the cells are able to satisfy their energy requirement by fermentation. Thus, *Saccharomyces cerevisiae* is an excellent model to study respiratory functions. Numerous yeast mutants have been isolated, and in many cases, a mutation displays pleiotropic effects. To characterize the respiratory defects in respiratory-deficient mutants, we use several protocols on purified mitochondria to evaluate the activity of the different complexes of the respiratory chain *in vitro*. Some of these tests can be performed on complexes separated by blue native gel electrophoresis (BN–PAGE), which can then be enzymatically digested for a subsequent direct analysis by mass spectrometry.

Mitochondria are prepared routinely by differential centrifugations (*see* **Subheading 3.1.**) and for some instances further purified on sucrose gradient according to reference *(1)*. Analysis of their heme content can be performed by spectrometric analysis or heme stain after gel electrophoresis (*see* **Subheading 3.2.**). *In vitro* activity assays for complexes III, IV, and V are described in **Subheading 3.3**. The activity assay for complex II has been developed in reference *(2)*, and activities measuring the electron transfer from complexes II to III and NADHases to complex III can also be monitored *(3)*. Fractionation of mitochondria can be performed to separate soluble proteins and membrane proteins by alkali treatment (*see* **Subheading 3.4.**). Other methods have been devised to enable the separation of inner membrane from outer membrane and isolation of the mitoplasts *(4)*. Finally, analysis of each complex in its native form and of interactions between complexes can be performed using BN–PAGE (*see* **Subheading 3.6.**), a method initially established in reference *(5)*. In-gel activities can also be performed for complexes II, IV, and V (*see* **Subheading 3.6.4.**), whereas no test is yet available for complex III, the analysis of which requires the use of immunoblot with a specific antibody able to recognize the complex in its native form.

We have mainly developed these methods on *S. cerevisiae*, but for complex I that is absent in *S. cerevisiae*, we have used the filamentous fungi *Podospora anserina (6)*.

Mitochondrial proteins of *S. cerevisiae* separated by sodium dodecyl sulphate–polyacrylamide gel electrophoresis (SDS–PAGE) (*see* **Subheading 3.1.**) or respiratory complexes of *S. cerevisiae* separated by BN–PAGE (*see* **Subheading 3.6.**) can be further analyzed by LC-MS/MS *(7,8)* (Guillot A. et al., in preparation).

2. Materials

2.1. Cell Culture

The yeast media were described in reference *(9)*.

1. Respiratory-competent cells are grown in lactate medium: 0.75% (w/w) yeast extract, 0.75% (w/w) bactopeptone, 0.5% (v/v) lactic acid, and 0.02 mg/mL of adenine, pH adjusted to 4.5 with 3 M KOH.
2. Respiratory-deficient cells are grown in YPGAL medium: 1% (w/w) yeast extract, 1% (w/w) bactopeptone, 2% (w/w) galactose, and 0.1% (w/w) glucose. 20 µg/mL of adenine is then added after sterilization of the medium.

The cells are grown in liquid medium with shaking (130–180 rpm) at 28°C.

2.2. Preparation of Mitochondria by Differential Centrifugations

1. Sorbitol buffer A: 1.2 M sorbitol, 50 mM Tris–HCl, pH 7.5, 10 mM ethylenediaminetetraacetic acid (EDTA), and 0.3% (v/v) 2-mercaptoethanol. Stored at 4°C.
2. Sorbitol buffer B: 0.7 M sorbitol, 50 mM Tris–HCl, pH 7.5, and 0.2 mM EDTA. Stored at 4°C.
3. Zymolyase-100T (Seikagaku America, Rockville, MD).

2.3. In vitro Activity Assays of the Respiratory Chain Complexes

2.3.1. Ubiquinol c Oxido Reductase (Complex III) and Cytochrome c Oxidase (Complex IV) Activities

1. Phosphate buffer: 200 mL of 100 mM phosphate buffer, pH 7.4, is prepared by mixing 19 mL of 0.2 M KH_2PO_4 and 81 mL of 0.2 M K_2HPO_4 and finally bringing the volume to 200 mL with water.
2. Assay buffer: the phosphate buffer is diluted with an equal volume of H_2O and EDTA is added at a final concentration of 50 µM. This buffer must be filtered before use.
3. Oxidized cytochrome *c*: 50 mg of horse heart cytochrome *c* (Sigma) in oxidized form is prepared in 500 µL of assay buffer and stored at –20°C. The concentration is measured spectrophotometrically at 550 nm with an extinction coefficient ε_{ox} of 8/mM/cm *(10)* ($C = A_{550\ nm}/\varepsilon_{ox} \times l$). *l* is the path length.
4. Reduction of cytochrome *c*:
   a. 60 mg of horse heart cytochrome *c* prepared in 0.5 mL of assay buffer (as above) is reduced by addition of 5 mg of ascorbate. Mix by inverting the tubes.
   b. Reduced cytochrome *c* is loaded on a gel filtration chromatography column (Sephadex G25) equilibrated in assay buffer to eliminate ascorbate. Elution with assay buffer enables purification of a red-colored fraction containing cytochrome *c* (about 2 mL), which is stored in a dark bottle under argon at –70°C.

c. The level of reduction of cytochrome c is systematically checked by recording the spectra at 550 nm with (A^1) and without ascorbate (A^0). The concentration of cytochrome c in reduced form is calculated as follows:

$$C_{red}(mM) = \frac{A^0_{550nm} - \varepsilon_{ox} \frac{A^1_{550nm}}{\varepsilon_{red}}}{\Delta\varepsilon_{550}}, \tag{1}$$

where ε_{red} = 27.6/mM/cm, ε_{ox} = 8/mM/cm, and $\Delta\varepsilon_{550}$ = ε_{red} − ε_{ox} = 19.6/mM/cm (*10*).Finally, C_{red} (mM) = A^0_{550nm} − [8× (A^1_{550nm}/27.6]/19.6.

5. Quinone reduction: decylubiquinone (DB) (Sigma) is reduced chemically to decylubiquinol (DBH$_2$) as described in reference (*11*) and kept at –70°C in ethanol.

   a. Dissolve 25 mg of DB by adding directly 5 mL of ethanol to the flask containing the DB powder.
   b. In a screw tube, put 15 mL of filtered phosphate buffer containing 0.25 M sorbitol. Add the DB-containing ethanol solution.
   c. Add 2 mL of cyclohexane (work under a fume hood).
   d. Close the tube and shake it slightly. DB goes into the upper organic phase, cyclohexane.
   e. Add a few milligrams of potassium borohydride to reduce quinones. Shake well. Open the cap gently, so that gas can be released. There is a change of color from yellow to color-free.
   f. Decant the organic phase in a second tube containing 15 mL of phosphate buffer containing 0.25 M sorbitol.
   g. Further add 2 mL of cyclohexane. Mix and allow to settle for a few minutes.
   h. Transfer the organic phase in a light-tight tube and bubble argon through the solution to evaporate cyclohexane.
   i. Add 1 mL of ethanol and acidify with two drops of 0.1 M HCl to stabilize the quinones.
   j. Keep at –70°C under argon.

6. Antimycin solution (*see* **Note 1**): 6 mg of antimycin A (Sigma) is dissolved in 11.4 mL of ethanol to prepare a 1 mM solution that is kept at –20°C. A 20 μM antimycin final concentration is used to inhibit complex III activity.
7. Potassium Cyanide (KCN) solution (*see* **Note 2**): 1 M solution is prepared by dissolving 65 mg of KCN (Aldrich) in 700 μL of distilled water. 4 mM KCN final concentration is used to inhibit cytochrome c oxidase activity.

2.3.2. ATPase (Complex V) Activity

1. Reaction buffer: 0.2 M KCl, 3 mM MgCl$_2$, and 10 mM Tris–HCl, pH 8.4.
2. Assay solution (*see* **Note 3**):

   a. Prepare 600 mL of 0.37 M H$_2$SO$_4$.

    b. Dissolve 3.3 g of ammonium molybdate in 100 mL of the 0.37 M sulfuric acid solution.

    c. Dissolve 4 g of ferrous sulfate in 100 mL of the 0.37 M sulfuric acid solution.

    d. Dilute the ammonium molybdate up to 380 mL with the 0.37 M sulfuric acid solution.

    e. Add the ferrous sulfate solution and complete to 500 mL with the 0.37 M sulfuric acid solution.

3. Oligomycin solution: Stock solution at 0.6 mg/mL of oligomycin (Sigma) is prepared in ethanol and can be stored at –20°C. Oligomycin is used at a final concentration of 10 μg/mg protein to inhibit ATPase activity.

2.4. Isolation of Membrane and Soluble Proteins from Mitochondria (Alkali Treatment)

1. Extraction buffer: 0.1 M Na_2CO_3, pH 11.5, 5 mM dithiothreitol (DTT), 1 mM phenylmethylsulfonyl fluoride (PMSF), 1 μg/mL pepstatin, 10 μg/mL chymostatin, 10 μg/mL antipain, and 1 μg/mL leupeptin (see **Subheading 2.4.2**). This buffer is made extemporaneously and the pH adjusted with NaOH to pH 11.5.

2. Protease inhibitor stock solutions are prepared as follows: PMSF (Sigma) is dissolved in ethanol at a concentration of 100 mM and stored at 25°C for up to 9 months; pepstatin (Sigma) is dissolved in ethanol at 1 mg/mL and stored in aliquots at –20°C for up to 1 month; antipain (Sigma) is dissolved in H_2O at 1 mg/mL and stored in aliquots at –20°C for up to 1 month; and leupeptin (Sigma) is dissolved in H_2O at 10 mg/mL and stored in aliquots at –20°C for up to 6 months.

3. Solubilization solution: 2% (w/v) SDS, 0.8% (v/v) β-mercaptoethanol, 20% (v/v) glycerol, 50 mM Tris–HCl, pH 6.8, and 0.02% (w/v) bromophenol blue.

2.5. Immunoblot Analysis

1. Transfer buffer: 50 mM Tris, 386 mM glycine, 0.1% (w/v) SDS, and 20% (v/v) ethanol.

2. Reinforced nitrocellulose membrane from Schleicher and Schuell: 0.45 μm and 3 MM Chr chromatography paper from Whatman.

3. Phosphate-buffered saline (PBS) solution (10×): 1.37 M NaCl, 26.8 mM KCl, 0.1 M Na_2HPO_4, and 14.7 mM KH_2PO_4. Store at room temperature. Dilute 100 mL of this stock solution with 900 mL of water for use.

4. PBS-T solution: add 0.1% (w/v) of Tween-20 to the 1× PBS solution.

5. Blocking solution: add 3% (w/v) of no-fat dry milk to the PBS-T solution.

6. Secondary antibody: anti-mouse IgG or anti-rabbit IgG conjugated to horseradish peroxidase (Promega).

7. Enhanced chemiluminescent substrate: SuperSignal Substrate from Pierce.

8. Detection: place the blot against a film (Hyperfilm ECL; Amersham Biosciences) and expose or use an imaging system (LAS3000, Fuji).

2.6. Separation of Respiratory Complexes by Non-Denaturing Gels (BN–PAGE)

1. Acrylamide/Bis solution: 40% solution 29:1 (Bio-Rad).
2. Stock solutions: 2 M ε-amino-*n*-caproic acid (Sigma) and 500 mM Bis-Tris (Sigma) solutions that are each filtered on Millipore 0.45-μm filters.
3. Gel buffer (3×): 1.5 M ε-amino-*n*-caproic acid and 150 mM Bis-Tris. The pH is adjusted to 7 with HCl at 4°C (by placing the buffer on ice). The buffer is filtered on Millipore 0.45-μm filters.
4. BN sample buffer: 750 mM ε-amino-*n*-caproic acid, 50 mM Bis-Tris, and 0.5 mM EDTA. The buffer is filtered on Millipore 0.45-μm filters, then 5% (w/v) Serva Blue G-250 is added. The buffer is sonicated in a bath sonicator (Branson 2510) to solubilize the Serva Blue G-250.
5. Ammonium persulfate (APS): prepare a 10% (w/v) solution in water and store at 4°C for up to 1 month.
6. Gradient gel: volumes are indicated in **Table 1** for the casting of two mini-gels. The gradient gel is cast by mixing 4 mL of 5% acrylamide (or 6%) and 4.6 mL of 10% acrylamide (or 15%) using a gradient mixer.
7. Stacking gel (4%): Mix 0.5 mL of acrylamide/Bis solution, 1.67 mL of gel buffer (3×), 2.83 mL of H_2O, 55 μL of 10% (w/v) APS, and 5.5 μL of TEMED.
8. Anode buffer: 50 mM Bis-Tris. The pH is adjusted to 7.0 with 10 M HCl at 4°C. The buffer is filtered on Millipore 0.45-μm filters.
9. Cathode buffer A: 15 mM Bis-Tris and 50 mM Tricine. The pH is adjusted to 7.0 with a concentrated HCl solution at 4°C. The buffer is filtered on Millipore 0.45-μm filters, then 0.02% (w/v) Serva Blue G-250 is added and the buffer is sonicated as above.
10. Cathode buffer B: same as cathode buffer A except that no Serva Blue G-250 is added.
11. Molecular weight markers: High Molecular Weight Calibration Kit for Native Electrophoresis (Amersham Biosciences).

Table 1
Solutions to be Mixed for Casting BN–PAGE Gels

	5%	6%	10%	15%
Acrylamide/bis solution	0.63 mL	0.75 mL	1.25 mL	1.87 mL
Gel buffer (3×)	1.67 mL	1.67 mL	1.67 mL	1.67 mL
Glycerol	–	–	0.5 g	0.5 g
H_2O	2.7 mL	2.58 mL	1.79 mL	1.17 mL
Total volume	5 mL	5 mL	5 mL	5 mL
APS 10%	30 μL	27.5 μL	22.5 μL	15 μL
TEMED	3 μL	2.75 μL	2.25 μL	1.5 μL

2.7. In-Gel Activities

1. Complex II—reaction solution: 50 mM potassium phosphate buffer, pH 7.4, 84 mM succinic acid, 0.2 mM phenazine methasulfate (*see* **Note 4**), 2 mg/mL blue nitrotetrazolium, 4.5 mM EDTA, and 10 mM KCN (keep the solution in darkness).
2. Complex IV—reaction solution: in 9 mL phosphate buffer (0.05 M, pH 7.4) dissolve 5 mg of 3,3′-diaminobenzidine tetrahydrochloride (DAB) (*see* **Note 5**), 200 µL of catalase at a concentration of 20 µg/mL, 10 mg of cytochrome c, and 750 mg of sucrose.
3. Complex V—reaction solution: Prepare 100 mL of solution containing 50 mM glycine-NaOH, pH 8.4, 5 mM $MgCl_2$, 0.05% (w/v) lead nitrate, and 0.1% (w/v) Triton X-100.

2.8. Western Blotting of Blue Native Gels

Transfer buffer: 25 mM Tris, 192 mM glycine, 0.02% (w/v) SDS, and 20% (v/v) ethanol.

3. Methods

3.1. Preparation of Yeast Mitochondria by Differential Centrifugations

This protocol is adapted from reference (*12*).

1. Cells are harvested in mid-exponential phase, rinsed with 10 mM EDTA, and resuspended in sorbitol buffer A at 3 mL/g wet mass of cells.
2. The cell wall is digested enzymatically with Zymolyase-100T at 1 mg/g of cells at 37°C for 30 min.
3. The spheroplasts are harvested by centrifugation at 1800 *g* for 15 min at 4°C and resuspended in ice-cold sorbitol buffer B (15 mL when starting from a culture of 1 L).
4. Cell fragments are subsequently removed by centrifugation at 2500 *g* for 15 min at 4°C. Keep the supernatant.
5. Mitochondria are pelleted by centrifugation at 20,000 *g* for 15 min at 4°C.
6. Re-suspend the pellet of mitochondria in 4 mL of ice-cold sorbitol buffer B containing protease inhibitors (EDTA-free cocktail from Boehringer Mannheim). Centrifuge at 800 *g* for 5 min at 4°C. Decant and centrifuge the mitochondria-containing supernatant at 15,000 *g* for 15 min at 4°C. Re-suspend the pellet in 4 mL of ice-cold sorbitol buffer B. This cycle of two successive centrifugations is repeated twice to wash the mitochondrial sample.
7. Mitochondrial proteins are analyzed by standard SDS–PAGE or other methods (see following sections).

3.2. Characterization of Heme Content

3.2.1. Whole Cells Cytochrome Spectra

This method has been originally developed in reference (13).

1. Yeast cells are grown on plates, collected with a spatula, and dried between two filters (Durieux, n. 111, 90 m/m).
2. The cell paste is transferred to a home-made cuvette (1 mm width) [see reference (13)]. Spectra are then recorded on the dry cells at liquid nitrogen temperature (−196°C) between 490 and 630 nm using a Cary 400 spectrophotometer (Varian). Generally, we analyze the spectra after complete reduction of cytochromes. This is achieved by mixing about 3–5 mg of sodium dithionite into the paste. The α-peak of cytochromes c, c1, b, and aa3 are detected at 546, 552, 558, and 602 nm, respectively.

Cytochrome spectra of wild-type and mutant cells are shown in **Fig. 1A**.

3.2.2. In-Gel Detection of Type c Hemes

Mitochondrial proteins are separated by Lithium Dodecyl sulfate LDS–PAGE according to reference (14). Detection of c-type cytochromes is performed on the nitrocellulose membrane as described in reference (15) using the heme-dependent peroxidase activity.

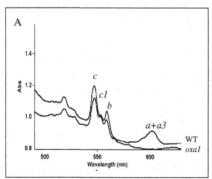 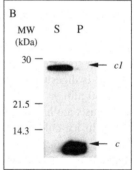

Fig. 1. Cytochrome spectra and in-gel detection of type-c hemes. (A) Low-temperature cytochrome absorption spectra of wild-type and *oxa1* cells grown on galactose medium. Absorption maxima of the α-bands of cytochromes c, c1, b, and aa3 are 546, 552, 558, and 602 nm, respectively. In the *oxa1* mutant (24), cytochrome aa3 is undetectable and cytochrome b is reduced. (B) Wild-type mitochondria were treated with sodium carbonate, and the supernatant (S) and pellet (P) were migrated separately on 12% LDS–PAGE gels and transferred to nitrocellulose. Detection of c-type cytochromes reveals cytochrome c in the soluble fraction and cytochrome c1 in the pellet.

1. Mitochondria proteins are resuspended in 2% LDS, 5% sucrose, and 0.02% bromophenol blue and incubated for 30 min at 4°C. LDS–PAGE is performed as a classical SDS–PAGE except that SDS is substituted by LDS.
2. After electrophoresis, the proteins are transferred onto a nitrocellulose membrane (*see* **Subheading 3.5.1.**).
3. After the transfer, the membrane is rinsed in PBS buffer for 5 min and kept wet until being incubated with the chemiluminescence western blotting detection system for 5–10 min (Super Signal West Pico Chemiluminescent Substrate from Pierce).

A result of direct detection of type *c* hemes in wild-type cells is shown in **Fig. 1B**.

3.3. In Vitro Activity of Respiratory Complexes

3.3.1. Ubiquinol c Oxido Reductase Activity

This protocol is adapted from reference *(16)*.

Ubiquinol *c* oxido reductase is assayed by the rate of reduction of cytochrome *c* by ubiquinol measured spectrophotometrically at 550 nm at 30°C.

1. Mitochondria are diluted in assay buffer (1–2 mg/mL mitochondrial proteins) and kept on ice.
2. In a cuvette, add assay buffer containing 4 mM KCN, 120 µM oxidized horse heart cytochrome *c*, and 10–20 µg of mitochondria.
3. Start the reaction by the addition of 120–160 µM DBH$_2$. The increase in absorbance at 550 nm is generally followed for 30 s, then antimycin 20 µM is added to test the specificity of the signal. The specific activity is expressed in nanomoles of cytochrome *c* in reduced form/minute/milligram of protein and is calculated from the Beer–Lambert law equation: $C = \Delta A_{550nm} /(\Delta \varepsilon_{550nm} \times l)$, where *C* is the concentration of reduced cytochrome *c*, ε the extinction coefficient, and *l* the path length. $\varepsilon_{550nm} = 19.6/\text{mM/cm}$ *(10)*.

3.3.2. Cytochrome c Oxidase Activity

Cytochrome *c* oxidation is followed by recording the absorbance at 550 nm at 30°C as described in reference *(17)*. The decrease in absorbance at 550 nm is generally followed for 30 s. The activity is expressed in nanomoles of cytochrome *c* in oxidized form/minute/milligram of protein and is calculated from the Beer–Lambert law equation: $C = -\Delta A_{550\ nm}/(-\Delta \varepsilon_{550nm} \times l)$, where *C* is the concentration of oxidized cytochrome *c*, ε is the extinction coefficient, and *l* is the path length. $\Delta \varepsilon_{550\ nm} = 19.6/\text{mM/cm}$.

1. Mitochondria are diluted in assay buffer (1–2 mg/mL mitochondrial proteins) and kept on ice.

2. In a cuvette, add assay buffer containing 30 μM reduced horse heart cytochrome *c* (final concentration).
3. Start the reaction by adding the mitochondria (10–20 μg). The enzymatic rate of cytochrome *c* oxidation is linear for generally 1 min in these conditions, and we check the specificity of the signal by adding 4 mM KCN final concentration that totally inhibits the reaction.

3.3.3. ATPase Activity

This protocol has been developed in the laboratory of Dr. J. Velours (I.B.G.C., Bordeaux, France) according to reference *(18)*. It is based on the measurement of the phosphate produced by hydrolysis of ATP through ATPase activity. The phosphate released by hydrolysis is measured by a colorimetric assay. Phosphate readily reacts with ammonium molybdate in the presence of suitable reducing agents to form a blue-color complex, the intensity of which is directly proportional to the concentration of phosphate in the solution. The mitochondrial ATPase activity is inhibited in the presence of oligomycin.

1. Put 900 μL of reaction buffer in microfuge tubes and incubate at 30°C (water bath).
2. Add 0.3 mg of mitochondria with or without oligomycin (10 μg of oligomycin per milligram of protein is sufficient to inhibit ATPase activity). Incubate for 2 min at 30°C.
3. At $t = 0$, add 50 μL of 0.1 M ATP, pH 7.0. The reaction is linear in these conditions for about 40 s.
4. At $t = 30$ s, stop the reaction by addition of 100 μL of 3 M Trichloroacetic acid (TCA). Incubate at 4°C.
5. After addition of 3 M TCA, the tubes are centrifuged for 10 min at 9000 *g* at 4°C.
6. Add 500 μL of the supernatant to 2.5 mL of assay solution.
7. Incubate for 15 min at ambient temperature.
8. Read the absorbance at 610 nm.
9. Prepare standard solutions of inorganic phosphate by introducing 0, 50, 100, 150, 200, or 300 μL of 1 mM potassium phosphate (KH_2PO_4) into tubes, which are completed with water up to a final volume of 500 μL. Add 2.5 mL of assay solution to the standard solutions and measure OD at 610 nm. The standard curve is used to determine the amount of phosphate released through ATP hydrolysis. The ATPase activity is expressed in micromoles of Pi/min/mg of proteins.
10. One tube containing 900 μL of reaction buffer, mitochondria, 100 μL of 3 M TCA, and 50 μL of 0.1 M ATP constitutes the zero point. This OD value will be deduced from the OD values given by the hydrolysis tubes to obtain the amount of phosphate released from ATP.

3.4. Isolation of Membrane and Soluble Mitochondrial Proteins from Mitochondria by Alkali Treatment

Numerous protocols have been described to separate extrinsic proteins from membranes. They are based on extraction by chaotropic agents, salts, or alkaline pH. These treatments are known to destabilize the hydrating shell of the proteins and diminish either hydrophobic (chaotropic agents) or electrostatic interactions (salts and alkaline pH). Here, we present the protocol using alkaline pH that we use, adapted from reference *(19)*.

1. 150 µg of mitochondria in sorbitol buffer B is centrifuged for 15 min at 10,000 g at 4°C.
2. The supernatant is discarded and the pellet is resuspended in 300 µL of extraction buffer.
3. The samples are incubated on ice for 30 min with occasional shaking by inverting the tubes four times.
4. Centrifuge for 1 h at 4°C at 100,000 g.
5. The supernatant (S1), containing soluble proteins and proteins loosely bound to the membranes, is transferred to a new tube and then centrifuged for 1 h at 100,000 g at 4°C. The pellet (P2) is discarded and the supernatant (S2) is precipitated with 10% TCA final concentration, with an incubation of 15 min at 4°C, and is then centrifuged at 20,000 g for 15 min at 4°C. The pellet is washed in 1 mL of cold acetone (pre-cooled at –20°C). After evaporation of acetone in free air, the pellet is resuspended in 30 µL of solubilization solution.
6. The pellet (P1), which contains the hydrophobic proteins, is resuspended in 300 µL of extraction buffer and centrifuged for 1 h at 100,000 g at 4°C. After centrifugation, the supernatant (S3) is discarded and the pellet (P3) is resuspended in 30 µL of solubilization solution.
7. Apply samples (S2 or P3) to SDS–PAGE or LDS–PAGE. Reveal either by western blot (SDS–PAGE) or by direct heme detection (LDS–PAGE).

Examples of alkaline treatments of wild-type or mutant mitochondria followed by either heme detection or immunodetection with specific antibodies are shown in **Figs. 1B and 2**, respectively.

3.5. Immunoblot Analysis

1. For western blotting, mini-gels (8.3 × 6 × 0.075 cm) are transferred at 150 mA constant for 1.5 h at 4°C. Then all the steps of immunodetection are realized on a rocking platform.
2. The nitrocellulose membrane is incubated in 30 mL of blocking buffer for 1 h at room temperature.
3. The blocking buffer is discarded and the primary antibody diluted in the blocking solution is added and generally incubated for 1 h at room temperature.

4. The primary antibody is then removed and the membrane washed once for 10 min and twice for 5 min with 30 mL of PBS-T solution each time.
5. The membrane is then incubated with the secondary antibody diluted in the blocking solution for 1 h at room temperature.
6. The secondary antibody is discarded and the membrane washed once for 10 min and twice for 5 min with 30 mL PBS-T solution each time.
7. The blot is then incubated in the Super Signal Substrate working solution (100 µL/cm^2) for 5 min (*see* **Note 6**). The blot must be completely wetted with substrate and is kept in the dark during incubation.
8. Remove the blot from the solution and place it in a plastic sheet taking care to remove all the bubbles between the blot and the plastic sheet.

Examples of immunoblots are presented in **Fig. 2**.

3.6. Isolation of Respiratory Complexes from Non-Denaturing Gels (BN–PAGE)

This method was initially developed in reference *(5)* and adapted to mini-gel system in reference *(20)*. It is illustrated in **Fig. 3**.

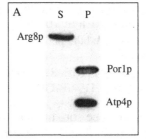
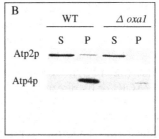

Fig. 2. Carbonate extraction of mitochondrial proteins analyzed by immunoblotting. Mitochondrial proteins were treated with sodium carbonate as in **Fig. 1B**. The mitochondrial proteins were detected by immunoblotting with specific antibodies. (**A**) The soluble matrix protein Arg8p was found in the supernatant (S), whereas the outer-membrane protein Por1p and the inner membrane protein Atp4p were found in the pellet (P). (**B**) In wild-type cells, the soluble F1 subunit of ATPase, Atp2p, is mainly present in the supernatant, whereas the F0 membrane-bound subunit, Atp4p, is found in the pellet. In the Δ*oxa1* mutant *(25)*, the steady state level of Atp4p is drastically decreased and Atp2p is only found in the supernatant. Antibodies are from T. D. Fox (Arg8p; Ithaca, NY), J. Velours (Atp2p and Atp4p; Bordeaux, France), and Molecular Probes, Eugene, Oregon, USA (Por1p).

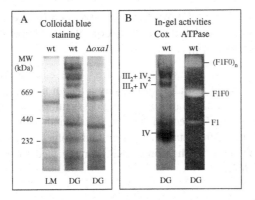

Fig. 3. BN–PAGE of mitochondrial proteins of wild-type and mutant strains. Wild-type and Δ*oxa1* mitochondrial extracts were separated on BN–PAGE after either digitonin (DG) or laurylmaltoside (LM) solubilization. (**A**) Complexes were detected with colloidal blue staining. Note the disappearance of high supramolecular complexes in the Δ*oxa1* mutant. (**B**) In-gel cytochrome *c* oxidase (Cox) and ATPase activities were used to localize the respective complexes in the wild-type strain. Cytochrome *c* oxidase was detected either in association with complex III or alone (IV$_2$ and III$_2$: dimers). ATPase activity was detected in the bands corresponding to oligomers (F1F0)n, monomer F1F0, or the F1 sector alone.

3.6.1. Solubilization of Mitochondria

Mitochondria are solubilized using detergents like laurylmaltoside or digitonin to analyze either the complexes individually *(5)* or the supra-molecular association, respectively *(21,22)*.

1. Resuspend mitochondria in 3× gel buffer containing 1 mM PMSF and 5 mM MgCl$_2$ at 1 mg of protein per mL.
2. Add a solution of DNase (Sigma) at a final concentration of 0.25 mg/mL. Incubate for 30 min at 25°C.
3. Centrifuge at 53,000 *g* for 30 min.
4. Discard the supernatant.
5. Add detergent to the pellet at a final concentration of 2%. Mix well and leave for 30 min at 4°C.
6. Centrifuge at 100,000 *g* for 15 min at 4°C.
7. Add BN sample buffer to the supernatant at a final concentration of 0.25%.
8. Apply samples to the BN–PAGE.

3.6.2. Preparation of BN-PAGE

To separate the respiratory chain complexes from *S. cerevisiae*, we use two types of gradient gels. A gradient gel of 5–10% acrylamide is routinely used to analyze high-molecular-weight complexes and a gradient gel of 6–15% acrylamide for the low-molecular-weight complexes (<200 kDa). Gel dimensions are 8.3 × 6 × 0.075 cm.

1. Gels are polymerized at room temperature. After casting the gradient gel, the stacking gel is poured just after or one day after. In this latter case, the gradient gel is overlaid by water, wrapped to avoid dessication, and kept at 4°C.
2. Electrophoresis is performed for 1 h at 100 V and then for 2 h at 200 V at 4°C. Gels are then either stained with colloidal blue, or transferred to nitrocellulose membrane (*see* **Subheading 3.6.3.**), or revealed for in-gel activities (*see* **Subheading 3.6.4.**).

3.6.3. Western Blotting

When BN–PAGE is used for immunoblot analysis, we change the cathode buffer A by cathode buffer B (without Serva Blue G-250) after 1 h at 200 V to remove excess dye. Then the BN–PAGE is transferred to nitrocellulose at 100 V for 1.5 h at 4°C.

3.6.4. In-Gel Activities of Complexes II, IV, and V

In-gel activities have been reported in reference *(20)* for complexes II and IV and in reference *(23)* for complex V.

1. Complex II: The gel is incubated in the dark in the reaction solution. A violet coloration will develop within 15–30 min of incubation. This coloration corresponds to the reduction of the tetrazolium salt that forms an insoluble violet precipitate.
2. Complex IV: The gel is incubated in the reaction solution for 1 h at room temperature. Then the reaction solution is discarded and substituted with water. The bands displaying complex IV activity appear brown colored as the oxidized diaminobenzidine precipitates (*see* **Fig. 3**). The coloration will intensify and appear with a better contrast if the gel is soaked in water during the night at 4°C.
3. Complex V: The gel is incubated for 15 min with 50 mL of reaction solution and then overnight at 4°C with 50 mL of the same solution containing 4 mM ATP. The inorganic phosphate obtained by the hydrolysis of ATP forms a white precipitate (*see* **Fig. 3**).

3.7. Conclusion

These various methods provide powerful tools to analyze the defect in respiratory complex assembly due to various mutations in the genes encoding the respiratory subunits or extrinsic assembly factors. We have shown that LC-MS/MS analysis can be performed on mitochondrial proteins separated on SDS–PAGE and BN–PAGE. In particular, the analysis from BN–PAGE allows the identification of subunit composition of super-complexes or sub-complexes that accumulate in assembly mutants (Guillot et al., in preparation).

4. Notes

1. Be careful, this product is extremely toxic.
2. Be careful, this product is extremely toxic. The aqueous solution is strongly alkaline and rapidly decomposes, so keep it only for 3 weeks when stored at 4°C.
3. This solution is unstable. Nevertheless, it is possible to keep it for some weeks at 4°C.
4. Note that phenazine methasulfate is light sensitive.
5. Be careful, DAB may act as a carcinogen.
6. For increased sensitivity, the substrate incubation time may be increased to 5–10 min.

Acknowledgments

We thank N. Lachacinski and S. Marsy for excellent technical assistance; Drs T.D. Fox and J. Velours for antibodies; and Dr. P. Hamel for critical reading of the manuscript and for looking over the English. This work was supported by grants from the "Association Française contre les Myopathies" and from the C.N.R.S. program "Protéomique et Génie des Protéines".

References

1. Meisinger, C., Sommer, T., and Pfanner, N. (2000) Purification of *Saccharomyces cerevisiae*. Mitochondria devoid of microsomal and cytosolic contaminations. *Anal. Biochem.* **287**, 339–342.
2. Hatefi, Y. and Stiggall, D. (1978) Preparation and properties of succinate: ubiquinone oxidoreductase (complex II). *Methods Enzymol.* **53**, 21–27.
3. Bousquet, I., Dujardin, G., and Slonimski, P. (1991) *ABC1*, a novel yeast nuclear gene has a dual function in mitochondria: it suppresses a cytochrome *b* mRNA translation defect and is essential for the electron transfer in the *bc1* complex. *EMBO J.* **10**, 2023–2031.
4. Ryan, M. T., Voos, W., and Pfanner, N. (2001) Assaying protein import into mitochondria. *Methods Cell Biol.* **65**, 189–215.
5. Schagger, H. and von Jagow, G. (1991) Blue native electrophoresis for isolation of membrane protein complexes in enzymatically active form. *Anal. Biochem.* **199**, 223–231.
6. Sellem, C. H., Lemaire, C., Lorin, S., Dujardin, G., and Sainsard-Chanet, A. (2005) Interaction between the *oxa1* and *rmp1* genes modulates respiratory complex assembly and life span in *Podospora anserina*. *Genetics* **169**, 1379–1389.

7. Pflieger, D., Le Caer, J. P., Lemaire, C., Bernard, B. A., Dujardin, G., and Rossier, J. (2002) Systematic identification of mitochondrial proteins by LC-MS/MS. *Anal. Chem.* **74**, 2400–2406.

8. Sickmann, A., Reinders, J., Wagner, Y., Joppich, C., Zahedi, R., Meyer, H. E., et al. (2003) The proteome of *Saccharomyces cerevisiae* mitochondria. *Proc. Natl. Acad. Sci. U.S.A.* **100**, 13207–13212.

9. Dujardin, G., Pajot, P., Groudinsky, O., and Slonimski, P. (1980) Long range control circuits within mitochondria and between nucleus and mitochondria. I. Methodology and phenomenology of suppressors. *Mol. Gen. Genet.* **179**, 469–482.

10. Yonetani, T. (1965) Studies on cytochrome c peroxydase II – stoichiometry between enzyme, H_2O_2, and ferricytochrome c and enzymic determination of extinction coefficients of cytochrome c. *J. Biol. Chem.* **240**, 4509–4513.

11. Rieske, J. S. (1967) Preparation and properties of a respiratory chain iron-protein. *Methods Enzymol.* **10**, 239–245.

12. Wallis, M., Groudinsky, O., Slonimski, P., and Dujardin, G. (1994) The NAM1 protein (NAM1p), which is selectively required for *cox1*, *cytb* and *atp6* transcript processing/stabilisation, is located in the yeast mitochondrial matrix. *Eur. J. Biochem.* **222**, 27–32.

13. Claisse, M. L., Pere-Aubert, G. A., Clavilier, L. P., and Slonimski, P. P. (1970) Method for the determination of cytochrome concentrations in whole yeast cells. *Eur. J. Biochem.* **16**, 430–438.

14. Dutta, C. and Henry, H. L. (1990) Detection of hemoprotein peroxidase activity on polyvinylidene difluoride membrane. *Anal. Biochem.* **184**, 96–99.

15. Vargas, C., McEwan, A. G., and Downie, J. A. (1993) Detection of c-type cytochromes using enhanced chemiluminescence. *Anal. Biochem.* **209**, 323–326.

16. Brasseur, G., Coppee, J. Y., Colson, A. M., and Brivet-Chevillotte, P. (1995) Structure-function relationships of the mitochondrial *bc1* complex in temperature-sensitive mutants of the cytochrome *b* gene, impaired in the catalytic center N. *J. Biol. Chem.* **270**, 29356–29364.

17. Pajot, P., Wambier-Kluppel, M., Kotylak, Z., and Slonimski, P. P. (1976) Regulation of cytochrome oxidase formation by mutations in a mitochondrial gene for cytochrome *b*. In *Genetics and Biogenesis of Chloroplasts and Mitochondria* (Bucher, T. H., Neupert, W., Sebald, W. and Werner, S., eds), Elsevier, North Holland, and Biochemical Press, Amsterdam, 443–451.

18. Somlo, M. (1968) Induction and repression of mitochondrial ATPase in yeast. *Eur. J. Biochem.* **5**, 276–284.

19. Kassenbrock, C., Cao, W., and Douglas, M. (1993) Genetic and biochemical characterization of ISP6, a small mitochondrial outer membrane protein associated with the protein translocation complex. *EMBO J.* **12**, 3023–3034.

20. Nijtmans, L., Henderson, N., and Holt, I. (2002) Blue native electrophoresis to study mitochondrial and other protein complexes. *Methods* **26**, 327–334.

21. Cruciat, C., Brunner, S., Baumann, F., Neupert, W., and Stuart, R. (2000) The cytochrome *bc1* and cytochrome *c* oxidase complexes associate to form a single supracomplex in yeast mitochondria. *J. Biol. Chem.* **275**, 18093–18098.

22. Schagger, H. and Pfeiffer, K. (2000) Supercomplexes in the respiratory chains of yeast and mammalian mitochondria. *EMBO J.* **19**, 1777–1783.

23. Soubannier, V., Vaillier, J., Paumard, P., Coulary, B., Schaeffer, J., and Velours, J. (2002) In the absence of the first membrane-spanning segment of subunit 4(b), the yeast ATP synthase is functional but does not dimerize or oligomerize. *J. Biol. Chem.* **277**, 10739–10745.

24. Lemaire, C., Guibet-Grandmougin, F., Angles, D., Dujardin, G., and Bonnefoy, N. (2004) A yeast mitochondrial membrane methyltransferase-like protein can compensate for *oxa1* mutations. *J. Biol. Chem.* **279**, 47464–47472.

25. Bonnefoy, N., Chalvet, F., Hamel, P., Slonimski, P., and Dujardin, G. (1994) *OXA1*, a *Saccharomyces cerevisiae* nuclear gene whose sequence is conserved from prokaryotes to eukaryotes controls cytochrome oxidase biogenesis. *J. Mol. Biol.* **239**, 201–212.

6

Mitochondrial Proteomics
Analysis of a Whole Mitochondrial Extract with Two-Dimensional Electrophoresis

Thierry Rabilloud

Summary

Mitochondria are complex organelles, and their proteomics analysis requires a combination of techniques. The emphasis in this chapter is made first on mitochondria preparation from cultured mammalian cells, then on the separation of the mitochondrial proteins with two-dimensional electrophoresis (2DE), showing some adjustment over the classical techniques to improve resolution of the mitochondrial proteins. This covers both the protein solubilization, the electrophoretic part per se, and the protein detection on the gels, which makes the interface with the protein identification part relying on mass spectrometry.

Key Words: Proteomics; mitochondria; two-dimensional electrophoresis; immobilized pH gradients; silver staining

1. Introduction

Mitochondria are among the most complex cell organelles and contain up to 10% of the cell protein content. Furthermore, they are among the few cell organelles that are separated from the bulk of the cytoplasm by a double membrane (i.e., a double lipid bilayer). As a matter of fact, this is probably linked with the functioning of the mitochondrial energy transducing machinery, which uses a proton gradient across the inner membrane. This proton gradient is built by the oxidative phosphorylation complexes, also named complexes I–IV or respiratory complexes, but it also requires a "tight" membrane, in the sense that it must be impermeant even to protons that must reenter in the

From: *Methods in Molecular Biology, vol. 432: Organelle Proteomics*
Edited by: D. Pflieger and J. Rossier © Humana Press, Totowa, NJ

mitochondrial matrix only through the ATP synthase (also named complex V) for an efficient ATP production. This implies in turn that this membrane is also impermeant to many other solutes, including those that must be present in the mitochondrial matrix for the various biochemical reactions occurring in the mitochondria. This further implies that a whole range of transporters is present in the inner membrane to allow the selective import of these substrates.

Mitochondria are also peculiar in the fact that they possess an autonomous genome. In mammals, this genome is almost vestigial and encodes only 13 protein subunits, all very hydrophobic. This implies in turn that a few hundreds of proteins present in the mitochondria are imported, even the mitochondrial ribosomal proteins and the mitochondrial RNA polymerase that are used to produce in situ the mitochondrially encoded proteins. This import mechanism is quite different from the one used for endoplasmic reticulum (ER)-derived organelles and has been reviewed elsewhere (1).

Thus, on a protein composition point of view, mitochondria are quite complex and encompass both very soluble proteins (present in the matrix and the intermembrane space) and very hydrophobic membrane proteins, plus membrane proteins of intermediate solubility, such as some subunits of the oxidative phosphorylation complexes or the outer membrane porins. This chemical heterogeneity is a real challenge for the proteomic analysis of mitochondria.

Dysfunction of mitochondria can lead to several disorders of varied severity, ranging from intolerance to an intense effort to perinatally fatal diseases. Progressive mitochondrial dysfunction has also been implicated in the aging process. This has led to interest in comparative mitochondrial proteomics. As many mitochondrial proteins are assembled into complexes of defined stoichiometry (e.g., the respiratory complexes whose structure is sometimes known) (2,3), it is interesting to reach a fine quantification level that allows to investigate mis-stoichiometries caused by deficient complex assembly. Not all proteomics techniques allow reaching this precision level, and two-dimensional electrophoresis (2DE) is among the few available choices nowadays. However, this technique is not without drawbacks, especially for hydrophobic proteins (4), and adequate protein solubilization conditions must be used to visualize at least part of the inner membrane-embedded proteins.

The methodological part of this chapter will therefore start with the biological sample (e.g., cultured cells) and detail the mitochondrial preparation, the protein solubilization, and the 2DE. The mass spectrometry techniques used are quite standard and can be found in any proteomics textbook. A brief outline only will be given in this chapter.

2. Materials

2.1. Mitochondria Preparation

1. Solution A: 10 mM Tris–HCl, pH 6.7, 10 mM KCl, and 15 mM $MgCl_2$. This solution is made fresh when needed from stock solutions of Tris buffer (usually 1 M Tris–HCl, pH 7.6), 1 M KCl, and 1 M $MgCl_2$. The stock solutions are stable for months at room temperature.
2. Solution B: 2 M sucrose. This solution is prepared with gentle warming (up to 60°C) to help dissolution. Once made, it is kept at 4°C for a few months. Degradation of the solution is indicated by mold or bacterial growth, which is fairly visible on swirling the bottle containing the solution.
3. Solution C: 0.25 M sucrose, 10 mM Tris–HCl, pH 6.7 (at 25°C), and $1.5 \times 10\text{–}4$ M MgCl2. This solution is prepared on the day of use from solution B and from the stock solutions used for the preparation of solution A.
4. Solution D: 0.25 M sucrose, 10 mM Tris–HCl, pH 7.6, and 10 mM ethylenedi-aminetetraacetic acid (EDTA). This solution is prepared on the day of use from solution B, from 1 M Tris buffer, and from stock 0.5 M EDTA–NaOH, pH 8.0. The latter solution is prepared by suspending EDTA disodium salt and adding concentrated NaOH (10 M) up to the desired pH, which is close to the dissolution point. Thus, care must be taken not to add too much sodium hydroxide. This EDTA stock solution is stable for months at room temperature.
5. Solution E: 0.225 M sucrose, 75 mM mannitol, 10 mM Tris–HCl, pH 7.6, and 1 mM EDTA.

2.2. Mitochondrial Proteins Solubilization

1. Solution F: 8.75 M urea, 2.5 M thiourea, 6 mM Tris carboxyethyl phosphine, and 0.5% (v/v) 3–10 carrier ampholytes (all from Fluka). This solution is difficult to prepare, as water occupies less than 50% of the volume. The most convenient way is to place the capped tube in a bath sonicator and to let sonicate until complete dissolution (occasional tube inverting speeds up the process). Once made, this solution is stored in aliquots at –20°C for up to 1 year.

2.3. Two-Dimensional Gel Electrophoresis

1. Orange G solution: 2 mg orange G (Fluka) per mL of water. Stable at room temperature for months.
2. Dithiodiethanol (from Fluka): used as supplied, stable for months at room temperature. A slight yellow color may develop and does not prevent its use.
3. Solution G: 6 M urea, 30% (v/v) glycerol, 2.5% (w/v) sodium dodecyl sulphate (SDS), and 0.125 M Tris–HCl, pH 7.5.
4. Solution H: 130g/L Tris and 0.6 M HCl.
5. Solution I: 150 g/L Tris and 0.6M HCl.

6. Solution J: 30% acrylamide and 0.8% methylene Bis acrylamide (to be stored at 4°C). This solution can be purchased ready-made, and this is recommended to avoid handling of monomer powders.

7. Ammonium persulfate solution: 10% (w/v) ammonium persulfate in water, stable for 1 week at room temperature.

8. Solution K: 2% (w/v) low-melting agarose in 0.125 M Tris–HCl, pH 7.5, 0.4% (w/v) SDS, and 0.002% (w/v) bromophenol blue.

9. Solution L: 6 g/L Tris, 30 g/L glycine, and 1 g/L SDS.

10. Solution M: 3 g/L Tris, 1 g/L SDS, and 25 g/L Taurine.

11. Solution N: The silver–ammonia solution is prepared as follows—for ca. 500 mL of staining solution, 475 mL of water is placed in a flask with strong magnetic stirring. First, 7 mL of 1 N sodium hydroxide is added, followed first by 7.5 mL of 5 N ammonium hydroxide (Aldrich) and then by 12 mL of 1 N silver nitrate. A transient brown precipitate forms during silver nitrate addition. It should disappear in a few seconds after the end of silver addition. Persistence of a brown precipitate or color indicates exhaustion of the stock ammonium hydroxide solution. Attempts to correct the problem by adding more ammonium hydroxide solution generally lead to poorer sensitivity.

    The ammonia–silver ratio is a critical parameter for good sensitivity *(5)*. The above proportions give a ratio of 3.1, which is one of the lowest practicable ratios. This ensures highest sensitivity and good reproducibility control of the ammonia concentration through silver hydroxide precipitation. This solution should be prepared at most 30 min before use.

    Flasks used for preparation of silver–ammonia complexes and silver–ammonia solutions must not be left to dry out, as explosive silver azide may form. Flasks must be rinsed at once with distilled water, while used silver solutions should be put in a dedicated waste vessel containing either sodium chloride or a reducer (e.g., ascorbic acid) to precipitate silver.

12. Solution O: 2% (w/v) phosphoric acid, 15% (v/v) ethanol, and 12% (w/v) ammonium sulfate. Phosphoric acid and ammonium sulfate are added to water (70% of the final volume). Ethanol is added once the salt is dissolved, and the volume is adjusted with water.

2.4. Protein Digestion and Analysis by Mass Spectrometry

1. 25 mM ammonium bicarbonate.
2. HPLC grade acetonitrile and formic acid.
3. 10 mM dithiothreitol.
4. 55 mM iodoacetamide.
5. Sequencing-grade bovine trypsin.
5. α-Cyano-4-hydroxycinnamic acid.
6. Mass spectrometry measurements are carried out on an ULTRAFLEX™ matrix-assisted laser desorption ionization time of flight (MALDI-TOF)/TOF mass spectrometer (Bruker-Daltonik GmbH, Bremen, Germany).

3. Methods

3.1. Mitochondria Preparation

The preparation starts from a cell pellet. As small scale preparations lead to more severe losses and also to lesser mitochondrial purity, it is recommended to start from a billion cells, leading to a few milligrams of mitochondrial proteins. This amount is sufficient to carry out a complete set of comparative mitochondrial proteomics, including several replicate gels and preparative gels for the identification of minor-abundance proteins.

1. After isolation, wash the cells in a standard saline solution (e.g., PBS).
2. Swell the cells in solution A (at least 10 mL of solution A per gram of cell pellet) for 5 min on ice.
3. Break the cells with a motor-driven Potter-Elvejehm homogenizer set at 80–100 rpm. Ten strokes are generally needed to break >80% of the cells, but this may depend on the cell type. This step is carried out in a cold room to limit proteolysis. Protease and phosphatase inhibitors are not used in the lysis buffer because mitochondria are tight organelles, which means that the interior of intact mitochondria is protected from what happens outside. Moreover, many inhibitors do not enter in the mitochondria.
4. Measure the volume of the homogenate and add 1/7th of this volume of cold (4°C) solution B (*see* **Note 1**).
5. Centrifuge this homogenate at 1200 *g* for 5 min at 4°C to get rid of unbroken cells, large debris, and nuclei.
6. Collect the mitochondria by centrifugation at 8000 *g* for 10 min at 4°C.
7. Resuspend the mitochondrial pellet (by homogenization with 10 strokes of a hand-driven Potter-Elvejem homogenizer) in 20–50 times its volume of solution C.
8. Centrifuge at 1200 *g* for 5 min at 4°C, save the supernatant and centrifuge at 8000 *g* for 10 min at 4°C.
9. Save the pellet and wash again once by the same procedure but using solution D (*see* **Note 2**).
10. Store the final mitochondrial pellet in aliquots at −80°C as a concentrated suspension in solution E (estimate the volume of the final pellet and use at most five times this volume of solution E to prepare the suspension) (*see* **Note 3**). Protein concentration is estimated by a standard protein assay (BCA or Bradford type).

3.2. Mitochondrial Protein Solubilization

Protein solubilization for 2D gel electrophoresis is carried out the day of use by mixing at room temperature 1 volume of mitochondrial suspension, 1 volume of detergent solution, and 8 volumes of solution F (*see* **Notes 4** and **5**). Extraction is carried out at room temperature for 0.5–3 h. The solution is then loaded on the isoelectric focusing strip.

3.3. Two-Dimensional Gel Electrophoresis

3.3.1. Isoelectric Focusing

Because of their simplicity of use and because of their high performance in the analysis of basic proteins (mitochondria are quite rich in basic proteins), the use of immobilized pH gradients (IPG) strips is strongly recommended. Strips of various pH ranges are commercially available (e.g., from GE Healthcare or from Bio-Rad). Otherwise, immobilized pH gradient plates can be prepared in the laboratory and cut into strips of required width with a paper cutter. This home-made pH gradient preparation is, however, beyond the scope of this chapter and can be found in adequate textbooks *(6)*. Nevertheless, commercial or home-made IPG strips are handled the same way. The strips are reswollen in the adequate solution and the protein sample is applied either at this reswelling stage or after reswelling in a sample cup. As a rule of thumb, application by reswelling is preferred, except when it leads to poor resolution, that is, for basic gradients (e.g., 6–10 and 7–11 ranges). However, sample application by reswelling is adequate for wide gradients even if they extend into the basic pH (e.g., 3–10 and 4–12). Both sample application procedures are presented here.

3.3.1.1. SAMPLE APPLICATION BY RESWELLING

The total amount of solution needed for complete reswelling (i.e., including the sample solution volume) depends on the size of the strip. Commercial strips are 3.3 mm wide and cast as 0.5-mm thick gels. Thus, the strip gel volume in microliters is 1.65 × strip gel length (in mm). However, best results are obtained when the gel is reswollen to 1.25- to 1.3-fold over their initial volume *(7)*. Thus, the reswelling volume in microliters is 2 × strip gel length (in mm).

1. Practically, prepare a sample dilution solution on the day of use by mixing 8 volumes of solution F, 1 volume of water, and 1 volume of 20% detergent solution.
2. Once the required reswelling volume and the required sample volume are known, dilute the sample up to the reswelling volume with this dilution solution.
3. To this reswelling sample solution, add (1) 1 µL of orange G solution and (2) 0.1 volume of dithiodiethanol (*see* **Note 6**).
4. Place this complete, colored rehydration solution in the grooved rehydration chamber or in the strip holder, depending on the system used, and let rehydration take place overnight at room temperature, the whole strip plus solution being covered by mineral oil to prevent evaporation.
5. Then place the reswollen gel in the IEF apparatus that is ready for running.

3.3.1.2. SAMPLE APPLICATION BY CUP LOADING

In this case, the ideal gel rehydration volume is the initial one, that is, (in µL) 1.65 × strip gel length (in mm). The rehydration solution is made by mixing

8 volumes of solution F, 1 volume of dithiodiethanol, 1 volume of 20% detergent solution, and 2 μL of orange G solution per milliliter of rehydration solution. Rehydration takes place overnight at room temperature in the chamber provided by the IEF apparatus supplier. On the day of use, apply the sample at the anodic side of the gel on a plastic cup or in the molded chamber depending on the apparatus used.

3.3.1.3. Isoelectric Focusing and Equilibration

1. Place the rehydrated strip in the strip holder and apply the sample anodically if necessary. This is required when alkaline pH gradients (e.g., 6–12 and 7–11) are used.
2. Cover the strip and the sample with mineral oil and connect the power supply.
3. It is advisable to use a thermostated IEF apparatus to guarantee the constancy of the spot position in the 2D pattern *(8)*. The strips can be run at any temperature above 10°C to avoid urea crystallization. The strips are usually run at 22°C to avoid any precipitation of urea-detergent complex.
4. To avoid any overheating, even local ones, it is recommended to use a voltage-controlled migration program. For a wide pH gradient (3 pH units or more) that is 16–20 cm long, the following program is used: 100 V for 1 h, 300 V for 3 h, 1000 V for 1 h, and 3500 V for 18 h or more. To adjust the migration for each condition, the following rule of thumb can be applied: most proteins have reached their steady-state position after 100 Vh/cm^2, where the cm^2 means the square of the strip length in cm. For example, a 20-cm-long strip needs at least 20 × 20 × 100 Vh, that is, 40,000 Vh. However, as most gradients are stable over time, more Vh can be applied without any problem.
5. After the IEF migration has been completed, remove the mineral oil and equilibrate the strips for 20 min in solution G. 5–10 mL of solution G is used per strip. The equilibrated strips can then be used immediately for the second dimension or frozen in the equilibration solution at –20°C. Frozen strips are stable for a few weeks at –20°C. They are thawed in a water bath at 20–30°C and used when necessary.

3.3.2. SDS Electrophoresis

3.3.2.1. Gel Casting

In addition to being rich in basic proteins, mitochondria are rich in low-molecular-weight proteins, and many subunits of the respiratory complexes are below 15 kDa. This makes the standard Laemmli system not optimal for the resolution of mitochondrial proteins, and it is recommended to use the recently introduced Tris-Taurine system *(9)* (*see* **Note 7**).

1. 1.5-mm-thick gels are routinely used. For a 160 × 200 × 1.5 mm gel, 60 mL of gel mix is prepared. This gel mix is optimized for the molecular weight range that needs to be investigated.

2. For a 20–200 kDa range, the gel mix is made of 10 mL of solution H, 20 mL of solution J, and 30 mL of water. For a 5–200 kDa range, which provides resolution of the low-molecular-weight proteins at the expense of the compression of proteins above 35 kDa, the gel mix is composed of 10 mL of solution I, 22.5 mL of solution J, and 27.5 mL of water.

3. Initiate polymerization by the sequential addition of 20 μL of TEMED (tetramethyl ethylene diamine) and 400 μL of ammonium persulfate solution (*see* **Note 8**).

4. Cast the gels between the plates (5 mm free of gel mix are left at the top of the gel cassette) and overlay with 0.8 mL of water-saturated 2-butanol. Polymerization should occur within 30 min.

5. It is recommended to cast the second dimension gels the day before their use for a complete and uniform polymerization. Once polymerized, remove the gels from the casting chamber, remove the butanol and replace with water, and store the gels assemblies in a closed polyethylene box. To avoid glass plate sticking, separate each gel assembly from its neighbors by a plastic sheet (polycarbonate plastic sheets from Bio-Rad).

3.3.2.2. STRIP TRANSFER AND GEL RUNNING

1. Place the second dimension gel assembly on its stand.

2. Catch the equilibrated strip with tweezers at one end. The use of inverted tweezers that hold the strip without hand pressure is quite convenient.

3. Clip the excess plastic and gel at the free end (the one not covered by the tweezers) with scissors.

4. Pour 0.8 mL of molten agarose (solution K) on the top of the second dimension gel and put the strip in place (clipping of the excess plastic and gel at the site of the tweezers releases the strip in place). Care must be taken to eliminate any bubble between the top of the second dimension gel and the strip.

5. Allow 10 min for the agarose to set (*see* **Note 9**) and secure the gel in the gel tank.

6. Fill the lower chamber of the tank with buffer L and the upper chamber with buffer M (*see* **Note 10**).

7. Run the second dimension gels at 10°C (thermostated) for 1 h at 25 V, then at 12 W/gel until the bromophenol blue front reaches the bottom of the gel.

3.3.3. Spot Visualization

Two main types of spot visualization are used in such proteomics experiments. Silver staining is used in the initial phases of the study, for example, to set the conditions and to perform comparative experiments. The rationale for using silver staining is based on its sensitivity, as a gel showing more than 1000 protein spots can be obtained with 0.1 mg of total mitochondrial proteins. Such "analytical" gels can be used for image analysis but also for spot excision and subsequent protein identification with mass spectrometry. However, silver-stained gels can be deceptive in spot identification because (1) small and weak silver-stained spots contain very small amounts of proteins and (2) the silver staining process results in peptide losses in the mass spectrometry identification process *(10)*.

Two main processes can be used for silver staining. The silver nitrate process works well for acidic proteins but less well for basic proteins. In addition, peptide losses are often important. The silver–ammonia process works nicely for basic proteins but frequently gives artifacts (weak or hollow or negative spots) in the acidic range. However, the peptide losses are generally lesser with this process. For optimal performance, silver–ammonia staining requires home-made gels cast with sodium thiosulfate (*see* **Note 10**). Both processes are given below.

3.3.3.1. FAST SILVER NITRATE STAINING (*SEE* **NOTE 11**)

1. Fix the gels (1 h + overnight) in acetic acid/ethanol/water, 5/30/65 (v/v/v).
2. Rinse in water four times for 10 min.
3. To sensitize, soak gels for 1 min (1 gel at a time) in 0.8 mM sodium thiosulfate.
4. Rinse twice for 1 min in water.
5. Impregnate for 30–60 min in 12 mM silver nitrate (0.2 g/L). The gels may become yellowish at this stage.
6. Rinse in water for 5–15 s.
7. Develop image (10–20 min) in 3% (w/v) potassium carbonate containing 250 µL of formalin and 125 µL of 10% (w/v) sodium thiosulfate per liter.
8. Stop development (30–60 min) in a solution containing 40 g of Tris and 20 mL of acetic acid per liter.
9. Rinse with water (several changes) prior to drying or densitometry.

3.3.3.2. AMMONIACAL SILVER STAINING (*SEE* **NOTE 11**)

Thiosulfate is added to the gel during polymerization (*see* **Note 10**). After electrophoresis, proceed as follows for silver staining:

1. Fix in acetic acid/ethanol/water, 5/30/65 (v/v/v), containing 0.05% (w/v) 2–7 naphtalene disulfonate (Acros) for 1 h.
2. Fix overnight in the same solution.
3. Rinse six times for 10 min in water.
4. Impregnate for 30–60 min in the ammoniacal silver solution (solution N).
5. Rinse thrice for 5 min in water.
6. Develop image (5–10 min) in 350 µM citric acid containing 1 mL formalin per liter.
7. Stop development in acetic acid/ethanolamine/water, 2/0.5/97.5 (v/v/v). Leave in this solution for 30–60 min.
8. Rinse with water (several changes) prior to drying or densitometry.

3.3.3.3. Protein Spots Destaining

When protein identification is planned on spots stained with silver, a destaining step is required for better results. To maximize peptide recovery by in-gel proteolytic digestion, this destaining step should be performed the same day as silver staining *(10)*.

Destaining proceeds as follows *(11)*:

1. Prepare a stock solution of potassium ferricyanide (30 mM in water) and a solution of sodium thiosulfate (100 mM in water). Just before use, mix equal volumes of the two solutions and cover the spots with the resulting mix.
2. Destain the spots for 5 min at room temperature.
3. Remove the destaining solution, rinse thrice for 5 min with water.
4. Soak the spots for 20 min in ammonium bicarbonate (200 mM).
5. Remove the bicarbonate solution and rinse thrice for 5 min in water.

For the less abundant protein spots whose identification fails from silver-stained gels or when maximal sequence coverage is desired (e.g., for assignment of modification sites), more heavily loaded gels are needed (0.5–1 mg protein loaded on the strip). These gels are usually stained with colloidal Coomassie blue, which is far less sensitive than silver staining but gives a better sequence coverage.

3.3.3.4. Colloidal Coomassie Blue Staining

1. After electrophoresis, fix the gels thrice for 30 min in ethanol/water, 30/70 (v/v), containing 1.7% (w/v) phosphoric acid. This fixation can also proceed with a 1-h bath followed by an overnight bath.
2. Rinse thrice for 20 min in 1.7% (w/v) phosphoric acid.
3. Equilibrate for 30 min in solution O.
4. Without removing the solution surrounding the gels, add 1% (v/v) of a solution containing 20 g of Brilliant Blue G per liter (dissolved in hot water, stable for months at room temperature). Let the staining proceed for 24–72 h.
5. If needed, destain the background with water. Avoid alcohol-containing solutions.

3.4. Mass Spectrometry

The details of mass spectrometry analysis are not fully in the scope of this chapter, as rather classical procedures are used. The detailed procedures used are mentioned for information.

3.4.1. In-Gel Digestion

1. Wash the spots (a robotic device can be used) with 0.1 mL of 25 mM ammonium bicarbonate for 8 min.
2. Remove the bicarbonate and shrink the gel pieces thrice for 8 min with pure acetonitrile.
3. Remove the acetonitrile and dry completely the gel pieces in a vacuum centrifugal concentrator (e.g., a SpeedVac).
4. Cover the gel pieces with 0.1 mL of 10 mM dithiothreitol in 25 mM ammonium bicarbonate, and break disulfide bridges by incubating at 50°C for 1 h.
5. Add 0.1 mL of 55 mM iodoacetamide in 25 mM ammonium bicarbonate, and alkylate the thiol groups at room temperature for 1 day (in the dark).
6. Wash alternatively with 25 mM ammonium bicarbonate and acetonitrile (5 min each). Repeat this double washing three times.
7. Dry completely the gel pieces in a vacuum centrifugal concentrator (e.g., a SpeedVac).
8. Estimate the dried gel volume and add three times this volume of trypsin solution (12.5 mg/L in 25 mM ammonium bicarbonate). Let digestion proceed overnight at 35°C.
9. Dry partially the gel pieces for 5 min in a vacuum centrifugal concentrator (e.g., a SpeedVac).
10. Add 5 µL of water/acetonitrile/formic acid, 35/60/5 (v/v/v), and extract the peptides by sonication (bath sonicator) for 5 min. Recover the liquid phase and repeat this extraction once.

3.4.2. Mass Spectrometry Measurements

The MALDI-TOF/TOF instrument is used at a maximum accelerating potential of 20 kV and operated in reflector positive mode. Sample preparation is performed with the dried droplet method using a mixture of 0.5 µL of sample with 0.5 µL of matrix solution. The matrix solution is prepared from a saturated solution of α-cyano-4-hydroxycinnamic acid in H_2O/acetonitrile, 50/50 (v/v/v), diluted three times in H_2O/acetonitrile, 50/50 (v/v). Internal calibration is performed with tryptic peptides resulting from autodigestion of trypsin (monoisotopic masses at m/z = 842.51 Da, m/z = 1045.56 Da, and m/z = 2211.11 Da).

Monoisotopic peptide masses are assigned and used for database searches using the search engines MASCOT (Matrix Science, London, UK) and Aldente (www.expasy.org). All proteins present in Swiss-Prot are used without any pI and Mr restrictions. The error on peptide mass measurement is limited to 50 ppm, one possible cleavage site missed by trypsin is accepted.

3.5. A Few Results

To illustrate the methods presented in this chapter, some results obtained on mitochondria prepared from human cultured cells are shown. **Figure 1** shows the resolving power of 2DE of mitochondria prepared from HeLa cells. The resolution in the low-molecular-weight region, as well as in the basic range, should be noted. These methodological improvements have allowed a better coverage of the mitochondrial proteome *(12,13)*.

Figure 2 shows a typical comparative experiment between mitochondria extracted from normal cells (gel A) and mitochondria extracted from cells devoid of mitochondrial DNA (gel B). Some of the differentially expressed spots have been excised from the silver-stained gels, digested, and analyzed by mass spectrometry to obtain their peptide mass fingerprint. The results in terms of protein identifications are shown in **Table 1**. More complete results can be found in the literature *(14)*.

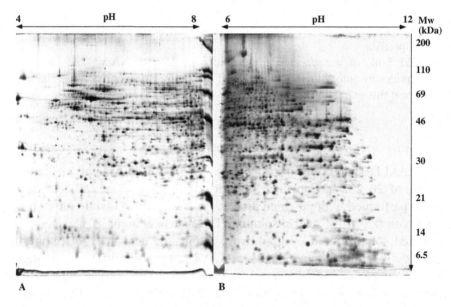

Fig. 1. Two-dimensional separation of mitochondrial proteins. Mitochondrial proteins obtained from HeLa cells are separated on the gels. 0.12 mg of total mitochondrial protein has been loaded on each gel. The second dimension gel is a 11.5% gel cast with the alkaline gel buffer (solution I). (**A**) pH 4–8 linear pH gradient in the first dimension. Sample application by in-gel rehydration. Spot visualization with silver nitrate staining. (**B**) pH 6–12 linear pH gradient in the first dimension. Sample application by cup loading. Spot visualization with silver–ammonia staining.

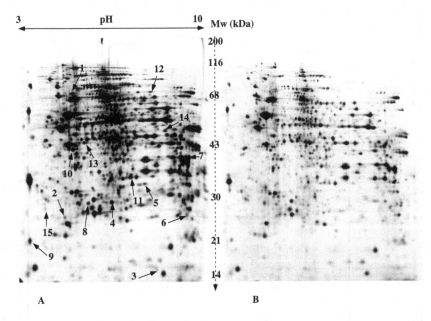

Fig. 2. Comparative proteomics on mitochondrial proteins. Mitochondrial proteins obtained from 143B cells or Rho0 cells are separated on the gels. 0.12 mg of total mitochondrial protein has been loaded on each gel. pH gradient: linear 3–10. Sample application by in-gel rehydration. The second dimension gel is a 10% gel cast with the standard gel buffer (solution H). Spot visualization by silver nitrate staining. (**A**) Mitochondrial proteins extracted from 143B cells (normal mitochondrial DNA). (**B**) Mitochondrial proteins extracted from 143B Rho0 cells (cells devoid of mitochondrial DNA). Some differentially expressed spots have been excised, destained, and submitted to protein identification by mass spectrometry (peptide mass fingerprinting). The results are shown in **Table 1**.

4. Notes

1. The addition of solution B is intended to restore the osmolarity of the solution to a level close to the one present in cells. This limits mitochondria and nuclei breakage in the following steps, thereby increasing the yield and purity of mitochondria. However, solution B is dense and viscous, and thorough mixing is needed. This is usually achieved by inverting the capped tube and/or by gentle vortex mixing.

2. The washing steps are critical for increasing the purity of the mitochondria. Pure mitochondria are best obtained by density gradient centrifugation or free-flow electrophoresis (*15*), but these procedures have low yields and are best suited for large samples (e.g. bovine tissues or yeast cultures). The low/high speed procedure presented here is a good compromise between yield and purity, and is

Table 1
Protein Identification Results

Number on gels	Accession number	Protein name	pI	Mw (kDa)	Seq cov.	x-fold decrease in Rho0 versus 143B
1	P28331	NADH-ubiquinone oxidoreductase 75-kDa subunit	5.42	77	42	6
2	O00217	NADH-ubiquinone oxidoreductase 23-kDa subunit	5.10	20.3	42	21
3	Q9P0J0	NADH-ubiquinone oxidoreductase B16.6 subunit	8.24	16.5	20	5.4
4	P47985	Ubiquinol-cytochrome c reductase iron-sulfur subunit	6.3	21.6	31	9
5	Q9Y399	Mitochondrial 28S ribosomal protein S2	7.35	28.3	41	20
6	Q9Y3D9	Mitochondrial ribosomal protein S23	8.94	21.7	55	2.3
7	Q9UGM6	Tryptophanyl-tRNA synthetase	8.99/8.29	37.9/35.1	31	2.9
8	O75431	Metaxin 2	5.9	29.7	35	2.1
9	Q9NS69	Mitochondrial import receptor subunit TOM22	4.27	15.4	77	2.3
10	Q9UJZ1	Stomatin-like protein 2	5.26	35.0	64	1.9
11	P13804	Electron transfer flavoprotein alpha-subunit	7.1	32.9	52	2.4
12	P49748	Acyl-CoA dehydrogenase, very-long-chain specific	7.74	66.1	52	1.74
13	Q9P2R7	Succinyl-CoA ligase (ADP-forming) beta-chain	5.64	43.6	36	8.6
14	Q6YN16	Hydroxysteroid dehydrogenase	6.31	42.5	41	4.45
15	Q96EH3	Chromosome 7 open reading frame 30	4.87	21.6	27	7

The protein identification results obtained from the spots excised from the gels of **Fig. 2** are shown. The quantitative variations between 143B and Rho0 cells have been obtained by quantitative image analysis, using the Melanie software and the normalization of protein intensities in ppm of the total adequate for mitochondria isolated from cultured cells. The first wash in solution C removes the last contaminating nuclei, and the second wash in solution D eliminates most of the contaminating ribosomes. The first wash in solution C must not be skipped, as any nuclei remaining in the pellet suspended in solution D will burst because of the presence of EDTA. Exploded nuclei may lead to considerable increase in viscosity and inefficient washing, leading to heavily contaminated mitochondria.

3. Mitochondrial suspensions are more stable when stored concentrated. A good procedure is to resuspend the final pellet in twice its estimated volume of solution E. Suspending is achieved by vortex mixing and pipetting. The bulk of the suspension is saved in a tube, and an equal volume of solution E is added to the tube containing the remnants of the pellet. Thorough suspending is carried out again, and this new suspension is combined to the previous one. This process is carried out in a cold room.

4. There is no single detergent allowing the optimal extraction and thus visualization of all classes of proteins under the conditions prevailing in isoelectric focusing. Soluble proteins are best analyzed with CHAPS, while membrane proteins are poorly if at all solubilized by this detergent. Analysis of membrane proteins requires other detergents such as Brij 56 or dodecyl maltoside which are not equivalent in their solubilization patterns *(16)*. The choice of detergent will then depend on the focus of the study or on the amount of sample, which will allow or not series of experiments to be carried out with different detergents.

5. In some cases, this procedure will lead to too dilute a protein solution (e.g. for heavily loaded preparative gels or for cup loading). If this is the case, collect first the mitochondria in an Eppendorf-type tube by centrifugation (10,000g for 10 min at 4°C), and then suspend the pellet in an equal volume of 10% detergent solution plus four volumes of solution F.

6. The orange G is used as a tracker dye to check for any lack of electrical contact which would prevent protein migration at the isoelectric focusing stage. The dye must migrate to the anode and collect in a small zone close to the anode. Dye remaining over a large portion of the strip indicates a migration problem (electrical contact problem or too high salt concentration).

   Dithiodiethanol (used as supplied) has been shown to increase resolution in the basic portion of the IPG gels **(17)** and simplifies equilibration between the IPG and SDS dimension. It can be used with any pH range.

7. The Tris-Taurine system uses both the pH and the acrylamide concentration to control the speed of the moving boundary (visualized as the bromophenol blue front). This speed controls in turn the speed of the proteins that can comigrate with this dye front, and thus the lower limit (in molecular weight) of the proteins that can be resolved in the gel. This strong dependency of the boundary speed upon pH explains the way of preparing the gel buffer. These recipes have been empirically optimized, and their way of preparation is designed to give the best reproducibility over time.

8. For optimal results, alterations must be made at the level of gel casting when silver ammonia staining is to be performed. Thiosulfate is added at the gel polymerization step. Practically, the initiating system is composed of 1 μL of TEMED, 7 μL of 10% (w/v) sodium thiosulfate solution and 8 μL of 10% (w/v) ammonium persulfate solution per mL of gel mix. This ensures correct gel formation and gives minimal background upon staining.

9. Low melting agarose is used as it will leave more time to put the strip in place before the agarose gel sets. This can be a problem in summer in warm labs, where the temperature is close to the setting temperature of low melting agarose. If this is the case, the most convenient way is to place the gel assemblies in the fridge for 1 h prior to strip transfer. This will secure agarose gel setting.

10. Two different buffers can be used in gel tanks where the cathode and anode chambers do not communicate on a fluidic point of view (e.g. Bio-Rad Protean chambers). In this case, the lower chamber is usually much bigger than the upper one, and a cheaper Tris glycine buffer can be used in the lower chamber. Taurine is needed only in the cathode (upper) chamber for the system to operate. In "submarine" type gel tanks (e.g. Bio-Rad dodeca cell) the Taurine buffer (solution M) must be the only electrode buffer used.

11. General practice for silver staining. Batches of gels (up to five gels per box) can be stained. For a batch of three to five medium-sized gels (e.g. 160 × 200 × 1.5 mm), 1 L of the required solution is used, which corresponds to a solution/gel volume ratio of 5 or more; 500 mL of solution is used for one or two gels. Batch processing can be used for every step longer than 5 min, except for image development, where one gel per box is required. For steps shorter than 5 min, the gels should be dipped individually in the corresponding solution.

For changing solutions, the best way is to use a plastic sheet. The sheet is pressed on the pile of gels with the aid of a gloved hand. Inclining the entire setup allows the emptying of the box while keeping the gels in it. The next solution is poured with the plastic sheet in place, which prevents the solution flow from breaking the gels. The plastic sheet is removed after the solution change and kept in a separate box filled with water until the next solution change. This water is changed after each complete round of silver staining. The above statements are not true when gels supported by a plastic film are stained. In this case, only one gel per dish is required. A setup for multiple staining of supported gels has been described elsewhere *(18)*.

When gels must be handled individually, they are manipulated with gloved hands. The use of powder-free, nitrile gloves is strongly recommended, as powdered latex gloves are often the cause of pressure marks. Except for development or short steps, where occasional hand agitation of the staining vessel is convenient, constant agitation is required for all the steps. A reciprocal ("ping-pong") shaker is used at 30-40 strokes per min.

Dishes used for silver staining can be made of glass or plastic. It is very important to avoid scratches in the inner surface of the dishes, as scratches promote silver reduction and thus artifacts. Cleaning is best achieved by wiping with a tissue soaked with ethanol. If this is not sufficient, use instantly prepared Farmer's reducer (50 mM ammonia, 0.3% potassium ferricyanide, 0.6% sodim thiosulfate). Let the yellow-green solution dissolve any trace of silver, discard, rinse thoroughly with water (until the yellow color is no longer visible), then rinse with 95% ethanol and wipe.

Formalin stands for 37% formaldehyde. It is stable for months at room temperature. However, solutions containing a thick layer of polymerized formaldehyde must not be used. Never put formalin in the fridge, as this promotes polymerization. 95% ethanol can be use instead of absolute ethanol. Do not use denatured alcohol. It is possible to purchase 1 M silver nitrate ready-made. The solution is cheaper than solid silver nitrate on a silver weight basis. It is stable for months in the fridge.

Last, but not least, the quality of water is critical. Best results are obtained with water treated with ion exchange resins (resistivity higher than 15 MΩ/cm). Distilled water gives more erratic results.

References

1. Mokranjac, D., Neupert, W. (2005) Protein import into mitochondria. *Biochem. Soc. Trans.* **33**, 1019–1023.
2. Iwata, S., Lee, J. W., Okada, K., Lee, J. K., Iwata, M., Rasmussen, B., et al. (1998) Complete structure of the 11-subunit bovine mitochondrial cytochrome bc1 complex. *Science* **281**, 64–71.
3. Tsukihara, T., Aoyama, H., Yamashita, E., Tomizaki, T., Yamaguchi, H., Shinzawa-Itoh, K., et al. (1996) The whole structure of the 13-subunit oxidized cytochrome c oxidase at 2.8 A. *Science* **272**, 1136–1144.
4. Santoni, V., Molloy, M. P., Rabilloud, T. (2000) Membrane proteins and proteomics: un amour impossible? *Electrophoresis* **21**, 1054–1070.
5. Eschenbruch, M., Bürk, R. R. (1982) Experimentally improved reliability of ultrasensitive silver staining of protein in polyacrylamide gels. *Anal. Biochem.* **125**, 96–99.
6. Righetti, P. G. (1990) *Immobilized pH gradients. Theory and methodology.* Elsevier, Amsterdam.
7. Sanchez, J. C., Rouge, V., Pisteur, M., Ravier, F., Tonella, L., Moosmayer, M., et al. (1997) Improved and simplified in-gel sample application using reswelling of dry immobilized pH gradients. *Electrophoresis* **18**, 324–327.
8. Gorg, A., Postel, W., Friedrich, C., Kuick, R., Strahler, J. R., Hanash, S. M. (1991) Temperature-dependent spot positional variability in two-dimensional polypeptide patterns. *Electrophoresis* **12**, 653–658.
9. Tastet, C., Lescuyer, P., Diemer, H., Luche, S., van Dorsselaer, A., Rabilloud, T. (2003) A versatile electrophoresis system for the analysis of high- and low-molecular-weight proteins. *Electrophoresis* **24,** 1787–1794.

10. Richert, S., Luche, S., Chevallet, M., Van Dorsselaer, A., Leize-Wagner, E., Rabilloud, T. (2004) About the mechanism of interference of silver staining with peptide mass spectrometry. *Proteomics* **4**, 909–916.
11. Gharahdaghi, F., Weinberg, C. R., Meagher, D. A., Imai, B. S., Mische, S. M. (1999) Mass spectrometric identification of proteins from silver-stained polyacrylamide gel: a method for the removal of silver ions to enhance sensitivity. *Electrophoresis* **20**, 601–605.
12. Rabilloud, T., Kieffer, S., Procaccio, V., Louwagie, M., Courchesne, P. L., Patterson, S. D., et al. (1998) Two-dimensional electrophoresis of human placental mitochondria and protein identification by mass spectrometry: toward a human mitochondrial proteome. *Electrophoresis* **19**, 1006–1014.
13. Lescuyer, P., Strub, J. M., Luche, S., Diemer, H., Martinez, P., Van Dorsselaer, A., et al. (2003) Progress in the definition of a reference human mitochondrial proteome. *Proteomics* **3**, 157–167.
14. Chevallet, M., Lescuyer, P., Diemer, H., van Dorsselaer, A., Leize-Wagner, E., Rabilloud T. (2006) Alterations of the mitochondrial proteome caused by the absence of mitochondrial DNA: a proteomic view. *Electrophoresis* **27**, 1574–1583.
15. Zischka, H., Weber, G., Weber, P. J., Posch, A., Braun, R. J., Buhringer, D., et al. (2003) Improved proteome analysis of Saccharomyces cerevisiae mitochondria by free-flow electrophoresis. *Proteomics* **3**, 906–916.
16. Luche, S., Santoni, V., Rabilloud, T. (2003) Evaluation of nonionic and zwitterionic detergents as membrane protein solubilizers in two-dimensional electrophoresis. *Proteomics* **3**, 249–253.
17. Luche, S., Diemer, H., Tastet, C., Chevallet, M., Van Dorsselaer, A., Leize-Wagner, E., et al. (2004) About thiol derivatization and resolution of basic proteins in two-dimensional electrophoresis. *Proteomics* **4**, 551–561.
18. Granier, F., De Vienne, D. (1986) Silver staining of proteins: standardized procedure for two-dimensional gels bound to polyester sheets. *Anal. Biochem.* **155**, 45–50.

7

Purification and Proteomic Analysis of the Mouse Liver Mitochondrial Inner Membrane

Sandrine Da Cruz and Jean-Claude Martinou

Summary

Mitochondria are key organelles that play a crucial role in cellular homeostasis. Dysfunction of these organelles is associated with a wide range of human diseases. Therefore, mapping the different components of mitochondria would provide invaluable information to gain further understanding of mitochondrial functions and mitochondria-associated diseases. The mitochondrial inner membrane (MIM) contains a variety of proteins that are still unknown at their molecular level but are thought to play an essential role in several cellular processes including oxidative stress, cell death and transport of ions or metabolites. Here, we have used a new proteomics-based approach to establish a proteome of the MIM. This approach combines the use of highly purified mouse liver MIM, extraction of membrane proteins with organic acid and two-dimensional liquid chromatography coupled to mass spectrometry. This procedure allowed us to identify 182 different proteins that are involved in several biochemical processes, such as the electron transport, protein import, metabolism and ion or metabolite transport. The full range of isoelectric points, molecular masses and hydrophobicity values were represented in our list of proteins. Amongst the 182 proteins identified, 20 were unknown or had never previously been associated with the MIM. Altogether, this study demonstrates that the proteomics-based approach we have used is a powerful technique to identify new mitochondrial membrane proteins.

Key Words: Mitochondria; proteomics; LC-MS/MS; mitochondrial inner membrane.

1. Introduction

Mitochondria are eukaryotic organelles that play a crucial role in several cellular processes including energy production, fatty acid metabolism, oxidative

From: *Methods in Molecular Biology, vol. 432: Organelle Proteomics*
Edited by: D. Pflieger and J. Rossier © Humana Press, Totowa, NJ

stress, ion homeostasis and programmed cell death. The importance of mitochondria in cellular homeostasis is further underlined by the wide range of pathologies involving mitochondrial dysfunction that include cancer, myopathies, diabetes, obesity, aging and neurodegenerative diseases (1). These complex organelles are composed of an outer and inner membrane (MOM and MIM, respectively) defining the intermembrane space (IMS), the cristae and the matrix where the mitochondrial genome is confined (2). In mammals, the mitochondrial DNA (mtDNA) is organized into several hundreds of nucleoids that contain an average of two to eight copies of the mitochondrial genome (3) encoding 2 rRNAs, 22 tRNAs and 13 polypeptides, all of which are components of the respiratory chain. The low complexity of the mitochondrial genome indicates that the vast majority of the mitochondrial proteins (estimated to be ~1000) (4) are encoded by the nuclear genome. The MIM is of particular interest because, besides the well-known components of the respiratory chain complexes, it contains several ion channels and a variety of carrier proteins that certainly play a key role in mitochondrial homeostasis but whose molecular identity remains unknown. To address the latter issue, we used a proteomics-based approach whereby membrane proteins were extracted from highly purified mouse liver MIM with organic acid, and their tryptic peptides were analysed by two-dimensional liquid chromatography coupled to tandem mass spectrometry (2DLC-MS/MS). This procedure allowed us to identify 182 proteins, of which 20 were unknown or had never previously been associated with mitochondria.

2. Materials

2.1. Isolation of MIM

1. Glass homogenizer (30 mL).
2. MB buffer: 210 mM mannitol, 70 mM sucrose, 10 mM Hepes–NaOH, pH 7.5, and 1 mM ethylenediaminetetraacetic acid (EDTA).
3. Nycodenz (Sigma, St. Louis, MO, USA).
4. Ultracentrifuges and appropriate rotors.
5. 0.1 M Na_2CO_3, pH 11.5.
6. Peristaltic pump (Bio-Rad, Hercules, CA, USA).
7. Plastic gradient pourer.
8. Bradford assay (Bio-Rad).

2.2. Sodium Dodecyl Sulphate–Polyacrylamide Gel Electrophoresis

1. Separating buffer for two gels at 15% acrylamide: 15% acrylamide/bisacrylamide (Sigma), 400 mM Trizma-Base, pH 8.8 (adjusted with HCl), 0.1% (w/v) sodium

    dodecyl sulphate (SDS), 0.1% (w/v) ammonium persulfate (APS), and 0.1% (v/v) TEMED.

2. Stacking buffer at 5% acrylamide: 5% acrylamide/bisacrylamide (Sigma), 125 mM Trizma-Base, pH 6.8 (adjusted with HCl), 0.1% (w/v) SDS, 0.1% (w/v) APS, and 0.1% (v/v) TEMED.
3. Running Buffer (10x): 250 mM Trizma-Base, pH 8.3 (adjusted with HCl), 2.5 M glycine, and 35 mM SDS.
4. Sample buffer (3x): 50 mM dithiothreitol (DTT), 150 mM Trizma-Base, pH 6.8 (adjusted with HCl), 6% (w/v) SDS, 40% (v/v) glycerol, and 0.015% (w/v) bromophenol blue.
5. Pre-stained molecular weight markers of mass range 6–180 kDa (Invitrogen, Carlsbad, CA, USA).

2.3. Western Blot Analysis

1. Transfer Buffer (1x): 190 mM glycine, 40 mM Trizma-Base, and 20% (v/v) methanol.
2. Nitrocellulose membrane (Protran, Florham Park, NJ, USA) and 3 MM sheets (Whatman, Florham Park, NJ, USA).
3. PBST: phosphate-buffered saline (1x PBS)—137 mM NaCl, 2.7 mM KCl, 1.4 mM KH_2PO_4, 4.3 mM Na_2HPO_4, pH 7.4 with 0.1% (v/v) Tween 20.
4. Blocking solution: 5% (w/v) milk in PBST.
5. Antibody dilution buffer: PBST supplemented with 2.5% (w/v) milk.
6. Monoclonal antibodies: anti-prohibitin (Neomarkers, Fremont, CA, USA), anti-calnexin (Neomarkers, Fremont, CA, USA), anti-COXIV (Molecular Probes, Eugene, OR, USA), anti-mHSP70 (ABR, Golden, CO, USA), and anti-Bcl-xL (Pharmingen, San Diego, CA, USA). Polyclonal antibodies: anti-ANT (kindly provided by Dr. Stepien) and anti-SKL (Zymed Laboratories, San Francisco, CA, USA). The secondary antibodies were from DakoCytomation (Carpinteira, CA, USA) and Uptima Interchim (Montluçon, France).
7. Enhanced chemiluminescence (ECL) reagents from Amersham BioSciences, (Piscataway, NJ, USA).

2.4. Solubilization and Digestion of MIM Proteins

1. Formic acid (FA) 90% (*see* **Note 1**).
2. 3.3 M Na_2CO_3.
3. 8 M urea.
4. 1 mM DTT.
5. 10 mM iodoacetamide (IA).
6. 0.1 M $NaHCO_3$, pH 8.5.
7. Trypsin, sequencing grade (Promega, Madison, WI, USA).
8. SPEC-PLUS C18 columns (Ansys, Canonsburg, PA, USA).
9. Acetonitrile containing 0.1% (v/v) acetic acid.

2.5. Strong Cation Exchange Chromatography

1. Polysulfoethyl A column (2.1 mm × 20 cm) from PolyLC Inc. connected to a HPLC (binary pump system, such as Waters 1525 Binary HPLC Pump connected to a Waters Model 2487 Dual Absorbance detector Firmware Version 1.02 (Millford, MA, USA). Separations are monitored by Millennium 32 V 4.0 Chromatography management system).
2. Mobile phase (buffer A): 10 mM KH_2PO_4 in water/acetonitrile, 75/25 (v/v).
3. Mobile phase (buffer B): 10 mM KH_2PO_4 and 500 mM KCl in water/acetonitrile, 75/25 (v/v).
4. SPEC-PLUS C18 columns (Ansys).

2.6. Electrospray LC-MS/MS

1. Q-Tof Ultima mass spectrometer (hybrid quadrupole orthogonal acceleration time-of-flight mass spectrometer) fitted with a nanoflow electrospray source, online coupled to an analytical CapLC system (Waters, Milford, MA, USA).
2. Trapping column packed with 5 μm Symmetry C18 stationary phase (300 μm inner diameter × 5 mm).
3. Analytical column packed with PepMap C18 material (150 mm × 75 μm).
4. Buffer A: 0.1% (v/v) FA in H_2O/acetonitrile, 95/5 (v/v).
5. Buffer B: 0.1% (v/v) FA in acetonitrile.

2.7. Analysis of MS/MS Data

1. Mascot program (www.matrixscience.com).
2. Mus musculus database.

3. Methods

The study of the MIM proteome involves several critical steps:

1. Isolation and purification of mitochondria so as to reduce the possibility of identifying contaminating proteins from other cellular organelles (*see* **Scheme 1**).
2. Isolation and purification of the MIM to enrich for MIM proteins (*see* **Scheme 2**).
3. Extraction and solubilization of the membrane proteins (*see* **Scheme 3**).
4. Development of the appropriate proteomics-based method to identify MIM proteins (*see* **Scheme 3**).

3.1. Purification of the Mouse Liver MIM

3.1.1. Isolation and Purification of Mitochondria

1. Kill four female mice by exposure to CO_2. The livers (three lobes per mouse) are dissected out, rinsed twice with ice-cold MB buffer, and finely chopped with

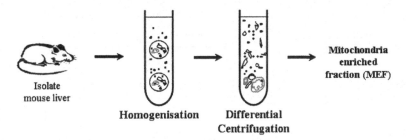

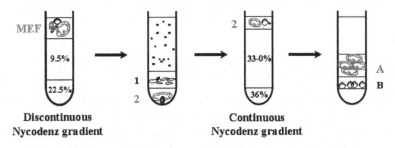

Scheme. 1. Schematic representation of the method for the purification of mitochondria from mouse liver.

a scalpel blade. The minced livers are then hand-homogenized in a total volume of 60 mL of MB buffer using a glass homogenizer kept on ice.

2. The resulting homogenate is centrifuged at 2000 g for 5 min at 4°C to remove the nuclei as well as non-homogenized tissue.

3. The 2000 g supernatant is centrifuged twice at 11,700 g for 10 min at 4°C to pellet a mitochondria-enriched fraction (MEF) (*see* **Scheme 1**).

4. The mitochondria-enriched pellet (*see* **Note 2**) is gently resuspended in 50 mL of MB buffer kept on ice, using a 10-mL pipette (*see* **Note 3**), to be purified on a discontinuous Nycodenz gradient *(5)*.

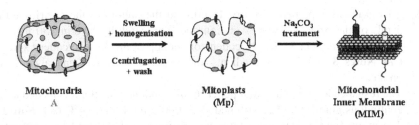

Scheme. 2. Schematic representation of the method for the purification of mitochondrial inner membranes.

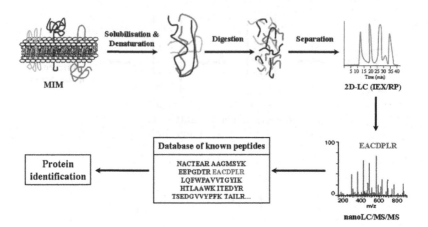

Scheme. 3. Schematic representation of the proteomic method.

5. The discontinuous gradient (*see* **Note 4**) is prepared by layering 10 mL of a 22.5% (v/v) Nycodenz solution at the bottom of a 35-mL centrifuge tube. 17 mL of a 9.5% (v/v) Nycodenz solution is carefully layered on the top of the 22.5% Nycodenz solution. The 22.5 and 9.5% Nycodenz solutions are obtained by dilution in MB buffer of a stock solution at 36% Nycodenz (v/v). Six discontinuous gradients are prepared separately.

6. The 50 mL of MEF is layered carefully (drop-wise) per 8-mL fractions on the top of each discontinuous Nycodenz gradient kept on ice to be centrifuged at 141,000 g for 1 h at 4°C.

7. The obtained pellet (0.5–1) with a light brown colour (fraction 2 on **Scheme 1** and **Fig. 1**), which contains the mitochondria, is resuspended in 20 mL of MB buffer on ice and further centrifuged twice at 11,700 g for 5 min to wash the organelles. It is noteworthy that fraction 1 (**Scheme 1** and **Fig. 1**) that is collected at the bottom of the 9.5% Nycodenz layer mainly contains proteins from the endoplasmic reticulum (ER) (*see* **Note 5**).

8. The mitochondrial pellet thus obtained is resuspended in 6 mL of MB buffer (*see* **Note 6**) to be further purified on a continuous Nycodenz gradient (*see* **Note 7**).

9. This gradient is prepared by first layering at the bottom of the centrifuge tube 5 mL of the 36% Nycodenz solution and then carefully layering 30 mL of a continuous gradient 33–0% Nycodenz on the top of the 36% Nycodenz solution (*see* **Notes 8** and **9**). Six gradients are prepared separately and always kept on ice.

10. The resuspended mitochondrial pellet (6 mL) obtained in **step 8** is loaded on the continuous gradient and centrifuged at 39,000 g for 1.5 h at 4°C (*see* **Note 10**). The major light-brown band above the interface of the 36 and 33% Nycodenz solutions is collected, diluted with 3 volumes of MB buffer and centrifuged twice at 11,700 g for 10 min. The obtained pellet, which contains the mitochondria

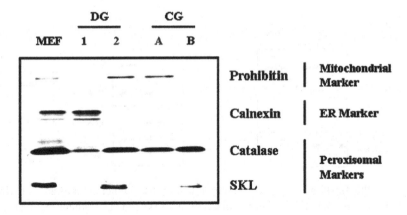

DG: Discontinuous gradient
CG: Continuous gradient

Fig. 1. Purification of mouse liver mitochondria. 25 µg of proteins from the mitochondrial fraction after differential centrifugation (MEF); the fractions obtained after a discontinuous Nycodenz gradient (DG) [endoplasmic reticulum (1) and mitochondria (2)] and the fractions obtained after a continuous Nycodenz gradient (CG) [mitochondria (A) and peroxisomes (B)] were analyzed by SDS–PAGE and western blot using antibodies against markers of the mitochondria (prohibitin), the endoplasmic reticulum (calnexin) and peroxisomes (catalase, SKL).

(fraction A, **Scheme 1** and **Fig. 1**), is resuspended in 2 mL of MB buffer (*see* **Note 11**). Of note, just below the band containing the mitochondria, there is a band containing peroxisomal proteins (fraction B, **Scheme 1** and **Fig. 1**) (*see* **Note 12**).

11. The protein concentration of the different obtained fractions (MEF, fractions 1, 2, A and B) is further assessed using a Bradford assay according to the manufacturer's instructions.

3.1.2. Isolation and Purification of the MIM

1. The purified mitochondria are resuspended in H_2O at a concentration of 5 mg/mL and stirred for 20 min on ice.
2. The mixture is further hand-homogenized 20 times using a glass homogenizer on ice and centrifuged at 12,000 g for 5 min at 4°C. The obtained pellet is resuspended in MB buffer and centrifuged again at 12,000 g for 5 min at 4°C. The supernatant is discarded, and the obtained pellet contains the purified mitoplasts (Mp, **Scheme 2** and **Fig. 2A and B**) (*see* **Notes 13, 14** and **15**).
3. The mitoplasts are resuspended in 0.1 M Na_2CO_3, at a final concentration of 0.5 mg/mL, and incubated on ice for 20 min. The preparation is then ultra-centrifuged at 100,000 g for 30 min at 4°C. The supernatant is removed, and the remaining pellet represents the purified MIM (*see* **Scheme 2** and **Fig. 2A and B**) (*see* **Notes 16, 17** and **18**).

A)   B)

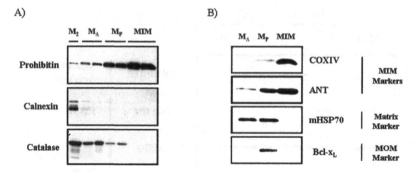

Fig. 2. Purification of mouse liver mitochondrial inner membranes (MIMs). (A) Samples of mitochondria after the discontinuous Nycodenz gradient (M_2), after the two consecutive gradients (M_A), mitoplasts (Mp) and MIM corresponding to 5 μg of total protein each were analyzed by SDS–PAGE and western blot using antibodies against markers of the mitochondria (prohibitin), the endoplasmic reticulum (calnexin) and peroxisomes (catalase). (B) 5 μg of mitochondria after the two consecutive Nycodenz gradients (M_A), mitoplasts (Mp) and MIM were analyzed by SDS–PAGE and western blot using antibodies against markers of the MIM (prohibitin, adenine nucleotide translocator, ANT; cytochrome c oxidase subunit IV, COXIV), the matrix (mitochondrial heat shock protein 70, mHSP70) and the mitochondrial outer membrane (Bcl-X_L).

3.1.3. Preparation of Samples to Assess the Purity of the Mitochondria and the MIM

1. 8.35 μL of 3× loading buffer is added to 25 μg (*see* **Fig. 1**) or 5 μg (*see* **Fig. 2A and B**) of each fraction obtained during the different purification steps, and the sample is diluted with MB buffer to a final volume of 25 μL.
2. The samples are incubated at 95°C for 5 min and centrifuged for 2 min at maximum speed in a tabletop centrifuge to spin down the drops of condensation. The samples are now ready to be separated by SDS–polyacrylamide gel electrophoresis (PAGE).

3.1.4. SDS–PAGE

1. The system that was used for the migration and transfer of proteins is from Hoefer (Mighty Small SE 250/260 gel system, San Franciso, CA, USA).
2. Prepare a 1.5-mm thick, 15% separating gel as described in **Subheading 2.2** using a 10.1 × 8.3 cm glass and silica plates that have both been thoroughly cleaned and rinsed first with distilled water and then with ethanol. TEMED and APS are added last to initiate the process of polymerization. Pour the gel, leaving space for a stacking gel, and overlay with a few drops of water-saturated isobutanol: the gel should polymerize in about 45 min.
3. Pour off the isobutanol and rinse abundantly the top of the gel with water to eliminate traces of isobutanol.

4. Prepare the stacking gel as described in **Subheading 2.2**, pour the stacking solution onto the polymerized separating gel and very rapidly insert the comb.

5. Prepare the 1× running buffer by diluting 50 mL of 10× running buffer with 450 mL of water in a measuring cylinder. Cover with parafilm and invert to mix.

6. Once the stacking gel has polymerized, carefully remove the comb and rinse the wells with distilled water.

7. Place the polymerized gel on the Hoefer gel migrating apparatus. Add the running buffer in the small chamber of the gel until it reaches the top of the gel and then add running buffer to the big chamber so that about 3 cm of the gel bath in the buffer. Rinse the wells with running buffer using a syringe.

8. Load the samples (25 μg in 25 μL) and the pre-stained molecular weight marker (10 μL).

9. Complete the assembly of the gel unit and connect to a power supply. The gel can be run at 180 V for ~45 min. The dye front (blue) can be run off the gel if preferred.

3.1.5. Western Blot Analysis of the Mitochondrial and the MIM Fractions

1. The samples that were separated by SDS–PAGE are transferred onto nitrocellulose membranes by electrophoresis. Prepare two sheets of 3 MM and a sheet of nitrocellulose membrane cut to the size of the separating gel. The nitrocellulose membrane is soaked in transfer buffer for at least 5 min before use to hydrate it properly. The 3 MM, nitrocellulose membrane and the pieces of foam are then left in transfer buffer until use.

2. The gel unit is disconnected from the power supply and disassembled. The stacking gel is removed and discarded.

3. The separating gel is then laid on one sheet of wet 3 MM paper, which is itself on a piece of foam. Then the nitrocellulose membrane is carefully laid on the top of the gel, ensuring that no bubbles remain trapped. To complete the sandwich, add the other sheet of 3 MM paper on the top of the membrane and finally, the other foam. Assembling the sandwich should be done in a tray containing transfer buffer to keep all the components soaked in buffer. Put the sandwich in the transfer cassette and place the cassette in the transfer tank so that the gel faces the negative pole while the nitrocellulose membrane faces the positive pole. The orientation of the sandwich in the tank is crucial as the proteins will migrate towards the positive pole.

4. The refrigerated/circulating water bath is switched on and a magnetic stir bar is activated in the tank.

5. The lid is put on the tank and the power supply switched on. The transfer is done at 80 V during 90 min or overnight at 20 V.

6. Once the transfer is finished, the cassette is taken out of the tank and disassembled to recover the nitrocellulose membrane. The coloured molecular weight markers should be clearly visible on the membrane.

7. The nitrocellulose membrane is then placed in a plastic box containing 10 mL of blocking solution and incubated for 1 h at room temperature on a rocking platform.

8. The blocking solution is discarded and the membrane is incubated in the appropriate antibody dilution buffer, overnight at 4°C on a rocking platform. The anti-prohibitin, anti-calnexin, anti-COXIV, anti-mHSP70 and anti-Bcl-xL were diluted at 1:1000, whereas the anti-ANT was 1:500-fold diluted and anti-catalase was at a 1:20,000 dilution. These antibodies were all diluted in blocking buffer.

9. The primary antibody is then removed (*see* **Note 19**), and the membrane is washed three times for 10 min with 20 mL of PBST at room temperature.

10. The secondary antibody is freshly prepared at a 1:2000 dilution in blocking solution and added to the membrane for 1 h at room temperature on a rocking platform.

11. The secondary antibody is discarded and the membrane washed three times for 10 min in PBST at room temperature on a rocking platform. Finally, the membrane is rinsed with 1× PBS.

12. Before removing the 1× PBS, prepare the ECL reagents according to the manufacturer's instructions and incubate the nitrocellulose membrane in the ECL solution for about 1 min, making sure that all parts of the membrane are appropriately soaked in ECL solution.

13. The blot is removed from the ECL reagents and placed between two sheets of acetate in an X-ray film cassette. The cassette is closed (*see* **Note 20**). This step has to be done very rapidly to avoid the decrease of luminescence due to the light.

14. In a dark room, open the cassette and place a film in the X-ray film cassette for a suitable exposure time (usually start with 1 min and then adapt the exposure time depending on the signal intensity). An example of the results obtained for the purification of the mitochondria and the MIM is shown in **Figs. 1, 2A and B**. The enrichment of the MIM proteins, such as prohibitin, COXIV or ANT, that is obtained with the purification method described above is between 10- and 20-fold compared with purified intact mitochondria. Furthermore, neither catalase nor calnexin are detected in the MIM fraction, indicating that the contamination of the mitochondrial preparation by peroxisomes and ER is negligible.

3.2. Proteomic Analysis of the MIM Proteins

To analyze the proteome of the MIM, we use an approach that has previously been shown to be very efficient to identify highly basic proteins and proteins that contain several transmembrane domains in yeast (*6*). This method is based on the tryptic digestion of the solubilized proteins followed by a peptide analysis by two-dimensional liquid chromatography coupled to tandem mass spectrometry (2D-LC/MS/MS) (*see* **Scheme 3**). These separation techniques reduce the complexity of the peptide mixture, according to their isoelectric point and their hydrophobicity, before sequencing by mass spectrometry.

3.2.1. Solubilization and Digestion of MIM Proteins

1. 500 µg of MIM is solubilized in 100 µL of a 90% FA solution (*see* **Notes 1** and **21**) for 10 min at room temperature on a rotating wheel.
2. Then, 340 µL of 3.3 M Na_2CO_3 is added to adjust the pH to 8.5 (*see* **Note 22**).
3. The sample is then transferred to an Eppendorf tube already containing 0.21 g of urea and incubated for 30 min at room temperature on a rotating wheel so as to denature the proteins (*see* **Note 23**).
4. To reduce the proteins, the sample is added to an Eppendorf tube already containing 1.5 mg of DTT and incubated for 30 min at room temperature on a rotating wheel (*see* **Note 23**).
5. Then the reduced sample is transferred to an Eppendorf tube already containing 15 mg of IA and incubated for 30 min at room temperature on a rotating wheel in the dark (*see* **Note 23**).
6. The alkylated solution is further diluted with 2 mL of 0.1 M $NaHCO_3$, pH 8.5. This step is important to dilute the concentration of urea to 2 M to be compatible with the tryptic digestion.
7. Add trypsin at the 1:20 dilution and incubate for 24 h at 37°C on a rotating wheel. Finally, the digested proteins are centrifuged at 100,000 g for 30 min at 4°C to remove undigested or unsolubilized proteins, which will be found in the pellet.
8. The supernatant is further applied to a prepared SPEC PLUS C18 column (*see* **Note 24**) using a method of solid phase extraction to concentrate the sample. Then add 500 µL of H_2O and let dry for 5 min, following manufacturer's instructions. Then the sample is eluted in 2× 500 µL of acetonitrile containing 0.1% (v/v) acetic acid. Finally, the eluted fraction is dried under N_2.

3.2.2. Strong Cation Exchange Chromatography

1. The strong cation exchange (SCX) column is connected to the HPLC system and washed first with 20 mL of H_2O and then with 20 mL of mobile phase B. Finally, the column is conditioned with 20 mL of mobile phase A at 0.2 mL/min.
2. The fraction dried with the SPEC column is resuspended in 200 µL of buffer A (*see* **Subheading 2.5.**), and 100 µL of the sample is loaded onto the SCX column.
3. The tryptic peptides are separated at a flow rate of 0.2 mL/min using the following step elution: 0–105 min, 0% of B; 105–155 min, 50% of B; 155–210 min, 70% of B; 210–265 min, 100% of B; and 265–300 min 0% of B. The UV detection system is set at a wavelength of 214 nm and 23 fractions are collected in 2-mL tubes.
4. Each collected fraction is desalted and concentrated using the SPEC PLUS C18 columns as described above.

3.2.3. Electrospray Nano-LC-MS/MS

1. The mass spectrometer conditions are as follows: positive ion mode, capillary voltage of 3.9 kV, cone voltage of 160 V, and source temperature 80°C. Tandem mass spectra are collected in one full MS scan (*m/z* range = 400–1500 Th)

followed by MS/MS spectra of the three ions detected with the highest intensities in the previous MS scan. MS/MS data are acquired over an *m/z* range of 50–2500 Th. The MS scan time is 1 s and the inter scan delay 0.1 s. Each MS/MS spectrum is acquired for 3 s before switching back to MS scan.

2. Each fraction collected from the SCX column and desalted is resuspended in 20 μL of 1% FA in H_2O.

3. Peptide mixtures are loaded onto the C18 trapping column at a flow rate of 30 μL/min to remove salts. After washing with the auxiliary pump with $FA/H_2O/ACN$, 0.1/95/5 (v/v/v), for 6 min, the trapped peptides are back-washed from the pre-column onto the analytical column. A linear gradient is used to elute the peptide mixture from mobile phase A ($FA/H_2O/ACN$, 0.1/95/5, v/v/v) to mobile phase B [0.1% (v/v) FA in ACN] at a flow rate of 5 μL/min before the splitter; this corresponds to a gradient flow rate through the analytical column of about 200 nL/min. The gradients that are used are the following: 6–90 min, 5–40% of B; 90–95 min, 40–60% of B and 95–96 min, 5% of B until 120 min. The mobile phases are degassed before use. The injector syringe is washed with degassed mobile phase A, and the injection volume is set at 6 μL. It is critical to maintain a steady flow without air bubbles in all CapLC operations and a stable ion beam is required.

4. Using this method, 21,000 MS/MS spectra were obtained, over all the SCX fractions.

3.2.4. Analysis of MS/MS Data

1. The list of spectra is compared to a mus musculus protein database using either Mascot algorithms or an algorithm similar to SEQUEST *(7)*. The tolerance on peptide mass measurement used with Mascot is 0.25 Da.

2. Using these algorithms with a score threshold of 30, about 30% of the MS/MS spectra were assigned. The sequences of the remaining spectra could not be identified mainly because of very low signal to noise ratio, chemical noise or co-fragmentation of two co-eluting species of close masses. Final assessment of each protein identified was confirmed by manual analysis of the corresponding spectra.

3. The proteomic analysis of the mouse liver MIM lead to the identification of 182 different proteins. As shown in **Fig. 3**, proteins representative of a variety of functions in which mitochondria are involved were identified. For example, 30% of the proteins are members of the respiratory transport chain, 13% are transporters or carriers and 8% of the proteins are part of the mitochondrial import machinery. Interestingly, 10% are unknown proteins or proteins that have never previously been associated with mitochondria.

About 3.5% of the identified proteins are known to reside outside the mitochondria, suggesting that these proteins could either be minor contaminants or could have a dual localization as it has recently been shown for the perox-isomal catalase *(8)*. In addition, the MIM preparation appears to be slightly

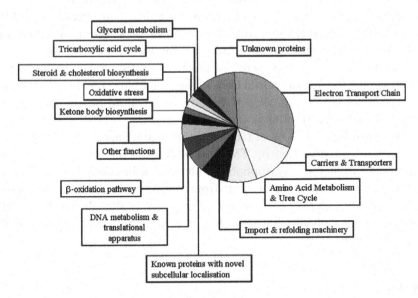

Fig. 3. Functional classification of the 182 different proteins identified in the mitochondrial inner membrane (MIM) proteome.

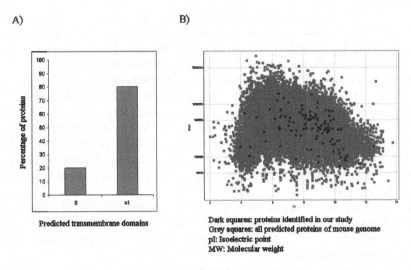

Fig. 4. Distribution of the identified proteins according to their predicted transmembrane domains (**A**), their pI and molecular weight (**B**).

(5%) contaminated by proteins known to be present at MOM. However, most of these proteins have been shown to localize at contact sites between the MIM and the MOM, thus suggesting that these contact sites were co-purified with the MIM. Finally, a number of matrix proteins (14%) are also found but in their precursor form. Their presence in the MIM proteome can be due to their association with the mitochondrial translocase membrane complex (TOM/TIM) as they have not yet been processed by the mitochondrial matrix peptidases to remove their mitochondrial targeting signals.

As shown in **Fig. 4A**, 80% of the proteins identified harbor 1–16 transmembrane domains (as predicted by TMpred), indicating that the procedure used is appropriate to identify hydrophobic proteins. Furthermore, this approach is very efficient to characterize proteins from a wide range of pI, as 50% of the identified proteins display a pI over 9 or below 4. It also allows the identification of proteins with very low (less than 10 kDa) or high (more than 100 kDa) molecular weight.

4. Notes

1. To prepare the solution of 90% FA, add 9 mL of FA onto 1 mL of ultra-pure H_2O.
2. An average of 150–200 mg of protein sample enriched in mitochondria is obtained in the MEF from four mice livers.
3. A small aliquot of the MEF is kept to assess the levels of purity by western blot analysis.
4. It is important to prepare the discontinuous Nycodenz gradients just before starting the preparation of the MEF and to keep them on ice. Indeed, as soon as the MEF is obtained, it is better to rapidly perform the discontinuous gradient to avoid degradation or modifications of the mitochondrial proteins.
5. The discontinuous Nycodenz gradient is performed mainly to remove ER, which is one of the major contaminating organelle of the mitochondrial fraction. This is possibly because of the existence of close contacts between the two organelles.
6. An average of 100 mg of protein sample enriched in mitochondria is obtained after the first Nycodenz gradient with four mice livers as starting material.
7. A small aliquot of the mitochondria- and ER-enriched fractions obtained after discontinuous Nycodenz gradient is kept to assess the levels of purity by western blot analysis.
8. It is important to prepare the continuous Nycodenz gradients while the discontinuous gradient is still centrifuging and to keep them on ice. Thus, once the mitochondrial fractions are obtained, they are immediately loaded on the continuous gradients.
9. To prepare the continuous Nycodenz gradients, it is necessary to use a peristaltic pump at a slow flow to avoid any disturbances. The pump is connected to a gradient pourer that contains the 33% Nycodenz solution in one cylinder and the 0% Nycodenz solution (only MB buffer) in the other cylinder.

10. The continuous Nycodenz gradient is performed mainly to remove peroxisomes from the mitochondrial preparation.

11. An average of 50 mg of mitochondria is obtained after the second Nycodenz gradient with four mice livers as starting material.

12. A small aliquot of the mitochondrial and peroxisomal fractions obtained after the continuous Nycodenz gradient is kept to assess the levels of purity by western blot analysis.

13. These steps of swelling and homogenization are crucial to eliminate proteins from the MOM and the IMS sub-compartment.

14. An average of 10–15 mg of mitoplasts is obtained after swelling and homogenizing mitochondria purified with the two consecutive Nycodenz gradients.

15. An aliquot of the obtained mitoplast-containing fraction is kept to assess the levels of purity by western blot analysis.

16. The sodium carbonate treatment is important to eliminate peripherally associated membrane proteins and proteins from the matrix.

17. An average of 2–4 mg of MIM is obtained after the sodium carbonate treatment.

18. It is important to keep an aliquot of the pellet obtained and further resuspend it in MB buffer to perform a Bradford assay and quantify the amount of obtained proteins.

19. The primary antibody can be saved for subsequent experiments by storage at –20°C. Some antibodies can be used up to 10 times over several months. However, with time, it is necessary to increase the time of exposure of the film with ECL reagents.

20. To be able to obtain the exact alignment of the subsequent film with the nitrocellulose membrane, it is useful to have a luminescent strip at the edge of the acetate sheet.

21. MIM proteins are extracted with organic acid because this method has previously been shown by Washburn and co-workers *(6)* to be very efficient to extract highly hydrophobic proteins from membranes and to be compatible with mass spectrometry.

22. Adjusting the pH is essential for the subsequent tryptic digestion.

23. Denaturation, reduction and alkykation steps are required to improve the efficacy of the digestion with trypsin. It is important to break the disulphide bonds and alkylate them to maintain proteins in an unfolded state.

24. SPEC PLUS C18 3-mL microcolumns are first rinsed with 500 μL of methanol and then 500 μL of H_2O before adding the sample. It is important not to dry the column before loading the sample onto the column.

References

1. Wallace, D. C. (1999) Mitochondrial diseases in man and mouse. *Science* **283**, 1482–1488.
2. Frey, T. G. and Mannella, C. A. (2000) The internal structure of mitochondria. *Trends Biochem. Sci.* **25**, 319–324.

3. Legros, F., Malka, F., Frachon, P., Lombes, A., and Rojo, M. (2004) Organization and dynamics of human mitochondrial DNA. *J. Cell Sci.* **117**, 2653–2662.

4. Scharfe, C., Zaccaria, P., Hoertnagel, K., Jaksch, M., Klopstock, T., Dembowski, M., et al. (2000) MITOP, the mitochondrial proteome database: 2000 update. *Nucleic Acids Res.* **28**, 155–158.

5. Stromhaug, P. E., Berg, T. O., Fengsrud, M., and Seglen, P. O. (1998) Purification and characterization of autophagosomes from rat hepatocytes. *Biochem. J.* **335**, 217–224.

6. Washburn, M. P., Wolters, D., and Yates, J. R., III. (2001) Large-scale analysis of the yeast proteome by multidimensional protein identification technology. *Nat. Biotechnol.* **19**, 242–247.

7. Da Cruz, S., Xenarios, I., Langridge, J., Vilbois, F., Parone, P. A., and Martinou, J. C. (2003) Proteomic analysis of the mouse liver mitochondrial inner membrane. *J. Biol. Chem.* **278**, 41566–41571.

8. Jiang, X. S., Dai, J., Sheng, Q. H., Zhang, L., Xia, Q. C., Wu, J. R., et al. (2005) A comparative proteomic strategy for subcellular proteome research: ICAT approach coupled with bioinformatics prediction to ascertain rat liver mitochondrial proteins and indication of mitochondrial localization for catalase. *Mol. Cell. Proteomics* **4**, 12–34.

8

Subcellular Fractionation and Proteomics of Nuclear Envelopes

Laurence Florens, Nadia Korfali, and Eric C. Schirmer

Summary

Because of its many connections to other cell systems, the nuclear envelope (NE) is essentially impossible to purify to homogeneity. To circumvent these problems, we developed a subtractive proteomics approach in which the fraction of interest and a fraction known to contaminate the fraction of interest are separately analyzed, and proteins identified in both fractions are subtracted from the data set. This requires that the contaminating fraction can be purified to homogeneity. In this case, microsomal membranes (MMs) are used to represent endoplasmic reticulum contamination, allowing the identification of transmembrane proteins specific to the NE. To circumvent problems commonly associated with analyzing membrane proteins, the multidimensional protein identification technology (MudPIT) proteomics methodology is employed.

Key Words: MudPIT; liquid chromatography; tandem mass spectrometry; nuclear envelope; microsomal membranes; transmembrane proteins; inner nuclear membrane; nuclear lamina.

1. Introduction

The nuclear envelope (NE) is comprised of a double membrane, its integral proteins, and the intermediate filament lamin polymer that underlies the membrane and binds several of its integral proteins. While the inner nuclear membrane is clearly a subdomain of the nucleus, the outer nuclear membrane is both the outermost layer of the nucleus and a subdomain of the endoplasmic reticulum with which it is continuous. As a result, the NE is highly complexed with both cytoplasmic membranes and filament systems

From: *Methods in Molecular Biology, vol. 432: Organelle Proteomics*
Edited by: D. Pflieger and J. Rossier © Humana Press, Totowa, NJ

as well as chromatin *(1–3)*. These many interconnections render biochemical isolation of pure NEs effectively impossible. To circumvent this, we developed a subtractive proteomics approach *(4)*. Two fractions are separately analyzed: (1) a NE-enriched fraction and (2) a fraction known to contaminate NEs, but which can be purified completely clean from nuclei. In this case, microsomal membranes (MMs) were chosen as the contaminating fraction because they can be cleanly separated from intact nuclei at an early stage in the preparation *(5)*. To purify NEs, nuclei are first isolated taking advantage of their mass and density *(6)* though many other cytoplasmic membranes and filament systems copurify. A large percentage of the contaminating membranes are removed by floating them on sucrose cushions that are penetrated by the dense nuclei. Finally, the nucleoplasmic contents are removed by swelling nuclei, digesting chromatin, and washing away released chromatin again by pelleting the larger/heavier NEs through sucrose *(7,8)*. MMs are purified by breaking the plasma membrane and vesiculating ER with dounce homogenization, removing nuclei and mitochondria by pelleting and floating MMs on dense sucrose *(5,9,10)*. Because of the complexity of the NE, subsequent extraction steps may be employed depending on the goals of a particular study; however, it should be noted that all additional extractions carry with them the possible loss of true NE proteins. Some subsequent purification procedures are based on the biochemical properties of the nuclear lamin polymer, which, together with associated transmembrane proteins, remains insoluble in the presence of the relatively high concentrations of salt and detergent used here [for review on lamin properties see *(11)*]. Other procedures rely on the solubility properties of membrane proteins, and here, we describe alkali extraction to enrich for transmembrane proteins.

Multidimensional protein identification technology (MudPIT) *(12–14)* was used to circumvent several inherent difficulties of working with membrane proteins. In shotgun proteomics, a complex protein mixture is digested into an even more complex peptide mixture. This step may seem counter-intuitive, but in actuality, peptides' physicochemical properties are more homogenous than proteins'. In particular, peptides can readily be separated by simple reverse-phase chromatography techniques, and their molecular weights are ideal to be analyzed by mass spectrometry. In addition, with the advent of tandem mass spectrometry, amino acid sequences can be deduced from peptide fragmentation patterns using software such as SEQUEST *(15)*. SEQUEST peptide level information is reassembled into protein lists using software such as DTASelect *(16)*, whereas multiple protein lists are compared using software such as CONTRAST *(16)*. Proteins appearing in both NE and MMs fractions are subtracted from the data set, leaving an in silico "purified" NE-specific data set ("subtractive proteomics"). This NE-specific list of proteins can then be queried for the presence of transmembrane proteins. All 13 previously known NE integral proteins (NETs) were readily identified, as well as all known components of the nuclear pore complex. An additional 67 hypothetical transmembrane proteins were identified. All eight of these novel putative NETs originally tested targeted to the NE *(4)* while subsequent characterizations have thus far confirmed the localization of nearly two dozen more NETs *(17,18)* (W. E. Powell, V. Lazou, D. Kavanagh, P. Malik, N. Korfali, and E. Schirmer, unpublished results).

2. Materials

2.1. Preparation of Tissue (Rodent Livers)

1. Volumes in the protocol are given based on grams of liver or OD of nuclei. To estimate how many animals to use ~5 g of liver can be obtained from one rat and ~1.25 g of liver can be obtained from one mouse. We generally produce 1000–2000 OD of nuclei (1 OD = 3,000,000 nuclei) from the livers of 10 rats (*see* **Note 1**).
2. 6- to 8-week-old rats (e.g., Sprague–Dawly or equivalent) or mice (e.g., CB6F1/J or equivalent) (*see* **Note 2**).
3. Guillotine or equivalent local method for euthanizing animals.
4. Dissection scissors, scalpel, and forceps/ tweezers.
5. Two beakers on ice, one with 200 mL of double distilled H2O and the other with 200 mL of 0.25 M SHKM buffer (*see* **Note 3**).
6. Appropriate materials for covering surfaces during procedure and for cleaning and waste disposal.
7. 0.25 M SHKM: 50 mM HEPES–KOH, pH 7.4, 25 mM KCl, 5 mM MgCl2, 250 mM sucrose, and freshly added 2 mM dithiothreitol (DTT) and 1 mM phenylmethylsulfonyl fluoride (PMSF; from a 100 mM solution in ethanol) (*see* **Note 4**).

2.2. Preparation of Nuclear Envelopes

2.2.1. Hardware

1. Potter–Elvehjem homogenizer with a motor-driven Teflon pestle providing 0.1–0.15 mm clearance and the drive motor capable of 1500 rotations per minute [e.g., Potter S Homogenizer catalogue numbers 853 3032 (motor), 854 2600 (60-mL homogenizer cylinder), and 854 3003 (Plunger made of PFTE) from Sartorius or equivalent].
2. Loose-fitting (Wheaton type B pestle) glass dounce homogenizer with clearance of between ~0.1 and 0.15 mm.
3. Swinging bucket rotor with a tube capacity of at least 200 mL if processing 10 rats (e.g., Beckman-Coulter SW28 rotor with Beckman-Coulter 344058 Ultra-Clear 25 × 89 mm centrifuge tubes).
4. Local standard light microscope, glass slides, and coverslips.
5. Assorted beakers, 2 funnels, and several spatulas.

6. Sterile cheesecloth (*see* **Note 5**).
7. Large bore luer lock stainless steel needles (e.g., 14 gauge) of greater length than centrifuge tubes and glass luer lock syringes.

2.2.2. Solutions

Solution names include the initials for the primary components: S for sucrose, H for HEPES–KOH, K for KCl, and M for MgCl2 (*see* **Note 6**).

1. DNase (e.g., Sigma DNase I D4527) resuspended at 10 U/ µL in H2O.
2. RNase (e.g., Sigma RNase A R4875) resuspended in H2O at 10 mg/ mL and boiled for 20 min.
3. Protease inhibitors (*see* **Note 7**): all solutions require freshly added 1 mM PMSF, 1 µg/mL aprotinin (from a 1 mg/mL stock in H2O), 1 µM pepstatin A [from a 1 mM stock in dimethyl sulfoxide (DMSO)], and 10 µM leupeptin hemisulfate (from a 10 mM stock in H2O) (*see* **Note 8**).
4. 0.25 M SHKM: 50 mM HEPES–KOH, pH 7.4, 25 mM KCl, 5 mM MgCl2, 250 mM sucrose, and freshly added 2 mM DTT (from a 1 M solution in H2O) and protease inhibitors. This is the same solution used to wash the freshly isolated livers, except that additional protease inhibitors are added to the fresh buffer in which homogenization occurs.
5. 2.3 M SHKM: 2.3 M sucrose, 50 mM HEPES–KOH, pH 7.4, 25 mM KCl, 5 mM MgCl2, and freshly added 2 mM DTT (*see* **Note 9**).
6. 30% SHKM: 0.9 M sucrose, 50 mM HEPES–KOH, pH 7.4, 25 mM KCl, 5 mM MgCl2, and freshly added 2 mM DTT and protease inhibitors (*see* **Note 10**).
7. 30% SHM buffer: 0.9 M sucrose, 10 mM HEPES–KOH, pH 7.4, 5 mM MgCl2, and freshly added 2 mM DTT and protease inhibitors.
8. 10% SHM buffer: 0.3 M sucrose, 10 mM HEPES–KOH, pH 7.4, 5 mM MgCl2, and freshly added 2 mM DTT and protease inhibitors.

2.3. Preparation of Microsomal Membranes

1. The same hardware is required as for "Preparation of Nuclear Envelopes".
2. A Ti45 fixed angle rotor or equivalent that can provide 150,000 g and matching tubes.
3. The same sucrose solutions used for the preparation of NE can be used in preparing microsomes. In particular, 2.3 M SHKM, 0.25 M SHKM, and a mixture of the two to 1.9 M sucrose will be required.

2.4. Extraction of Fractions

1. TLA100.3 rotor for table-top ultracentrifuge or equivalent and corresponding tubes (e.g., Beckman-Coulter 343778 polycarbonate 11 × 34 mm tubes).
2. Salt/ detergent extraction: Octyl β-D-glucopyranoside (also called *n*-octyl glucoside; Sigma O9882) resuspended at 1% (w/v) in a solution containing 25 mM HEPES–KOH, pH 7.5, and 400 mM NaCl.
3. Alkaline extraction: 0.1 N NaOH, 1 mM DTT.

2.5. Determining the Purity Quality of Fractions

1. Standard labware for western blotting and antibodies to lamins characterized integral NE and endoplasmic reticulum proteins.
2. Electron microscope facility and glutaraldehyde for fixation.

2.6. Preparation and Digestion of Proteins for MudPIT

1. pH indicator strips, 7.5 to 14 (EMD Chemicals, San Diego, CA, USA, Part # EM-9587-3; http://www.emdchemicals.com/).
2. Eppendorf Thermomixer and Thermomixer block for 1.5-mL tubes (Eppendorf, Hamburg, Germany, Part # 5355 000.011; http://www.eppendorf.com/).
3. Ammonium bicarbonate (Sigma-Aldrich, Saint Louis, MO, USA, Part # A6141; http://www.sigma-aldrich.com) as a 1 M solution in double distilled H2O stored at 4°C.
4. Urea, solid (Sigma-Aldrich, Part # U 1250).
5. 90% formic acid (J.T. Baker, Phillipsburg, NJ, USA, Part # JT0129-1; http://www.jtbaker.com).
6. Cyanogen bromide (Sigma-Aldrich, Part # 48,143-2).
7. Ammonium hydroxide solution, NH4OH (Sigma-Aldrich, Part # A6899), in H2O at 0.9 g/mL density).
8. Tris(2-carboxylethyl)-phosphine hydrochloride, TCEP (Pierce, Rockford, IL, USA, Part # 20490; http://www.piercenet.com/), as a 1 M stock in HPLC grade H2O stored at –20°C.
9. Iodoacetamide, IAM (Sigma-Aldrich, Part # I 1149), made fresh weekly as a 500 mM stock in double distilled H2O and stored at –20°C.
10. Endoproteinase LysC, sequencing grade (Roche Applied Science, Indianapolis, IN, USA, Part # 11047825001; http://www.roche-applied-science.com/), as a 1 μg/μL stock in double distilled H2O and stored at –20°C.
11. Calcium chloride (EMD Chemicals, Part # EM-3000) as a 500 mM stock in double distilled H2O and stored at room temperature.
12. Poroszyme bulk immobilized trypsin, bulk media (Applied Biosystems, Foster City, CA, USA, Part # 2-3127-00; http://www.appliedbiosystems.com/), stored at –20°C.

2.7. Packing and Loading of Microcapillary Column

2.7.1. Hardware

1. SPEC-PLUS PTC18 cartridges (Varian/Ansys Technologies Inc., Palo Alto, CA, USA, Part # 572-03; http://www.varianinc.com/)
2. 1-mL syringes (Becton Dickinson and Co, Franklin Lakes, NJ, USA, Part # BD309602; http://www.bd.com).
3. SpeedVac concentrator (Thermo Electron, San Jose, CA, USA, Part # SPD111V; http:/www.thermo.com/).
4. Micropipette laser-based puller (Sutter Instrument Co, Novato, CA, USA, Part # P-2000; http://www.sutter.com/).

5. Polyimide-coated fused silica capillary, 100 μm ID × 365 μm OD (Polymicro Technologies, Phoenix, AZ, USA, Part # TSP 100375; http://www.polymicro.com/).

6. Column scribe (Chromatography Research Supplies, Louisville, KY, USA, Part # 205312; http://www.chromres.com/crs/).

7. Helium pressure cell (custom-made, MTA for blueprints available by request from John Yates, Scripps Research Institute, La Jolla, CA; or Brechbuehler, Inc., Houtson, TX, USA, Part # 1 100 110; http://www.brechbuehler.com/).

2.7.2. Solutions

1. HPLC grade methanol (EMD Chemicals, Part # EM-MX0488-6).
2. HPLC grade acetonitrile (EMD Chemicals, Part # EM-AX0142-6).
3. Glacial acetic acid (Sigma-Aldrich, Part# A 6283).
4. 90% formic acid (J.T. Baker, Part # JT0129-1).
5. Ammonium acetate (J.T. Baker, Part # JT0599-8).
6. 5-μm Polaris C18A reversed-phase column (Varian/Metachem Technologies, Torrance, CA, USA, Part # 2000-250X046; *http://www.varianinc.com/*) (*see* **Note 11**).
7. 5-μm Partisphere strong cation exchange column (Whatman, Florham Park, NJ, USA, Part # WC4621-1507; http://www.whatman.com/) (*see* **Note 11**).
8. Buffer A: acetonitrile/formic acid/double distilled H2O, 5/0.1/95 (v/v/v).

2.8. Liquid Chromatography In-Line with Tandem Mass Spectrometry

2.8.1. Hardware

1. Agilent 1100 series G1379A degasser, G1311A quaternary pump, and G1323B controller (Agilent Technologies, Palo Alto, CA, USA; http://we.home.agilent.com/).
2. LCQ DECA tandem mass spectrometer (Thermo Electron).
3. Nano electrospray stage (custom-made, MTA for blueprints available by request from John Yates, Scripps Research Institute, La Jolla, CA; or Thermo Electron Nanospray II ion source; or Brechbuehler, Inc., Part # 1 2000 1000).
4. MicroTight Cross (UpChurch Scientific, Oak Harbor, WA, USA, Part # P-777; http://www.upchurch. com/).
5. Micro Ferrule for 360 μm o.d. tubing (UpChurch Scientific, Part # F-152).
6. Polyimide-coated fused silica, 50 μm ID × 365 μm OD (Polymicro Technologies, http://www.polymicro.com/; Part # TSP 050375).
7. Gold wire 0.025" diameter (Scientific Instrument Services, Ringoes, NJ, USA, http://www.sisweb.com/; Part # W352).

2.8.2. Solutions

1. Buffer A: acetonitrile/formic acid/double distilled H2O, 5/0.1/95 (v/v/v).
2. Buffer B: acetonitrile/formic acid/double distilled H2O, 80/0.1/20 (v/v/v).
3. Buffer C: 500 mM ammonium acetate in buffer A, filtered.

2.9. Data Analysis

1. SEQUEST™ *(15)* (Thermo Electron and John Yates, Scripps Research Institute, La Jolla, CA) and/or PEP_PROBE *(19)*.
2. DTASelect/CONTRAST *(16)* (available by request from John Yates, Scripps Research Institute, La Jolla, CA).
3. TMpred *(20)* (http://www.ch.embnet.org/software/TMPRED_form.html).

3. Methods

The first step in NE enrichment is the isolation of nuclei. Critical to this step and all subsequent steps is the fact that nuclei from different tissues have distinct densities, thus the concentration of sucrose in buffers may need to be altered or centrifugation steps lengthened if NEs are to be isolated from tissues other than liver (*see* **Note 12**).

3.1. Preparation of Tissue (Rodent Livers)

Most of the NE preparation procedures can be efficiently performed with one individual; however, euthanizing and dissecting the animals should be done quickly, and it is very helpful to have assistance at this point.

1. Overnight—fast the animals the night before procedure (*see* **Note 13**).
2. Euthanize rats or mice according to local animal protocols.
3. Immediately pull up on the ventral skin to isolate from the abdominal cavity. Make an incision anterior to posterior and two perpendicular incisions above the thoracic vertebrae and below the abdomen. Peel the skin back to access the liver.
4. Remove the liver with a scalpel being extremely careful to avoid the yellowish tube directly behind as it is protease rich (*see* **Note 14**).
5. While clasped in the forceps, rinse fresh livers quickly in the beaker on ice containing H2O and then immediately place in the pre-weighed beaker on ice containing 0.25 M SHKM buffer with freshly added PMSF.
6. Return to the laboratory with livers as soon as local animal protocols have been satisfied. If possible, have two people working at this point so that one can begin processing the material while the other deals with disposal and clean-up.
7. Weigh livers to determine the volume of buffer to be added for homogenization.

3.2. Preparation of Nuclear Envelopes

1. Pour off the buffer and resuspend the livers in fresh, ice-cold 0.25 M SHKM with freshly added protease inhibitors (*see* **Note 15**) at 2 mL of buffer for every gram of liver, for example, 50 g of livers should be resuspended in 100 mL of buffer (*see* **Note 16**).
2. Use scissors to chop livers into small pieces in the beaker with buffer.

3. Pour into a 55-mL or larger Potter–Elvehjem homogenizer and homogenize at 1500 rpm in the cold, bringing the pestle to the bottom three times (*see* **Note 17**). No more than 20 g of livers can be homogenized in a 55-mL homogenizer at a time.

4. Rinse the homogenizer with 1/10th volume of buffer and add to the homogenate. More buffer can be used as the nuclei will be pelleted in **step 7**, but the total volume should be kept at least 20% lower than the volume of the tubes used in **step 7** because an additional wash can be applied in **step 6**.

5. In a cold room, fold cheesecloth over four times and lay in funnel. Pour ~40 mL of crude homogenate through the cheesecloth (*see* **Note 18**).

6. As the flow slows, fold the cheesecloth over and roll a sterile pipette along the outside from top to bottom to squeeze the fluid out. A wash with buffer poured into the central cavity formed by the cheesecloth may increase yield slightly.

7. Remove to 250-mL round-bottom centrifuge tubes (*see* **Note 19**) and underlay with a 10-mL cushion of 30% SHKM using a 14 gauge needle and syringe. Pellet nuclei at 1000 g in a swinging bucket rotor (e.g., 2000 rpm in a Beckman-Coulter J6MI floor model centrifuge) for 10 min at 4°C.

8. Remove the supernatant carefully as the pellets are very soft. The pellets mostly contain intact nuclei, but contaminating membranes and collapsed cytoplasmic filaments will also be present in the pellet. Keep this supernatant if MMs are going to be prepared at the same time from the same tissue as they will be in this fraction.

9. Resuspend pellets in 2.3 M SHKM (roughly 20 mL for every 10 g of starting liver mass) and homogenize in Potter–Elvehjem homogenizer with three more strokes of the pestle at 1500 rpm. Rinse with the same buffer, accrue and dilute with 0.25 M SHKM to a concentration of 1.9 M sucrose (roughly 5 mL of 0.25 M SHKM for every 20 mL of 2.3 M SHKM).

10. Aliquot by 25 mL into each SW28 ultracentrifuge tube and underlay with 5 mL of 2.3 M SHKM using a 14 gauge needle in a luer lock syringe (*see* **Note 20**).

11. Balance the tubes by adding buffer to the top and spin in an SW28 rotor for 60 min at 82,000 g (25,000 rpm).

12. Move to cold room and scrape off the red layer at the top with a spatula. This lipid- and collagen-rich layer is thick and has a consistency that resembles rubber. Then pour off the rest of the supernatant by rapid inversion. Keep the tubes upside down in the cold for 10 min to drain them. Then gently wipe out the inside walls of the tubes with a folded kimwipe (or equivalent towel), being very careful not to touch the pellet.

13. Insert a clean dry spatula without touching the walls of the tube and scrape out the nuclear pellet. It is important that the spatula be dry so that the pellet will cling to it. Remove, again avoiding touching the walls of the tube, and resuspend each pellet in 2 mL of 0.25 M SHKM with freshly added 2 mM DTT and protease inhibitors (*see* **Note 21**).

14. Resuspend the pellet using a loose (Wheaton B-type pestle, ~0.1–0.15 mm clearance) dounce homogenizer until all aggregates are broken. Wash the homog-

enizer with an equal volume of the same buffer and remove to 40-mL round-bottom centrifuge tubes. Underlay with 5 mL of 30% SHKM with freshly added 2 mM DTT and protease inhibitors as in **step 7**. Pellet nuclei by centrifugation at 1000 g for 10 min. This pelleting step serves to further wash away collagen. Decant by inversion as the pellet should be compact (*see* **Note 22**).

15. Resuspend in 5 mL of 0.25 M SHKM for every 10 g of starting liver, dounce again, and take two small aliquot one for counting nuclei in a hemacytometer in **steps 6**, and the other for **step 17**. Then repeat pelleting as in **step 14**.

16. During centrifugation, count nuclei. The number of nuclei in the squares on the field should be multiplied by 10^4 (unless a different correction factor applies to your hemocytometer) and by the total volume (in mL) in which cells were resuspended for centrifugation in **step 15**. This number should be divided by 3 $\times 10^6$, which is the number of nuclei in an OD. The formula is

$$OD_{total} = \frac{\text{Number of nuclei on grid} \times 10^4 (\text{nuclei/mL}) \times \text{mL}}{3 \times 10^6 (\text{nuclei/OD})} \qquad (1)$$

17. Resuspend in 10% SHM with freshly added 2 mM DTT and protease inhibitors at 20 OD/mL. Take an aliquot to compare nuclei to the second aliquot saved in **step 15**. The cells should be observed to swell in the hypotonic SHM buffer.

18. Add 4 U/mL DNase and 1 μg/mL RNase and incubate at room temperature for 20 min. Observe digestion on the microscope in parallel. The phase grey of the nuclei should diminish slightly (*see* **Note 23**).

19. Underlay the solution with 30% SHM with freshly added DTT and protease inhibitors. Spin for 30 min at 6000 g using a swinging bucket rotor (e.g., 5000 rpm in a Beckman-Coulter floor model J6MI centrifuge) (*see* **Note 24**).

20. Carefully aspirate off the supernatant (do not decant by pouring) as the pellet will be very soft (*see* **Note 25**). The supernatant should contain a small fraction of the nuclear chromatin.

21. Resuspend the pellet at 50 OD/mL in 10% SHM. Add 20 U/mL DNase and 5 μg/mL RNase and incubate at room temperature, carefully following the digestion in an aliquot under the microscope. When 90% of nuclei are no longer phase-grey most of the nuclear chromatin will have been released; thus at this time, aliquot 4 mL of NE solution (200 OD) each to centrifuge tubes (chosen for desired storage method) and spin at 6000 g (5000 rpm) for 30 min (no cushion).

22. Carefully aspirate the supernatants and immediately flash-freeze the pellets in liquid nitrogen and store at −80°C.

3.3. Preparation of Microsomal Membranes

1. The supernatant after pelleting of nuclei (**step 7** in "Preparation of Nuclear Envelopes") is supplemented with 0.5 mM EDTA to inhibit metalloproteases and subjected to a subsequent centrifugation at 10,000 g (11,500 rpm in a Ti45 Beckman-Coulter ultracentrifuge rotor or 10,500 rpm in a JA25 Beckman-Coulter centrifuge rotor) for 15 min to remove mitochondria (*see* **Note 26**).

2. The post-mitochondrial supernatant is mixed with 5 volumes (e.g., 1 mL + 5 mL) of 2.3 M SHKM containing 0.5 mM EDTA, 2 mM DTT, and protease inhibitors to achieve a final sucrose concentration of just under 2 M sucrose.

3. The MMs are floated by pouring 35 mL of the diluted membranes in each Ti45 tube and overlayered with 9.5 mL of 1.9 M SHKM with 0.5 mM EDTA, 2 mM DTT, and protease inhibitors, and overlay this with 3 mL of 0.25 M SHKM with 0.5 mM EDTA, 2 mM DTT, and protease inhibitors. Centrifuge at 57,000 *g* (27,000 rpm) in Ti45 rotor for 5 h (*see* **Note 27**).

4. The MMs will be in the interphase between the 1.9 M sucrose phase and have a yellow-brown appearance (at least in liver). Above this band should be a clear phase on top of which will be a white flakey band. Below the 1.9 M to lower phase transition the media should begin taking on a pink shade. The yellow-brown MM band can be recovered by extraction with a syringe either through tube puncture from the side with a needle and or by inserting the needle through the upper phase (*see* **Note 28**).

5. The membranes are diluted with 4 volumes of 0.25 M SHKM containing 0.5 mM EDTA, freshly added DTT and protease inhibitors and are pelleted at 152,000 *g* (44,000 rpm in a type 45 Ti, 48,000 rpm in a type 50 Ti, or 60,000 rpm in a TLA100.3 rotor) for 75 min (*see* **Note 29**).

3.4. Extraction of Fractions

1. Alkaline extraction: resuspend 100 OD of NE or MM pellet on ice in 1 mL of 0.1 M NaOH. Immediately move to TLA100.3 ultracentrifuge tubes and pellet insoluble material at 104,000 *g* (35,000 rpm) for 35 min (*see* **Note 30**). Rinse the packed pellet rapidly with double distilled H2O and either freeze at –80°C or directly process to digest for mass spectrometry.

2. Extraction with salt and detergent: resuspend NE pellet in salt/detergent buffer and incubate on ice for 15 min, move to TLA100.3 ultracentrifuge tubes and pellet insoluble material at 104,000 *g* (35,000 rpm) for 35 min. Rinse the packed pellet with double distilled H2O and either freeze at –80°C or directly process to digest for mass spectrometry.

3.5. Determining the Purity Quality of Fractions

1. Western blotting, estimating a protein concentration of ~3 μg/OD, can be used to track the partitioning of known NE or MM proteins during cell fractionation (*see* **Fig. 1**) and in subsequent extraction of NEs. We generally use antibodies to nuclear lamins, the inner nuclear membrane proteins LAP1 and LAP2 (available from Babco/Covance, Richmond, CA, USA), and the endoplasmic reticulum marker RIC6 (anti-ribophorin).

2. Electron microscopy can also be used to determine the quality of NEs and MMs. For examples of clean preparations see reference *(7)*.

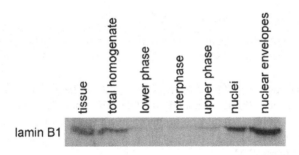

Fig. 1. Enrichment of rat liver NEs during the procedure. Different fractions were kept during the NE purification and concentrated so that the same percentage of each fraction was loaded in gel sample wells. Fractions were resolved by SDS–PAGE, transferred to PVDF membranes, and reacted with antibodies to lamin B1, a well-characterized NE protein. This NE marker was not notably lost during homogenization (compare "tissue" to "total homogenate") and chromatin extraction steps (compare "nuclei" to "nuclear envelopes"). Moreover, lamin B1 was largely absent from the three phases through which nuclei pelleted in the sucrose gradient step ("upper phase," "interphase," and "lower phase").

3.6. Preparation and Digestion of Proteins for MudPIT

Salt- and detergent-extracted NE proteins from mouse and rat livers ($\sim$225 µg and 375 µg proteins, respectively) are digested with endoproteinase Lys-C and trypsin while membrane pellets from NaOH-extracted mouse NEs and MMs are first solubilized in 90% formic acid and cyanogen bromide, before being further digested with endoproteinase Lys-C and trypsin.

3.6.1. Salt- and detergent-extracted NEs

1. Resuspend salt- and detergent-extracted NE pellets from mouse and rat livers in 200 µL of 0.1 M ammonium bicarbonate, pH 8.5, and 8 M urea.
2. To reduce disulfide bonds, add 1 M TCEP to 5 mM final concentration and incubate at room temperature for 30 min.
3. To carboxyamidomethylate free cysteines, add 500 mM IAM to 10 mM final concentration and incubate at room temperature for 30 min in the dark.
4. Add endoproteinase Lys-C (at 1 µg/µL) to a final substrate to enzyme ratio of 100:1 (w/w) and incubate at 37°C overnight with shaking.
5. Dilute sample to 2 M urea with 100 mM ammonium bicarbonate.
6. Add 1 M CaCl2 to a final concentration of 2 mM.
7. Add 10–15 µL of trypsin beads at 37°C (approximate substrate to enzyme ratio of 100:1) and incubate overnight with shaking.
8. Spin samples at 17,500 × g for 30 min.
9. Pull off supernatants (discard bead pellets).

3.6.2. NaOH-Extracted NEs and MMs

1. Under a fume hood, resuspend the membrane pellets in 100 μL of 90% formic acid and 500 mg/mL CNBr; mix well by pipetting and leave overnight in the dark.
2. Add concentrated NH4OH drop by drop on ice. Check pH using pH-indicator strips after every 100 μL of added NH4OH. The final pH should be around 8.5, and the final volume should be around 500 μL, corresponding to a 3- to 5-fold dilution (*see* **Note 31**).
3. Add solid urea to 8 M.
4. The next steps are as in **Subheading 3.6.1., steps 2** through **9**, with the exception that IAM is added to 20 mM final concentration.

3.7. Packing and Loading of Microcapillary Column

After digestion, the peptide mixtures are usually in large final volumes. In particular, the CNBr/trypsin digestion, with its multiple dilution steps, can reach volumes of well over 1 mL, which would take a while to load onto traditional 100-μm columns. To concentrate and desalt the samples, we use an off-line solid phase extraction on these peptide digests, before loading them onto custom-made microcapillary columns.

3.7.1. Solid Phase Extraction of Peptide Mixtures

1. Wash a SPEC-PLUS PTC18 cartridge with 50 μL of MeOH (push the solution through using a 1-mL syringe).
2. Wash the cartridge twice with 400 μL of 0.5% (v/v) acetic acid in water (after this step, do not allow the cartridge to dry).
3. Load the peptide mixture.
4. Wash the loaded cartridge four times with 400 μL of 0.5% acetic acid.
5. Elute peptides off with 50 μL of acetonitrile/acetic acid/water, 90/0.5/9.5 (v/v/v).
6. Speed-vac the samples down to dryness.
7. Add acetonitrile/acetic acid/water, 5/0.5/94.5 (v/v/v), to a final volume of 20 μL.
8. Store at –80°C or load directly onto a 100-μm microcapillary column.

3.7.2. Double-Phase Fused-Silica 100 μM Microcapillary Column

1. Place about 40 cm of 100 μm ID × 365 μm OD fused silica into P-2000 laser puller and use heating/pulling cycle settings (*see* **Note 32**) to pull the capillary to about a 5 μm opening.
2. Make slurries of 5-μm Polaris C18 Reverse Phase (RP) and of 5 μm strong cation exchange SCX material, both at about 15 mg/mL in 500 μL of methanol (*see* **Note 33**).

3. Pack fused silica column with 9–10 cm of 5 μm Polaris C18 RP using the high-pressure loading device (*see* **Fig. 2A**). Mark resin level in column with a marker (*see* **Note 34**).
4. Pack with 4–5 cm of 5 μm strong cation exchange material (*see* **Fig. 2B**).
5. Wash with methanol for at least 10 min.
6. Equilibrate in buffer A for at least 30 min.

3.7.3. Off-Line Loading and Desalting

1. Load the sample onto the microcapillary column by placing the sample-containing Eppendorf tube in the high-pressure device (*see* **Fig. 2C**).
2. Wash with buffer A for at least 30 min using the high-pressure device.

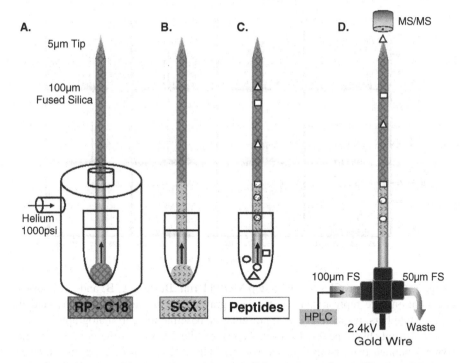

Fig. 2. Column packing, loading, and setup. (**A**) A 100-μm fused silica capillary with a pulled tip is inserted into a high-pressure device and packed using helium pressure with a slurry of Aqua C-18 RP in methanol. (**B**) The column is then packed with SCX material in a slurry, then washed with methanol and buffer A. (**C**) The complex peptide mixture is pressure loaded onto the column that is subsequently equilibrated in buffer A. (**D**) Loaded and washed column is installed in-line with a quaternary HPLC pump and a tandem mass spectrometer through a micro-cross.

3.8. Liquid Chromatography In-Line with Tandem Mass Spectrometry

3.8.1. Multidimensional Liquid Chromatography

1. Install the loaded and washed two-phase microcapillary column on the nanoelectrospray stage. Connect the microcapillary column, quaternary HPLC pump, and gold wire through which a 2.4 kV voltage is applied to the liquid phase and overflow tubing using a MicroTight Cross (*see* **Fig. 2D**).
2. Keep the HPLC flow rate constant at 0.1 mL/min throughout the chromatography. However, to achieve a slower flow rate at the tip of the column of about 200–300 nL/min, split the flow using a waste line consisting of 50-μm fused silica capillary cut to about 40 cm (i.e., back pressure of ~40 bar).
3. Run a 12-step chromatography (24 h) on samples with the gradient parameters described in **Fig. 3** The chromatography is set up through and controlled by the Xcalibur™ instrument software.

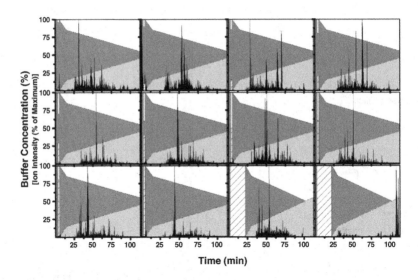

Fig. 3. Gradient profiles for a 12-step MudPIT run. Buffers A, B, and C of varying concentrations are represented by the dark gray areas, the light gray areas, and the striped white bars, respectively. A representative baseline ion chromatograph is shown in the forefront (black peaks) for each step (from the analysis of the complex peptide mixture obtained from mouse NaOH-extracted NEs). Each chromatographic step lasts for 112 min. The salt concentration is equal to 4, 10, 15, 20, 30, 40, 50, 60, 80, and 100% C in steps 1 through 10, respectively. In steps 1 through 10, the salt bump starts after 3 min and lasts for 2 min, while it lasts for 20 min in the last two chromatographic steps. In steps 1 through 10, a rapid increase from 0 to 15% buffer B occurs between 5 and 10 min, followed by a slow ramp to 45% B over 97 min. For steps 11 and 12, buffer B concentration increases rapidly to 15% between 22 and 30 min, followed by a slow ramp to 55% B over 82 min.

3.8.2. Tandem Mass Spectrometry

1. Using Xcalibur™, set up the collision energy at 35%.
2. Implement an acquisition scheme such that a cycle of one full MS scan (from 400 to 1600 m/z) followed by three MS/MS events on the top three most intense ions is repeated continuously throughout the chromatographic elution time.
3. To allow less intense ions to be analyzed, enable dynamic exclusion for 5 min.
4. Convert each RAW file (one per chromatographic step) into a DAT file using the XCalibur file converter function.

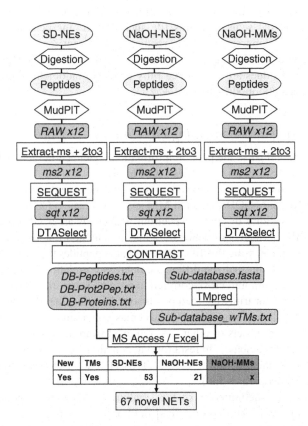

Fig. 4. Subtractive proteomics work flow and summary of results. Protein and peptide mixtures are circled in light gray ovals while processes are drawn within hexagons. Software are underlined in rectangles while file types are italicized and in gray boxes. A total of 67 uncharacterized proteins with at least one predicted transmembrane domain were found in at least one of the NEs preparations but not in the MMs runs. These are defined as putative nuclear envelope transmembrane proteins (NETs).

3.9. Data Analysis

3.9.1. Searching the MS/MS Data Set

1. To obtain the coordinates of the MS/MS spectra to be analyzed, convert each DAT file into a MS2 file using *extract-MS (21)* (*see* **Fig. 4**).
2. To remove spectra of poor quality and tentatively assign a charge state to precursor ions, apply the *2to3* software to MS2 files *(22)*.
3. To match MS/MS spectra to peptides, use SEQUEST™ *(15)* or PEP_PROBE *(19)* (*see* **Note 35**) against a database containing human, mouse, and rat protein sequences downloaded from National Center for Biotechnology Information (NCBI): this was 106,360 sequences on April 24, 2003, and was complemented with 172 sequences from usual contaminants (human keratins, IgGs, proteolytic enzymes). Set the sequest.params file such that (1) the peptide mass tolerance is 3, (2) no enzyme specificity is required, (3) parent ions are calculated with average masses, while fragment ions are modeled with monoisotopic masses, and (4) cysteine residues are considered fully carboxyamidomethylated and searched as bearing a static modification of +57 Da.

3.9.2. Assembling and Comparing Protein Lists

1. To assemble and parse the peptide information contained in the SEQUEST output files, run DTASelect on SQT files (*see* **Fig. 4**).
2. To compare the proteins detected in salt- and detergent-extracted NE, NaOH-extracted NE and NaOH-extracted MM samples, create a CONTRAST table (*see* **Fig. 4**). The validity of peptide/spectrum matches is assessed primarily with the PEP_PROBE-defined confidence for a match to be non-random (at least 85%). In addition, spectra passing this filter have to match peptides that are at least 7-amino-acid-long, with a normalized difference in cross-correlation scores (DeltCn) of at least 0.08 and minimum cross-correlation scores (Xcorr) of 1.8 for singly, 2.5 for doubly, and 3.5 for triply charged precursors. Proteins identified by single unique peptides are allowed (*see* **Note 36**), whereas contaminants, human keratins and proteins that are subsets of others are removed from the final list.
3. To create subset databases, in fasta format, containing only the proteins in the final list, use the "Database" utility of CONTRAST (add "Database" to the [Options] field of Contrast.params file).
4. To generate tab-delimited text files linking proteins to peptides, using the "−−DB" option in the [Criteria Sets] field of Contrast.params file (*see* **Fig. 4**).

3.9.3. Appending Transmembrane Domain Predictions

1. Predict the number of transmembrane segments using TMPred on a FASTA file containing the amino acid sequences for all detected proteins. Set a minimum score of 1000 in one orientation and 1900 in both (*see* **Note 37**).
2. Build a table of proteins and number of predicted transmembrane domains.

3.9.4. Generating a List of Putative Nuclear Envelope Transmembrane Proteins

1. Consolidate protein lists and TM table into a relational database using MSAccess.
2. Query for hypothetical proteins with transmembrane domains detected in either of the NE preparations but not the MM samples (*see* **Fig. 4**) (*see* **Note 38**).

4. Notes

1. As with most protocols there is an optimal middle ground with too little or too much starting material resulting in lower yields. In our hands, 10–12 rats produce optimal yields without saturating the six sucrose gradients in a Beckman-Coulter SW28 rotor.
2. This mouse strain was chosen to obtain more tissue per animal because it has much greater than average body weight (~35 g at 9 weeks). When choosing a heavier strain it is important to check that the mice do not have diabetes-related problems as this could bias results, some NE diseases having associated diabetes defects. The CB6F1/J strain is free of such defects.
3. It is recommended to pre-weigh the beaker containing the SHKM buffer as it simplifies measuring the weights of the livers.
4. This will also be needed in the "Preparation of Nuclear Envelopes" section, except that different and additional protease inhibitors should be added then. The PMSF is added here because additional proteases (mostly serine) may be released onto the liver tissue during dissection.
5. Muslin will also do if not chemically treated: make certain to ask the supplier.
6. MgCl2 concentration in the original procedure was 5 mM throughout; however, if NEs are being prepared for viewing by electron microscopy, dropping the concentration through most of the procedure to 0.1 mM will yield better structure. However, during DNase and RNase treatment, it is important to increase the MgCl2 concentration back to 5 mM.
7. The optimal protease inhibitors will vary according to the tissue being investigated; so it is important to investigate what proteases are present at high concentrations in the tissue of choice. The choice for liver focuses on inhibiting serine, trypsin, cysteine, and aspartic proteases present in this tissue.
8. If general protease cocktails are used, it is important to make certain that they do not contain EDTA as this is bad for NE preparation.
9. The solution can be prepared by adding 230 mL of an 85% (w/v) sucrose stock to 12.5 mL of 1 M HEPES-KOH, pH 7.4, 6.25 mL of 1 M KCl, and 1 mL of 1 M MgCl2, and freshly added 2 mM DTT and protease inhibitors. Other concentrations of sucrose can then be obtained by mixing the 2.3 M SHKM with 0.25 M SHKM.
10. The stock solution can be prepared by mixing 67 mL of 0.25 M SHKM to 33 mL of 2.3 M SHKM.
11. Bulk material is not available. The resin is extracted by cutting the HPLC column in half with a hacksaw, then washed with methanol, dried, and stored as a powder.

12. Chapters detailing modifications of the NE protocol for human blood lympho-cytes and rat muscle are being prepared for other volumes in the *Methods in Molecular Biology* series.

13. This has been experimentally shown to increase yields by 30–50% with this procedure and reduces the RNA/ DNA ratio *(6)*.

14. This tube is particularly rich in proteases, so it is important to avoid cutting it or mixing it with the isolated liver.

15. As many protease inhibitors are short-lived, it is important to add them fresh to buffers shortly before use throughout the procedure

16. One can avoid the weighing step and estimate 5 g per liver.

17. This requires a reasonable amount of physical strength, and one must take care to keep the homogenizer straight with the direction of the pestle, or the homogenizer can break. Also never stop the pestle rotation while it is inserted inside the homogenizer with liquid or this can become stuck or broken due to the vacuum produced during homogenization.

18. There is an enormous amount of collagen in rat liver, and this will clog the cheesecloth and reduce the yield if the cheesecloth is overloaded. Therefore, the homogenate from no more than three rats should be filtered through one four times-folded cheesecloth. Mice have much less collagen than rats and therefore roughly twice the amount of material can be used for the same amount of cheesecloth.

19. We prefer these because the pellet tends to distribute widely, but conical tubes could also be used.

20. Because of the high viscosity of the 2.3 M sucrose solution, it takes several minutes for each tube if an 18 gauge needle is used. In contrast, with the wide bore size of the 14 gauge, the same procedure can be performed in 30 s. It is important to use a luer lock syringe because the viscosity of the solution can produce high pressure on the connection.

21. It is critical to avoid touching the walls because they are rich with proteases that were associated with the floated membranes.

22. This and step 15 are done to wash away some of the high amount of contami-nating collagen prior to extraction of nucleoplasmic contents. These washes are necessary to prevent saturation of sucrose cushions when chromatin is released but are only necessary for liver preparations and other tissues that are particularly high in collagen.

23. Do not be concerned if the nuclei do not become phase-lucent at this point as this first treatment might be viewed as generally loosening rather than fully digesting the chromatin, and moreover, the digestion will continue during centrifugation.

24. It is very important to use a swinging bucket rotor when spinning through the sucrose cushion at this point to float any chromatin that is released away from the NEs.

25. The supernatant will appear cloudy, but this is mostly chromatin that has been ejected and should give a dark, worm-like appearance under the microscope quite distinct from NEs.

26. The pellet here may be larger than the nuclear pellet.
27. The membranes can be floated using either a fixed angle rotor or a swinging bucket rotor (e.g., SW28). Better separation can be achieved if a second SW28 rotor and ultracentrifuge are available when NEs and MMs are being prepared simultaneously. In this case, 20 mL of the diluted membranes are overlaid with 6 mL of the 1.9 M SHKM, and this is overlaid with 2.3 mL of the 0.25 M SHKM and centrifuged at 57,000 g (~21,000 rpm) for 4 h.
28. Extracting the MMs by side puncture is only possible if SW28 rotor and tubes are used, as the type 45 Ti rotor tubes are thick walled.
29. The type 45 Ti rotor tubes must be filled close to the top or they can collapse. There should be an orange pellet in the corner of the tube. Decant supernatant and freeze.
30. It is critical to get this spinning immediately as loss of membrane proteins was observed by western blot analysis and membrane vesicles looked very fragmented by electron microscopy even after just a 10-min sitting on ice prior to centrifugation (Tinglu Guan, personal communication).
31. Because adding the base to the acidic sample causes "bubbling," it is best to transfer the sample to a larger 15-mL conical tube in ice before neutralizing it, and then transfer it back to the smaller tube when the appropriate pH is reached.
32. A typical four-step parameter setup for pulling approx 3 to 5-μm tips from a 100 μm ID × 365 μm OD fused silica capillary is heat = 290, velocity = 40, and delay = 200; heat = 280, velocity = 30, and delay = 200; heat = 270, velocity = 25, and delay = 200; and heat = 260, velocity = 20, and delay = 200, with all other values set to zero.
33. This concentration roughly corresponds to an amount of resin powder covering the tip of a small spatula, about 2–3 mm^3.
34. Using a black brackground behind the capillary being packed helps seeing the RP and SCX resin levels inside the column.
35. PEP_PROBE is a modified version of SEQUEST, which can calculate the confidence for a match to be non-random based on an hypergeometric probability model. Alternatively, false discovery rates (FDRs) can be estimated by concatenating to the sequence database to be searched, randomized versions of each protein sequence (keeping the same amino acid composition and length). The theory is that for each spectrum matching a "shuffled" peptide (true negative), there should be a false positive in the "normal" peptide data set *(23)*. Xcorr and DeltaCn values are then set such as to obtain peptide FDRs of less than 1%.
36. Although a two-peptide per protein cut-off allows for higher confidence, single peptide identifications are reported in the final data set because some well-characterized NETs were detected by single peptides in previous studies, likely due to their lesser abundance.
37. These score limitations are based on an average to match those of the previously characterized NE proteins.
38. The selection criteria applied in this study *(4)* to discriminate true NET proteins from contaminating ER proteins is very conservative, as only proteins that

were not found in MM fractions were considered for further analysis. However, differences in sequence coverage (percentage of a protein sequence covered by identified peptides) *(24,25)* or spectral counts (number of spectra matching peptides from a protein) *(26)* could be used as semi-quantitative parameters to determine whether a particular protein is "enriched" in a fraction of interest.

Acknowledgments

The authors thank John R. Yates III in whose laboratory "shotgun" proteomics was developed and Tinglu Guan in the laboratory of Larry Gerace from whom the NE purification method was initially taught to Eric Schirmer. This work was supported by a Senior Research Fellowship to Eric Schirmer from the Wellcome Trust and by the Stowers Institute for Medical Research.

References

1. Schirmer, E. C. and Gerace, L. (2005) The nuclear membrane proteome: extending the envelope. *Trends Biochem. Sci.* **30**, 551–558.
2. D'Angelo, M. A. and Hetzer, M. W. (2006) The role of the nuclear envelope in cellular organization. *Cell Mol. Life Sci.* **63**, 316–332.
3. Starr, D. A. and Fischer, J. A. (2005) Kash 'n karry: The kash domain family of cargo-specific cytoskeletal adaptor proteins. *Bioessays* **27**, 1136–1146.
4. Schirmer, E. C., Florens, L., Guan, T., Yates, J. R., III, and Gerace, L. (2003) Nuclear membrane proteins with potential disease links found by subtractive proteomics. *Science* **301**, 1380–1382.
5. Bielinska, M., Rogers, G., Rucinsky, T., and Boime, I. (1979) Processing in vitro of placental peptide hormones by smooth microsomes. *Proc. Natl. Acad. Sci. U.S.A.* **76**, 6152–6156.
6. Blobel, G. and Potter, V. R. (1966) Nuclei from rat liver: isolation method that combines purity with high yield. *Science* **154**, 1662–1665.
7. Dwyer, N. and Blobel, G. (1976) A modified procedure for the isolation of a pore complex-lamina fraction from rat liver nuclei. *J. Cell Biol.* **70**, 581–591.
8. Gerace, L., Ottaviano, Y., and Kondor-Koch, C. (1982) Identification of a major polypeptide of the nuclear pore complex. *J. Cell Biol.* **95**, 826–837.
9. Scheele, G. (1983) Methods for the study of protein translocation across the rer membrane using the reticulocyte lysate translation system and canine pancreatic microsomal membranes. *Methods Enzymol.* **96**, 94–111.
10. Walter, P. and Blobel, G. (1983) Preparation of microsomal membranes for cotrans- lational protein translocation. *Methods Enzymol.* **96**, 84–93.
11. Stuurman, N., Heins, S., and Aebi, U. (1998) Nuclear lamins: Their structure, assembly, and interactions. *J. Struct. Biol.* **122**, 42–66.
12. Link, A. J., Eng, J., Schieltz, D. M., Carmack, E., Mize, G. J., Morris, D. R., et al. (1999) Direct analysis of protein complexes using mass spectrometry. *Nat. Biotechnol.* **17**, 676–682.

13. Washburn, M. P., Wolters, D., and Yates, J. R., 3rd (2001) Large-scale analysis of the yeast proteome by multidimensional protein identification technology. *Nat. Biotechnol.* **19**, 242–247.
14. Wolters, D. A., Washburn, M. P., and Yates, J. R., III (2001) An automated multidimensional protein identification technology for shotgun proteomics. *Anal. Chem.* **73**, 5683–5690.
15. Eng, J., McCormack, A. L., and Yates, J. R., III (1994) An approach to correlate tandem mass spectral data of peptides with amino acid sequences in a protein database. *J. Am. Mass Spectrom.* **5**, 976–989.
16. Tabb, D. L., McDonald, W. H., and Yates, J. R., 3rd (2002) Dtaselect and contrast: tools for assembling and comparing protein identifications from shotgun proteomics. *J. Proteome Res.* **1**, 21–26.
17. Brachner, A., Reipert, S., Foisner, R., and Gotzmann, J. (2005) Lem2 is a novel man1-related inner nuclear membrane protein associated with a-type lamins. *J. Cell Sci.* **118**, 5797–5810.
18. Wilhelmsen, K., Litjens, S. H., Kuikman, I., Tshimbalanga, N., Janssen, H., van den Bout, I., et al. (2005) Nesprin-3, a novel outer nuclear membrane protein, associates with the cytoskeletal linker protein plectin. *J. Cell Biol.* **171**, 799–810.
19. Sadygov, R. G. and Yates, J. R., III (2003) A hypergeometric probability model for protein identification and validation using tandem mass spectral data and protein sequence databases. *Anal. Chem.* **75**, 3792–3798.
20. Hofmann, K. and Stoffel, W. (1993) Tmbase - a database of membrane spanning proteins segments. *Biol. Chem. Hoppe-Seyler* **374**, 166.
21. McDonald, W. H., Tabb, D. L., Sadygov, R. G., MacCoss, M. J., Venable, J., Graumann, J., et al. (2004) Ms1, ms2, and sqt-three unified, compact, and easily parsed file formats for the storage of shotgun proteomic spectra and identifications. *Rapid Commun. Mass Spectrom.* **18**, 2162–2168.
22. Sadygov, R. G., Eng, J., Durr, E., Saraf, A., McDonald, H., MacCoss, M. J., et al. (2002) Code developments to improve the efficiency of automated ms/ms spectra interpretation. *J. Proteome Res.* **1**, 211–215.
23. Peng, J., Elias, J. E., Thoreen, C. C., Licklider, L. J., and Gygi, S. P. (2003) Evaluation of multidimensional chromatography coupled with tandem mass spectrometry (lc/lc-ms/ms) for large-scale protein analysis: the yeast proteome. *J. Proteome Res.* **2**, 43–50.
24. Florens, L., Liu, X., Wang, Y., Yang, S., Schwartz, O., Peglar, M., et al. (2004) Proteomics approach reveals novel proteins on the surface of malaria-infected erythrocytes. *Mol. Biochem. Parasitol.* **135**, 1–11.
25. Sam-Yellowe, T. Y., Florens, L., Wang, T., Raine, J. D., Carucci, D. J., Sinden, R., et al. (2004) Proteome analysis of rhoptry-enriched fractions isolated from plasmodium merozoites. *J. Proteome Res.* **3**, 995–1001.
26. Chu, D. S., Liu, H., Nix, P., Wu, T. F., Ralston, E. J., Yates, J. R., III, et al. (2006) Sperm chromatin proteomics identifies evolutionarily conserved fertility factors. *Nature* **443**, 101–105.

9

Purification and Proteomic Analysis of a Nuclear-Insoluble Protein Fraction

Tsuneyoshi Horigome, Kazuhiro Furukawa, and Kohei Ishii

Summary

We describe here a method for analyzing a rat liver nuclear-insoluble protein fraction to determine candidate proteins participating in nuclear architecture formation. Rat liver nuclei are purified by sucrose density gradient centrifugation. The purified nuclei are treated with DNase and RNase and then washed with high salt and detergent solutions. The residual nuclear-insoluble protein fraction is separated by reversed-phase high-performance liquid chromatography (HPLC) in 60% formic acid on a polystyrene resin column. This system allows good resolution and high recovery of most insoluble proteins, including intrinsic membrane proteins and even proteins larger than 140 kDa, with more than 70% recovery. The LC-fractionated proteins are further separated by sodium dodecyl sulphate–polyacrylamide gel electrophoresis (SDS–PAGE). Protein bands are excised, in-gel digested with trypsin, and then analyzed with a protein sequencer or mass spectrometer. Using this protocol, 138 were separated, 29 were identified, among which one appears as a novel nuclear constituent localized in the interchromatin space.

Key Words: Nuclear-insoluble protein; nuclear proteome; subcellular fractionation; polymer-based column; reversed-phase HPLC; membrane protein purification; perfusion chromatography.

1. Introduction

Recently, many integral nuclear membrane proteins have been found in proteome analyses, some of which are thought to be linked to a variety of dystrophies *(1)*. These findings suggest that the structure and function of the nuclear envelope are more complicated than previously expected from the limited set of known protein components. Many studies have shown that the cell

From: *Methods in Molecular Biology, vol. 432: Organelle Proteomics*
Edited by: D. Pflieger and J. Rossier © Humana Press, Totowa, NJ

nucleus contains distinct compartments including chromosome territories, the interchromatin space, and many kinds of discrete nuclear bodies. As it is the case for the nuclear envelope, it is likely that many unknown proteins participate in the maintenance of these compartments. Therefore, searches for novel nuclear structural proteins are necessary to gain further insights into the inner nuclear structure and higher-order nuclear functions.

With a view to identify candidate nuclear structural proteins, we have developed a procedure to analyze a rat liver nuclear-insoluble protein fraction. High resolution over a wide molecular weight range of insoluble proteins is achieved by two-dimensional separation involving polymer-based reversed-phase high-performance liquid chromatography (HPLC) in 60% formic acid and sodium dodecyl sulphate–polyacrylamide gel electrophoresis (SDS–PAGE). The separated proteins are identified by partial amino acid sequencing or mass spectrometry (MS). Using this approach, novel proteins along with many known structural proteins could be identified *(2)*.

2. Materials

2.1. Preparation of a Rat Liver Nuclear-Insoluble Fraction

1. Wister rats weighing 150–250 g.
2. Sucrose.
3. Buffer 1: 50 mM Tris–HCl (pH 7.5 at 4°C), 25 mM KCl, 5 mM $MgCl_2$, and 0.2 mM ethylenediaminetetraacetic acid (EDTA). Store at 4°C.
4. Proteinase inhibitor stock solutions: 1 M benzamidine-HCl in water, 10 mg/mL chymostatin in dimethylsulfoxide, 10 mg/mL leupeptin in water, 5 mg/mL pepstatin A in dimethylsulfoxide, and 5 mg/mL antipain in water. Store at – 20°C. 100 mM phenylmethylsulfonyl fluoride (PMSF) in isopropanol is freshly prepared daily. PMSF should be added to the working solution just before use or directly added to the sample solution because it easily decomposes in water. Final concentrations of inhibitors in the working solution are 1 mM PMSF, 2 mM benzamidine, 10 µg/mL chymostatin, 10 µg/mL leupeptin, 5 µg/mL pepstatin, and 5 µg/mL antipain.
5. Teflon pestle homogenizer.
6. Phosphate-buffered saline (PBS): 10 mM sodium phosphate buffer, pH 7.4, and 0.8% (w/v) NaCl. Store at 4°C.
7. Buffer 2: 50 mM Tris–HCl buffer, pH 7.4 at 4°C, and 5 mM $MgSO_4$. Store at 4°C.
8. Buffer 3: 10 mM Tris–HCl buffer, pH 7.4 at 4°C, and 0.2 mM $MgSO_4$. Store at 4°C.
9. Buffer 4: 40 mM MES–KOH, pH 6.0, 4% (w/v) Triton X-100, 0.6 M KCl, 20% (w/v) sucrose, and 4 mM EDTA. Store at 4°C.
10. DNase I [D-4513 (TypeII-S), DN-EP or higher grade] from Sigma, St. Louis, MO and RNase A (Sigma Type III-A).

2.2. Reversed-Phase HPLC in 60% Formic Acid for Insoluble Protein Purification

1. Poros 10R1 column: perfusion-type reversed-phase HPLC column packed with 10 µm polystyrene beads with 75 and 700 nm pores (Applied Biosystems, Foster City, CA).
2. 100 mM Tris–HCl, pH 9.0, and 5% (w/v) SDS. Store at room temperature.
3. Solvent A: formic acid/water, 60/40 (v/v). Prepare with reagent grade formic acid and filtrated water (0.45 µm). Store at room temperature.
4. Solvent B: *N*-butanol/formic acid/water, 33/60/7 (v/v/v). *N*-butanol and formic acid are reagent grade. Filtrated water (0.45 µm) is used. Store at room temperature.
5. 20 mM Tris–HCl, pH 9.0, and 10% (w/v) SDS. Store at room temperature.

2.3. SDS–PAGE

1. Sample buffer (5×): 250 mM Tris–HCl, pH 6.8, 10% (w/v) SDS, 10% (w/v) glycerol, 0.5% (w/v) bromophenol blue. Store at room temperature.
2. 1 M dithiothreitol: dissolve in water and freeze in single-use aliquots at –20°C.
3. 1 M iodoacetamide: freshly prepare every time.
4. 0.25% (w/v) Coomassie blue in methanol/acetic acid/water, 25/7/68 (v/v/v).
5. Methanol/acetic acid/water, 25/7/68 (v/v/v).

2.4. In-Gel Digestion of Proteins with Trypsin

1. 0.1 M NH$_4$HCO$_3$/acetonitrile, 50/50 (v/v). Acetonitrile is HPLC grade. Store at room temperature.
2. 0.1 M NH$_4$HCO$_3$, pH 8.4. Store at 4°C.
3. Trypsin: dissolve TPCK-treated trypsin (Sigma) in 1 mM HCl (0.1 mg/mL). It can be stored at –80°C in single-use aliquots for 3 weeks. Dilute with 0.1 M NH$_4$HCO$_3$ just before use.
4. Water: HPLC grade, Milli-Q (Millipore, Billerica, MA), or equivalent.

2.5. Protein Identification with a Protein Sequencer

1. Silica-based C8 HPLC column: Capcel Pack C8 column, 4.6 × 150 mm, 5-µm beads, and 30-nm pores (Shiseido, Tokyo, Japan).
2. Trifluoroacetic acid (TFA): amino acid sequencing grade.
3. Solvent C: acetonitrile/water/TFA, 5/95/0.1 (v/v/v). Store at room temperature.
4. Solvent D: acetonitrile/water/TFA, 75/25/0.1 (v/v/v). Store at room temperature.

2.6. Protein Identification by Mass Spectrometry

1. Solvent E: 0.1% (v/v) formic acid (sequencing grade) in water. Store at room temperature.

2. Solvent F: acetonitrile/water/formic acid, 90/10/0.1 (v/v/v). Store at room temperature.

3. Peptide mapping is performed with a model TSQ 700 triple stage quadrupole mass spectrometer equipped with an electrospray ionization source (Thermo Electron, Waltham, MA) coupled to a Gilson HPLC system (Gilson, Middleton, WI) equipped with an accurate splitter (1:100 v/v; LC Packings[]Please provide the city, state (if applicable) and the country of 'LC Packings'.) and a Magic C18 column (0.2 × 50 mm; Michrom BioResource, Auburn, CA).

4. An API QSTAR pulsar hybrid mass spectrometer system, involving nanoelectrospray ionization and quadrupole—time of flight analyzers (nano-ESI-Q-TOF), is used with a micro-liquid chromatography system (Magic 2002; Michrom BioResource, Auburn, CA) for MS/MS analysis to confirm the amino acid sequences of peptides.

3. Methods

The solubilization method used for "the nuclear-insoluble protein fraction" (which is insoluble in 10 mM Tris-HCl, pH 7.4, 0.5 M NaCl, 0.2 mM $MgSO_4$ and in 40 mM Mes-KOH, pH 6.0, 2% Triton X-100, 3 M KCl) is very important. Nuclear-insoluble proteins can be solubilized with urea, guanidine hydrochloride, SDS, and 60% formic acid and can be separated by ion exchange *(3)*, size exclusion *(4)*, ceramic hydroxyapatite *(3,5)*, or reversed-phase HPLC (RP-HPLC) *(2,6)*. The highest resolution over a wide molecular weight range of insoluble proteins is achieved by two-dimensional separation involving polymer-based RP-HPLC in 60% formic acid and SDS–PAGE *(6)*. The separated proteins can be identified by partial amino acid sequencing or MS after digestion with trypsin *(2)*.

3.1. Preparation of a Rat Liver Nuclear-Insoluble Fraction

1. Perform all experiments with Wister rats weighing 150–250 g (*see* **Note 1**). Rats fasted overnight are killed by decapitation, and then their livers are removed quickly and chilled immediately in several volumes of ice-cold 0.25 M sucrose in buffer 1. All subsequent operations are performed at 0–4°C.

2. Blot and weigh livers, mince with scissors, and then add to two volumes of ice-cold buffer 1 containing 0.25 M sucrose and proteinase inhibitors. Homogenize in a Teflon pestle homogenizer with 4 strokes at 600–800 rev/min. Filter the homogenate through four layers of cheese cloth and then centrifuge it at 800 *g* for 10 min.

3. Weigh the pelleted crude nuclear fraction and resuspend in 5.2 volumes (w/v) of buffer 1 containing 2.2 M sucrose and 0.5 mM PMSF (*see* **Note 2**) by homogenization (200–400 rev/min, 2 strokes).

4. After centrifugation for 90 min at 100,000 g at 0–4°C, pour off the supernatant. Remove cell debris adhering to the wall of the tube with a spatula, and then wipe the tube wall dry with tissue paper wrapped around a spatula.

5. Resuspend the white nuclear pellet in buffer 1 (0.15 mL/g liver) containing 0.25 M sucrose and proteinase inhibitors with a homogenizer and then pellet again by centrifugation at 1000 g for 15 min. Repeat the washing step once more to remove 2.2 M sucrose and contaminating cytosol proteins.

6. Resuspend the thus purified nuclear fraction in buffer 2 (0.15 mL/g liver) containing 0.25 M sucrose and proteinase inhibitors. Dissolve a 10-µL aliquot of the suspension in 3 mL of PBS containing 1% (w/v) SDS and then measure the absorbance at 260 nm to estimate the DNA concentration (*see* **Note 1**).

7. Add DNase I and RNase A (250 µg of each enzyme per mL) to the purified nuclei that should be at 200–250 U/mL in buffer 2 containing 0.25 M sucrose and proteinase inhibitors (*see* **Note 3**) to hydrolyze genomic DNA, mRNA, ribosome precursor rRNA and nuclear structural RNA. After incubation with occasional vortex-mixing for 60 min at 4°C, centrifuge the suspension at 800 g for 10 min to remove chromosomal proteins and most RNA-binding proteins.

8. Resuspend the pellet at 170 U/mL in buffer 3 containing 0.5 M NaCl, 1% (v/v) 2-mercaptoethanol, and proteinase inhibitors, and then let the suspension stand at 4°C for 15 min to dissociate proteins bound by ionic interactions.

9. Centrifuge the mixture at 10,000 g for 15 min (*see* **Note 4**). Wash the pellet with buffer 3 containing proteinase inhibitors, and then resuspend it in the same buffer at 500 U/mL. Add the suspension to an equal volume of buffer 4 containing protease inhibitors and then let stand the mixture at 4°C for 30 min to dissolve nuclear membrane proteins and lipids.

10. Centrifuge the suspension at 20,000 g for 30 min. The pellet is designated as the "nuclear-insoluble fraction." Resuspend the pellet in buffer 3 at about 700 U/mL by homogenization and then store the suspension at –80°C.

3.2. Reversed-Phase HPLC in 60% Formic Acid for Insoluble Protein Purification

1. Centrifuge the nuclear-insoluble fraction (21 mg) suspended in buffer 3 at 10,000 g for 10 min at 4°C and then discard the supernatant.

2. Add 3.0 mL of 100 mM Tris–HCl, pH 9.0, containing 5% (w/v) SDS and 0.15 mL of 2-mercaptoethanol to the precipitate, and then dissolve the precipitate by vibration with a micro-mixer for 15 min, standing overnight at room temperature and then heating at 100°C for 5 min.

3. After cooling, add 27 mL of formic acid to yield 90% formic acid to dissolve proteins completely. Then centrifuge the solution at 6500 g for 10 min.

4. Apply one-third of the supernatant (7 mg of protein) to a Poros 10R1 column (7.5 × 75 mm) equilibrated with solvent A, and then elute it with a 140-min linear gradient, 5–40% of solvent B at 1.5 mL/min (*see* **Note 5**). Next, bring the concentration of solvent B to 100% for 5 min. Elution is monitored by recording the absorbance at 280 nm (*2*), and fractions of 1.5 mL (1 min) are collected.

5. Carefully dry samples containing a high concentration of formic acid by centrifugal lyophilization to maintain the solubility according to the method of Heukeshoven and Dernick *(7)* as follows. Concentrate fractions (1.5 mL) to about 50 µL by centrifugal lyophilization, add 200 µL of water to each sample before drying, and then continue concentration. Repeat this dilution–concentration procedure three times. Finally, completely dry up the samples.

6. Solubilize the obtained pellets in 30 µL of 10% (w/v) SDS in 20 mM Tris–HCl (pH. 9.0) by vibration for 15 min and standing at room temperature overnight, and then store until use at –80°C (*see* **Note 6**).

7. Separate as above and at one time the remaining two-thirds (14 mg of protein) of the nuclear-insoluble fraction solubilized in 90% formic acid on the Poros 10R1 column (*see* **Note 5**) and dissolve each fraction in 40 µL of 10% (w/v) SDS, and 20 mM Tris–HCl, pH 9.0. Because capacity of the column is about 15 mg protein as mentioned in **Note 5**, the sample is divided into 7- and 14-mg fractions.

3.3. SDS–PAGE

1. Use one small aliquot of the obtained sample fractions to view the protein elution pattern by silver staining. Mix 0.3 µL out of the 30 µL of the above fractions (in some cases, several fractions are combined), 5 µL of 5× sample buffer, 1.0 µL of 1 M dithiothreitol (50 mM final concentration), and 14 µL of water, and then heat the mixture at 100°C for 5 min.

2. After cooling, add 3.6 µL of 1 M iodoacetamide (150 mM final concentration) to the sample to block sulfhydryl groups, and then incubate it at room temperature for 15 min in the dark.

3. Separate the alkylated samples by SDS–PAGE on 8.5% acrylamide uniform gels according to the method of Laemmli *(8)*. After electrophoresis, stain the gels with silver according to the method of Morrissey *(2,9)*. One hundred thirty-eight protein bands could be detected in our experiment.

4. For sequencing of the separated protein materials, treat in the same manner (**steps 1** and **2**) (28 µL out of the 40-µL fractions obtained in **Subheading 3.2., step 7**).

5. Stain this gel with 150 mL of 0.25% (w/v) Coomassie blue in methanol/acetic acid/water, 25/7/68 (v/v/v), for 1 h under gentle shaking. Then remove excess dye with methanol/acetic acid/water, 25/7/68 (v/v/v). Rotate the gels in water to remove methanol and acetic acid.

3.4. In-Gel Digestion of Proteins with Trypsin

1. Cut out protein bands of interest and cut them into 1 mm³ or smaller pieces. Place the pieces in microfuge tubes containing 0.5 mL of water, and wash them for 10 min at room temperature by shaking. Repeat the wash four more times by renewing the 0.5 mL of water (*see* **Note 7**).

2. To remove Coomassie blue: (1) dehydrate the gel pieces with 0.2 mL of 0.1 M NH_4HCO_3/acetonitrile, 50/50, (v/v) for 10 min, and then discard the solvent and (2) rehydrate the gel pieces with 0.5 mL of water at 37°C. Repeat procedures

(1) and (2) about six times until the blue color of the gel pieces has completely gone (*see* **Note 8**).

3. Incubate the gel pieces with 0.2 mL of 0.1 M NH_4HCO_3/acetonitrile, 50/50 (v/v), for 10 min, discard the solvent, then incubate with 0.2 mL of acetonitrile for 10 min, discard the solvent, and finally dry completely by centrifugal lyophilization for 10 min. The gel pieces should not stick to the walls of the microfuge tubes when completely dry.

4. Rehydrate the gel pieces by adding 10 μL of 0.1 M NH_4HCO_3 containing 0.5 μg of trypsin directly to the dried gel pieces on ice. After waiting for 10–20 min for absorption of the solution by the gel pieces, if necessary, repeat the above procedure to allow the gel pieces to fully swell.

5. Then add 0.1 M NH_4HCO_3 to fully immerse the gel pieces. Incubate the tube at 37°C for about 16 h.

6. Carefully remove the 0.1 M NH_4HCO_3 (now called the extract) and place it in a clean microfuge tube.

7. After addition of 30 μL of 25 mM NH_4HCO_3 to the gel pieces, rotate the tube for 10 min. Combine the supernatant with the above extract.

8. After addition of 30 μL of 20% (v/v) formic acid and rotation for 10 min, add 60 μL of acetonitrile to the gel piece and continue the rotation for 10 min. Combine the supernatant with the above extract.

9. Repeat **steps 7** and **8** once more.

10. After addition of 30 μL of water to the gel pieces, rotate the tube for 10 min. Combine the supernatant with the above extract.

11. Centrifuge the pooled extract at 10,000 *g* for 10 min, and then carefully remove the clarified supernatant and place it into a clean microfuge tube. Reduce the volume of the clarified extract by centrifugal lyophilization to less than 20 μL.

3.5. Protein Identification with a Protein Sequencer

Tryptic digests of selected protein gel bands derived from 21 mg of the nuclear-insoluble fraction are analyzed by partial amino acid sequencing with a protein sequencer or by MS.

1. Samples containing more than 10 μg of protein can be identified with the sequencer. The concentrated peptide sample is diluted to 100 μL with water/TFA, 98/2 (v/v), and then applied to a silica-based C8 column (4.6 × 150 mm) and eluted with solvents C and D, using a gradient program of 0.4% D/min for 150 min, 2% D/min for 20 min, and 100% solvent D applied for 20 min, at a flow rate of 0.4 mL/min. Elution of peptides is monitored at 216 nm.

2. The amino acid sequences of isolated peptides are determined with a Protein Sequencer 470A (Applied Biosystems). Partial amino acid sequences thus obtained are searched for using the program FASTA in the GENES, SWISS-PROT, PIR, PRF, and PDBSTR databases (http://fasta.bioch.virginia.edu/). Fifteen protein bands have been identified with this method (*2*).

3.6. Protein Identification by Mass Spectrometry

1. Peptides loaded on the C18 Magic column are eluted with solvents E and F, using a program of 5% solvent F for 5 min, a gradient of 1.05%/min for 90 min, and applying 100% solvent F for 5 min, at a flow rate of 2 μL/min provided by an accurate splitter.
2. Peptide mapping is obtained on a TSQ 700 instrument coupled and operated online with the C18 column. The MS conditions are as follows: ion spray voltage of 2.2–2.7 kV, electron multiplier voltage of 1000 eV, and acquisition mass range of 220–2500 Da (scans of 4 s).
3. Proteins are identified using the PROWL search engine (http://prowl.rockefeller. edu/) and the NCBI database.
4. For MS/MS analysis of peptides, the API QSTAR pulsar is used in the following conditions: ion spray voltage of 3.0–3.8 kV, electron multiplier voltage of 2200 V, 10 units of nitrogen curtain gas, 10 units of nitrogen collision gas, and collision energy of 20–55 eV.
5. Proteins are identified using the Mascot search engine (http://www.matrixscience. com/search_form_select.html) and the NCBInr database.

   Fourteen proteins including a novel interchromatin space protein of 36 kDa, ISP36, have been identified with finger printing and MS/MS methods *(2)*.

4. Notes

1. About 7 g of liver tissue is obtained from one rat. From 1 g of rat liver, about 1 mL of crude nuclear fraction, about 10 mg of protein (i.e., about 42 U) of the purified nuclear fraction, and about 0.3 mg of protein of the nuclear-insoluble fraction are obtained. One unit (1 U) of nuclei is defined as absorbance at 260 nm = 1; 1 mL of such a sample corresponds to about 3×10^6 nuclei. Usually, preparation starts with 70 g of rat liver tissue (10 rat livers), and 21 mg of the nuclear-insoluble fraction is obtained.
2. Precisely 5.2 mL of the buffer should be added per gram of pellet to bring the sucrose concentration to 1.85 M. If the concentration is lower than 1.85 M, the obtained nuclear fraction is indeed contaminated by other organelles.
3. The purity of DNase I is very important. Sigma D-4513 (TypeII-S), DN-EP, or higher grade of preparation should be used. DNase I should be dissolved freshly every time because the enzyme is not stable in solution. Purified RNase A (Sigma Type III-A) is used. RNase A is used after heating (100°C, 5 min) to inactivate possible contaminating proteinases because RNase A is very stable.
4. The amount of protein derived from 1 U nuclei is defined as 1 unit (1 U).
5. The Poros 10R1 column (7.5 × 75 mm) used in this study is a home-made one and 15 mg of protein can be applied at one time. A Poros 10R1 packed column (4.6 × 100 mm) is commercially available from Applied Biosystems. Five mg of protein may be applied to the column. A Resource RPC column from Pharmacia can be used instead of a Poros 10R1 one although the recovery of high-molecular-weight proteins is a little lower *(6)*. Silica-based columns, C18

and C4, have been used for such reversed-phase HPLC *(7)*. However, polystyrene resin columns, such as Poros 10R1, are useful for this purpose because handling is easy because of their stability in a wide pH range (pH 1–14), including the high pH values of the cleaning procedures. When crude membrane preparations and samples containing very hydrophobic and high-molecular-weight proteins are analyzed, some cleaning of the column is necessary. The hydrophobicity of Poros 10R1 corresponds to that of C4. So, high recoveries of higher molecular weight proteins are obtained *(6,10)*.

6. If a protein is directly dried from a solution at 60% formic acid, it forms a film and becomes very difficult to dissolve in a SDS-containing solution. The formic acid concentration should be lower than 1% before complete drying. It is still not easy to dissolve the thus dried protein in a SDS solution. Therefore, overnight treatment with SDS is necessary to dissolve the protein completely.

7. To minimize the loss of peptides through non-specific adsorption to their surfaces, microfuge tubes and pipettor-chips, for handling small amounts of peptides, are siliconized prior to use. They are treated with 2% dichlorodimethylsilane in CCl_4 for 5 min, rinsed with water, and then heated at 120°C for 1 h. Then they are rinsed with water and dried.

8. Complete removal of Coomassie blue stain is very important, because the dye inhibits trypsin activity.

References

1. Schirmer, E. C., Florens, L., Guan, T., Yates, J. R., III, and Gerace, L. (2003) Nuclear membrane proteins with potential disease links found by subtractive proteomics. *Science* **301**, 1380–1382.
2. Segawa, M., Niino, K., Mineki, R., Kaga, N., Murayama, K., Sugimoto, K., et al. (2005) Proteome analysis of a rat liver nuclear insoluble protein fraction and localization of a novel protein, ISP36, to compartments in the interchromatin space. *FEBS J.* **272**, 4327–4338.
3. Hiranuma, T., Horigome, T., and Sugano, H. (1990) Separation of membrane protein-sodium dodecyl sulfate complexes by high-performance liquid chromatography on hydroxyapatite. *J. Chromatogr.* **515**, 399–406.
4. Takagi, T. (1981) High-performance liquid chromatography of protein polypeptides on porous silica gel columns (TSK-GEL SW) in the presence of sodium dodecyl sulfate: comparison with SDS-polyacrylamide gel electrophoresis. *J. Chromatogr.* **219**, 123–127.
5. Horigome, T., Hiranuma, T., and Sugano, H. (1989) Ceramic hydroxyapatite high-performance liquid chromatography of complexes membrane protein and sodium dodecylsulfate. *Eur. J. Biochem.* **186**, 63–69.
6. Kikuchi, N., Yamakawa, Y., Ichimura, T., Omata, S., and Horigome, T. (1997) Reversed-phase high-performance liquid chromatography of membrane proteins on perfusion-type polystyrene resin columns in 60% formic acid. *Chromatography* **18**, 176–184.

7. Heukeshoven, J. and Dernick, R. (1985) Characterization of a solvent system for separation of water-insoluble poliovirus proteins by reversed-phase high-performance liquid chromatography. *J. Chromatogr.* **326**, 91–101.

8. Laemmli, U. K. (1970) Cleavage of structural proteins during the assembly of the head of bacteriophage T4. *Nature* **227**, 680–685.

9. Morrissey, J. H. (1981) Silver stain for proteins in polyacrylamide gels: a modified procedure with enhanced uniform sensitivity. *Anal. Biochem.* **117**, 307–310.

10. Afeyan, N. B., Fulton, S. P., and Regnier, F. E. (1991) Perfusion chromatography packing materials for proteins and peptides. *J. Chromatogr.* **544**, 267–279.

10

Preparation Methods of Human Metaphase Chromosomes for their Proteome Analysis

Kiichi Fukui, Hideaki Takata, and Susumu Uchiyama

Summary

Chromosomes are supermolecules that contain most of the DNA within a cell and are visible under optical and electron microscopes. Although they were observed at the earliest stage of genetics, their fundamental structure is not yet understood. The reasons for this are debated among researchers; however, it is clear that the accumulation of metaphase chromosomes for their biochemical analysis has been a significant challenge. In this chapter, a method is described for accumulating and preparing human metaphase chromosomes in sufficient amount to perform their proteome analysis. Preparation and separation methods of chromosome proteins are described, followed by a protocol for their identification by mass spectrometry.

Key Words: Human metaphase chromosomes; synchronization of cell cycle; preparation of chromosome proteins; gel electrophoresis; mass spectrometry.

1. Introduction

Chromosomes occupy a specific position among the organelles in a human cell, appearing only at a limited stage during cell cycle. As their formation accompanies the drastic event of nuclear disruption within a cell, they have been studied for three centuries. The major components of chromosomes are two DNA molecules and proteins, especially small basic histone proteins. Moreover, the complex formed of eight histone proteins and DNA, referred to as a nucleosome, has been detected as the common basic structure of chromosomal DNA associated to histones. Higher structures than nucleosomes or 30-nm

From: *Methods in Molecular Biology, vol. 432: Organelle Proteomics*
Edited by: D. Pflieger and J. Rossier © Humana Press, Totowa, NJ

fibers have not been clearly established yet although several models have been proposed by a number of researchers *(1)*.

The study of chromosome structure by biochemical means has been hampered by the difficulty of isolating and collecting metaphase chromosomes in sufficient amounts from human tissues. Thus, a proteome analysis of human chromosomes to comprehensively analyze chromosomal proteins, which may contribute to the elucidation of their higher order structure, has not been possible, except in a few exceptional cases *(2)*.

The development of a synchronization method for cultured cells combined with adequate chromosome isolation and protein extraction methods have enabled collection of chromosomal proteins from the massive metaphase chromosomes, in an amount suitable for their analysis by mass spectrometry *(3)*. Using these techniques, most of the human metaphase chromosome proteins that most likely have structural roles have been identified *(4–6)*. In this chapter, we describe the essential steps for synchronization of human cultured cells and isolation of high-quality human metaphase chromosomes. Preparation and separation methods of the chromosome proteins and their identification by mass spectrometry are also described.

2. Materials

2.1. Cell lines, cell culture, lysis, and chromosome harvesting

1. Cultured cell lines: BALL-1 and HeLa S3.
2. Culture medium: RPMI1640 with fetal bovine serum [FBS; 10% (w/v) for BALL-1 cells and 5% (w/v) for HeLa S3 cells].
3. Colcemid: 20 ng/mL final concentration for BALL-1 cells and 100 ng/mL for HeLa S3 cells.
4. Hypotonic solution: 75 mM KCl.
5. Ohnuki's buffer: 55 mM KCl, 55 mM $NaNO_3$, and 55 mM CH_3COONa, mixed in proportions 4:2:0.8 (v/v/v) *(7)*.
6. PA (polyamine) buffer: 15 mM Tris–HCl, pH 7.2, 2 mM ethylenediaminetetraacetic acid (EDTA), 80 mM KCl, 20 mM NaCl, 0.5 mM EGTA, 0.2 mM spermine, and 0.5 mM spermidine.
7. Modified PA buffer: PA buffer containing 1 mg/mL of digitonin, 14 mM 2-mercaptoethanol, and 0.1 mM phenylmethylsulfonyl fluoride (PMSF) *(8,9)*.
8. 1 μg/mL 4,6-diamidino-2-phenylindole (DAPI).
9. Sucrose solution for sucrose gradient (SG) chromosomes: 20% and 60% (w/v) sucrose in PA buffer.
10. Gradient fractionator (BioComp instrument Inc., Fredericton, NB, Canada).
11. Glycerol gradient solutions for PG chromosomes: 5 or 70% (v/v) glycerol in isolation buffer.

12. Wash buffer for PG chromosomes: 7.5 mM Tris–HCl, pH 7.4, 40 mM KCl, 1 mM EDTA, 0.1 mM spermine, 0.25 mM spermidine, 1% (v/v) thiodiglycol, and 0.1 mM PMSF.

13. Lysis buffer for PG chromosomes: 15 mM Tris–HCl, pH 7.4, 80 mM KCl, 2 mM EDTA, 0.2 mM spermine, 0.5 mM spermidine, 1% (v/v) thiodiglycol, 0.1 mM PMSF, and 0.1% Empigen.

14. Dounce homogenizer with B pestle.

15. Percoll solution for PG chromosomes: 89% (w/v) Percoll, 5 mM Tris–HCl, pH 7.4, 20 mM KCl, 20 mM EDTA, 0.8 mM spermine, 2.25 mM spermidine, 1% (v/v) thiodiglycol, 0.1 mM PMSF, and 0.1% Empigen.

16. Phosphate-buffered saline (10× PBS): 40 g of NaCl, 1.0 g of KCl, 14.5 g of $Na_2HPO_4 \cdot 12H_2O$, and 1.0 g of KH_2PO_4 in 500 mL of water.

17. JS-24.38 swinging rotor and JA-20 rotor (Beckman Instruments Inc., Fullerton, CA, USA).

18. Gradient maker (Bio-Rad Laboratories Inc., Hercules, CA, USA).

19. Isolation buffer for Percoll gradient (PG) chromosomes: 5 mM Tris–HCl, pH 7.4, 20 mM KCl, 20 mM EDTA, 0.25 mM spermidine, 1% (v/v) thiodiglycol, 0.1 mM PMSF, and 0.1% (v/v) Empigen (Calbiochem, La Jolla, CA, USA).

2.2. Isolation of Chromosome Proteins

1. 1 mM $MgCl_2$.
2. Acetic acid.
3. Protein extraction solution: 100 mM $MgCl_2$ in acetic acid/water, 66/34 (v/v).
4. Dialysis solution: 2% (v/v) acetic acid in water.
5. Spectra/P membrane (MWCO = 1000; Spectrum Medical Industries Inc., Houston, TX, USA).

2.3. One-Dimensional SDS–PAGE

1. SDS sample buffer (2×): 62.5 mM Tris–HCl, pH 6.8, 50 mM dithiothreitol, 5% (v/v) 2-mercaptoethanol, 2% (w/v) SDS, and 0.05% (w/v) bromophenol blue (BPB).

2. 40% acrylamide stock solution: 193.3 g of acrylamide and 6.7 g of N, N´-methylene-Bis(acrylamide) (abbreviated Bis) in 500 mL of water.

3. 10% (w/v) ammonium persulfate.

4. Stacking gel buffer (4×, for 100 mL): 0.4 g of SDS and 6.04 g of Tris in water. pH is adjusted to 6.8 with concentrated HCl solution.

5. Stacking gel solution for 1D SDS–PAGE (to obtain a 4% polyacrylamide gel). Add 40 µL of a 10% (w/v) ammonium persulfate stock solution and 5 µL of TEMED in 4 mL of 1× stacking gel buffer.

6. Separating gel buffer stock solution (4×, for 100 mL): 0.4 g of SDS and 18.2 g of Tris. pH is adjusted to 8.8 with concentrated HCl solution.

7. Separation gel solution for 1D SDS–PAGE: 6, 8, 12, or 15% polyacrylamide gel (*see* Note 1). Add 80 µL of a 10% (w/v) ammonium persulfate stock solution and 10 µL of TEMED to 8 mL of 1× separating gel buffer.

8. Running buffer: 14.4 g of glycine, 3.0 g of Tris, and 1.0 g of SDS in 1 L of water.

9. Staining solution: 1 tablet of PhastGel Blue R (GE Healthcare, Uppsala, Sweden) in methanol/acetic acid/water, 25/8/67 (v/v/v).
10. Destaining solution for PhastGel Blue R: methanol/acetic acid/water, 10/10/80 (v/v/v).

2.4. Two-Dimensional SDS–PAGE

1. MicroSpin G-25 columns (GE Healthcare).
2. Sample buffer for 2D SDS–PAGE: 7 M urea, 2 M thiourea, 65 mM dithiothreitol, and 2% (w/v) CHAPS.
3. Ampholine, pH 3.5–9.5 (GE Healthcare).
4. First dimension of separation: Immobiline DryStrip pH 4–7 and Immobiline DryStrip pH 6–11 (GE Healthcare).
5. Equilibration buffer stock solution: 50 mM Tris–HCl, pH 6.8, 6 M urea, 30% (v/v) glycerol, and 2% (w/v) SDS.
6. Equilibration buffer 1: 10 mL of equilibration buffer stock solution and 25 mg of dithiothreitol, add water to 100 mL.
7. Equilibration buffer 2: 10 mL of equilibration buffer stock solution, 0.45 g of iodoacetamide, and 1 mg of BPB, add water to 100 mL.
8. 0.5% agarose solution: 0.5% (w/v) agarose, 125 mM Tris–HCl, pH 6.8, and 0.1% (w/v) SDS.
9. Fixative solution: methanol/acetic acid/water, 20/8/72 (v/v/v).
10. Gel electrophoresis system: Ettan IPGphor (GE Healthcare).
11. Imaging system: ImageScanner (GE Healthcare) with ImageMaster 2D Elite version 4.01 (GE Healthcare).

2.5. Protein Digestion and Analysis by Matrix-Assisted Laser Desorption Ionization Time of Flight Mass Spectrometry

1. Destaining solution: 50 mM ammonium bicarbonate in methanol/water, 50/50 (v/v).
2. Reducing solution: 100 mM dithiothreitol and 100 mM ammonium bicarbonate.
3. Alkylating solution: 100 mM iodoacetamide and 100 mM ammonium bicarbonate.
4. Enzyme solution: 5 µg/mL of Lys-C or trypsin in 50 mM ammonium bicarbonate.
5. Peptide extraction solution: trifluoroacetic acid/acetonitrile/water, 0.1/50/50 (v/v/v).
6. ZipTipC$_{18}$ pipette tips (Millipore, Billerica, MA, USA).
7. Matrix solution: 10 mg/mL of a-cyano-4-hydroxycinnamic acid in trifluoroacetic acid/acetonitrile/water, 0.1/30/70 (v/v/v).
8. MTP 384 matrix-assisted laser desorption ionization (MALDI) target plate (Bruker Daltonics, Billerica, MA, USA).
9. Mass spectrometer: MALDI-time of flight mass spectrometry (TOF), Autoflex or Ultraflex (Bruker Daltonics).
10. Biotools 2.2 (Bruker Daltonics) employing the Mascot search engine (Matrix Science, www.matrixscience.com).

3. Methods

3.1. Cell Culture, Lysis, and Chromosome Harvesting

The two possible isolation methods for three different types of human chromosomes are summarized in **Fig. 1**.

3.1.1. Harvesting by Sucrose Gradient

1. Subculture the human cell lines, BALL-1 and HeLa S3, once in every 3 days in culture medium at 37°C under a 5% CO_2-containing atmosphere. At a cell

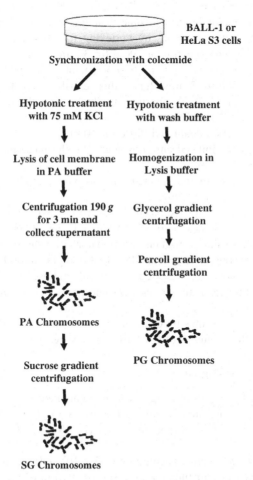

Fig. 1. Flow chart for preparation of three different types (PA, SG, and PG) of chromosomes for proteome analyses of chromosomal proteins. PA, SG, and PG indicate polyamine chromosomes, sucrose density gradient chromosome, and Percoll gradient chromosomes, respectively. See text for details.

concentration of 5×10^5 cells/mL, the cells are treated with colcemid for 12 h to arrest cell cycle at the M phase (*see* **Note 2**).

2. Collect synchronized cells from 400 mL of cell culture at 20°C by centrifugation at 440 *g* for 5 min and resuspend in 40 mL of hypotonic solution (75 mM KCl) or Ohnuki's buffer at 20°C. This treatment is required to swell the cells to lyse them efficiently in **step 5**.

3. After 30 min or 2 h of hypotonic treatment with KCl or Ohnuki's buffer, respectively, collect the cells at 20°C by centrifugation at 780 *g* for 10 min.

4. Hereafter, unless otherwise stated, all the procedures for chromosome preparation should be performed at 4°C. Resuspend the cell pellet in 15 mL of PA buffer for chromosome isolation. Polyamine molecules included in PA buffer are required to maintain chromosome structure stable in vitro.

5. After lysis of cell membranes by vortexing for 30 s, remove cell debris and intact nuclei by centrifugation at 190 *g* for 3 min.

6. Dilute the chromosome-containing supernatant in 15 mL of PA buffer and centrifuge at 190 *g* for 3 min. Repeat this centrifugation step once.

7. Carefully recover the final supernatant, which is the chromosome-rich fraction (C-fraction).

8. Centrifuge C-fraction again at 1750 *g* for 10 min.

9. Resuspend the precipitated chromosomes (PA chromosomes) in 1 mL of fresh PA buffer (*see* **Note 3**).

10. Confirm the nature of the isolated chromosomal sample by both optical and electron microscopy (*see* **Note 4**).

11. PA chromosomes are then subjected to sucrose density gradient centrifugation. Layer PA chromosomes onto 35 mL of a linear 20–60% (w/v) SG in PA buffer, prepared using a gradient fractionator (BioComp Instrument Inc.), and centrifuge at 2500 *g* for 15 min in a JS-24.38 swinging rotor. Fractions containing chromosomes should be assessed by optical microscopy after DAPI staining and combined together. In our case, chromosomes were obtained in the 40–50% sucrose fractions.

12. Collect SG chromosomes by centrifugation at 1000 *g* for 10 min after dilution of sucrose with 5 volumes of PA buffer (*see* **Note 5**).

3.1.2. Harvesting by Percoll gradient

1. Prepare a 5–70% exponential glycerol gradient made from 7 mL of the 70% (v/v) glycerol solution and 25 mL of the 5% (v/v) glycerol solution in 50 mL polycarbonate tubes. Use a gradient maker (Bio-Rad) *(5,10)*. Store the gradient on ice.

2. Harvest the cells at a cell concentration of 5×10^5 cells/mL from 400 mL of cell culture at 20°C by centrifugation at 1500 *g* for 10 min. Suspend them in wash buffer at 20°C and collect them at 1200 *g* for 5 min. Perform three such washes. The cells are thereby hypotonically swollen by wash buffer.

3. Lyse the cells on ice in 5 mL of lysis buffer with 10–12 pestle strokes (counting up and down as one stroke) using a dounce homogenizer B pestle (*see* **Note 6**).

4. Layer the lysate over the exponential glycerol gradient and spin in a JS-24.38 rotor for 5 min at 200 g followed by 15 min at 700 g. Chromosomes form a broad diffuse band in the glycerol gradient, between 10 and 25 mL from the top of the glycerol gradient.

5. Collect a 15-mL fraction containing chromosomes that form a broad band in **step 4**, place it on a 5-mL 70% (v/v) glycerol cushion, and then spin at 3000 g for 15 min to recover chromosomes in the glycerol cushion.

6. Combine 10 mL of 70% (v/v) glycerol cushion containing chromosomes with 10 mL of 89% Percoll solution and homogenize with 10 pestle strokes using a dounce homogenizer B pestle, followed by 30 pestle strokes after further addition of 15 mL of 89% Percoll solution. This step aims to further purify the chromosomes: the chromosomes isolated by glycerol gradient centrifugation indeed still contain many kinds of contaminants, such as mitochondrion and cytoskeletal proteins. To remove these proteins around chromosomes, chromosomes are mixed with Percoll solution and homogenized.

7. Centrifuge at 45,440 g for 30 min in a JA-20 rotor. Collect a band (a volume of about 5 mL) containing chromosomes located at a level that is one-fifth the length of the tube measured from the bottom and dilute three-fold in isolation buffer. The band of interest appears as a "haze layer" containing a lot of white small particles. Subsequently, centrifuge at 3000 g for 15 min to pellet PG chromosomes.

For PA chromosomes (obtained in **Subheading 3.1.1., step 9**) and PG chromosomes (obtained in **Subheading 3.1.2., step 7**) prepared from 2.8×10^8 cells, 500 µg and 44 µg of chromosome proteins were collected, respectively. PG chromosomes are much cleaner than PA chromosomes. By PG centrifugation, mitochondrion, cytoplasmic, and cytoskeletal proteins are efficiently removed from chromosomes. Note that PA chromosomes are cruder than the chromosomes obtained after glycerol gradient centrifugation.

3.2. Isolation of Chromosome Proteins and SDS–PAGE

1. Transfer the chromosomes isolated in PA buffer (PA or SG chromosomes) into microtubes and collect by centrifugation at 7700 g for 10 min at 4°C followed by resuspension in 1× PBS. 100 µL of PBS is added to the chromosomes isolated from 8.0×10^8 cells. For the already pelleted PG chromosomes, directly add PBS to the pellet.

2. Extract chromosome proteins by the acetic acid method *(11,12)* with minor modifications. 1/10th volume of 1 mM $MgCl_2$ and 2 volumes of acetic acid are added to the chromosome suspension, and the mixture is stirred for 1 h at 4°C with a microtube stirrer. Chromosome proteins are dissociated from chromosomes by decreasing pH with acetic acid.

3. After centrifugation at 17,400 g for 10 min, collect supernatants and add an equal volume of protein extraction solution to the precipitate, followed by stirring for 20 min at 4°C.

4. After centrifugation at 17,400 g for 10 min, collect this second supernatant and pool both supernatants obtained in **steps 3** and **4**.
5. Dialyze the supernatants against 2% (v/v) acetic acid in water with Spectra/P membrane for 2 h at 4°C to decrease the concentration of acetic acid. At least, three buffer changes are recommended.
6. Freeze the extracted protein solution by liquid nitrogen and lyophilize (*see* **Note 7**).
7. Clean glass plates, plastic combs, and spacers using 100% ethanol. Set up gel plates with plastic seal and paper clamps. Stand the plates on top of a sheet of aluminum foil. Pour ethanol to check that gel assembly does not leak. Discard the ethanol.
8. Prepare the separating gel solution and pour the gel solution into the gel assembly until it reaches about 2–3 cm below the top of the glass. Layer ethanol on top of the gel (squirt bottle). Wait for 20 min for the gel to polymerize.
9. Discard ethanol layer. Prepare stacking gel solution and pour the stacking gel solution on top of the polymerized separating gel.
10. Insert a dry and clean comb without making any air bubbles before the stacking gel polymerizes and wait for 15 min.
11. Remove the comb and spacer slowly, assemble the gel plate onto the electrophoresis tank. Fill the tank with running buffer.
12. Solubilize the lyophilized proteins into 1× SDS sample buffer and boil for 5 min. Load 5–30 µL (5–30 µg) of samples in different wells (the volume depends on the size of the well).
13. Run the electrophoresis at steady current (30–40 mA) for 1–1.5 h.
14. After electrophoresis, stain the gel with staining solution, followed by destaining with destaining solution.

3.3. Two-Dimensional SDS–PAGE

1. Solubilize 500 µg of lyophilized proteins into 50 µL of sample buffer. After centrifugation at 17,400 g for 10 min at 4°C, desalt supernatant using MicroSpin G-25 columns, because the salts present in the sample would perturb protein isoelectric focusing during the following electrophoresis step. Add 1 µL of ampholine, pH 3.5–9.5, into sample solutions for Immobiline DryStrip, pH 4–7 or pH 3–10 (*see* **Note 8**).
2. Set the Immobiline DryStrip on Ettan IPGphor and add sample solution. Run the electrophoresis. An example of electrophoresis conditions is as follows: 0 V for 2 h, 50 V for 8 h, 1000 V for 2 h, 1000 V for 1 h, 2000 V for 2 h, 5000 V for 2 h, and 8000 V for over 8 h.
3. After the first dimension of electrophoresis, shake the strip in equilibration buffer 1 for 15 min at room temperature, followed by in equilibration buffer 2 for 15 min at room temperature in the dark, respectively, to reduce and alkylate cysteines.
4. Prepare a 12% polyacrylamide gel as described in **Subheading 3.2**. Put the gel strip on the stacking gel and immobilize with 0.5% agarose solution.

5. Assemble the gel plate onto the electrophoresis tank. Fill the tank with running buffer. Run the electrophoresis at steady current (30 mA) for 4 h.
6. After electrophoresis, shake the gel in fixative solution for 1 h. Stain the gel in staining solution, followed by destaining in destaining solution.
7. Scan the gel images using ImageScanner. Analyze the images by ImageMaster 2D Elite ver. 4.01 and calculate isoelectric point, molecular weight, and amount of each spot.

3.4. Mass spectrometry

1. Excise the spots of interest from the gel. Cut the excised pieces into 1-mm cubes and transfer to a 0.5-mL microtube (*see* **Note 9**).
2. ncubate the gel pieces in destaining solution at 37°C. Mix by inverting the tube several times to facilitate the destaining step.
3. Remove the destaining solution and add enough acetonitrile to cover the gel pieces.
4. After the gel pieces have shrunk, remove the acetonitrile. Dry down gel pieces in a vacuum centrifuge.
5. For reduction and alkylation, swell the gel pieces in reducing solution and incubate for 30 min at 56°C, followed by replacement of the reducing solution with alkylating solution and incubation at 37°C for 30 min in the dark.
6. Remove the alkylation solution and add enough acetonitrile to cover the gel pieces.
7. After the gel pieces have shrunk, remove acetonitrile. Dry down gel pieces in a vacuum centrifuge.
8. Add just enough freshly prepared enzyme solution to cover the gel and incubate on ice for 30 min.
9. Remove excess of enzyme solution and add enough 25 mM ammonium bicarbonate to keep the gel wet overnight, but avoid excess liquid. Incubate at 37°C overnight.
10. Add 50 μL of peptide extraction solution and sonicate for 10 min to elute the peptides from the gel. Recover the supernatant. Repeat three times.
11. Pool the supernatants containing peptides and reduce the volume to 20 μL by vacuum centrifugation. Add 100 μL of 0.1% (v/v) trifluoroacetic acid in water and then concentrate to about 20 μL by further vacuum centrifugation.
12. To desalt the peptides, equilibrate ZipTipC$_{18}$ pipette tips with 0.1% (v/v) trifluoroacetic acid. After pipetting three times in the peptide solution to absorb the peptides, wash with 0.1% trifluoroacetic acid and elute peptides with 5 μL of peptide extraction solution into 0.5-mL microtubes.
13. Mix 1 μL of the desalted peptide solutions with an equal volume of matrix solution and deposit on a MALDI target plate to create a sample template for MALDI-TOF MS.
14. Measure peptide masses using MALDI-TOF mass spectrometer.

15. Identify proteins by peptide mass fingerprinting (MFP). Mass data extracted by the flex analysis are analyzed using Biotools 2.2 employing the Mascot search engine. Proteins are identified by matching the MFP results with the NCBInr database (*see* **Note 10**).

From PA and PG chromosomes isolated from 2.8×10^8 cells, about 200 and 100 proteins were identified, respectively *(4,5)*. In a 1D SDS–PAGE gel, about 50 bands were detected by Coomassie brilliant blue (CBB) staining, and three or four proteins were identified from several single bands. However, in general, it is difficult to identify more than three proteins from single bands by MALDI-TOF MS (MFP). If a single band contains several proteins, MS/MS analysis is recommended. Alternatively, although the separation of high-molecular-weight (>100 kDa) proteins is difficult, 2D gel separation is also useful for protein identification, because almost all spots contain one protein. In the case of chromosome proteins, more than 100 spots were detected by CBB staining. Furthermore, using 2D gel separation, we could estimate the amount of each protein after CBB staining.

4. Notes

1. The acrylamide percentage is adjusted according to the molecular weight range of the proteins to be detected: 6 or 8% PAGE is used for proteins above 100 kDa, 12% PAGE for proteins of MW 20–100 kDa, and 15% PAGE for proteins of MW 10–60 kDa.
2. For the isolation of chromosomes, 2 L of cell culture (at 7×10^5 cells/mL) can be handled easily in a single preparation. By this synchronization method, a mitotic index above 50% is routinely obtained.
3. The separation profiles between chromosomes and nuclei can be confirmed by flow cytometry (EPICS Elite, Beckman Coulter) after staining of DNA with 35 µg/mL propidium iodide (PI).
4. For observation by optical microscopy, isolated chromosomes are placed on glass slides and observed using a fluorescence microscope after staining with 1 µg/mL DAPI. For electron microscopy observation, isolated chromosomes are mounted on a plastic sheet that has been previously coated with poly-L-lysine and are fixed with 4% (*p*)-formaldehyde in PA buffer. After washing with the same solution, the specimens are dehydrated in an ethanol series and freeze-dried using the *t*-butyl alcohol drying method *(13)*. Finally, the specimens are osmium-coated with an osmium plasma coater and examined under the scanning electron microscope at 10 kV or 15 kV.
5. Isolated PA and SG chromosomes can be stored at –80°C in PA buffer containing 70% (v/v) glycerol until use. Glycerol is used for cryoprotection of chromosomes.

6. Homogenization should be performed gently to avoid damage to chromosomes. After cell lysis, all the procedures for PG chromosome isolation should be performed at 4°C.
7. Lyophilized proteins can be stored as powder material at –80°C.
8. We use 11-cm Immobiline DryStrip for 2D electrophoresis. Add 0.5 μL of IPG buffer, pH 6–11 (GE Healthcare), into sample solution if you wish to use Immobiline Drystrip, pH 6–11 (GE Healthcare).
9. Cut as close to the protein band as possible to reduce the amount of background gel. Excise a gel piece of roughly the same size from a non-protein-containing region of the gel for use as a control.
10. Values are compared with human protein databases at (±100–200 ppm) mass tolerance in MFP analysis.

Acknowledgements

The authors thank Profs Schihiro Matsunaga, and Shin'ichiro Kajiyama for their collaboration during the research on chromosome proteomics. This work was supported by Special Coordination Funds from the Ministry of Education, Culture, Sports, Science and Technology, Japan to K. F.

References

1. Marsden, M. P. and Laemmli, U. K. (1979) Metaphase chromosome structure: evidence for a radial loop model. *Cell* **17**, 849–858.
2. Gassmann, R., Henzing, A. J., and Earnshaw, W. C. (2004) Novel components of human mitotic chromosomes identified by proteomic analysis of the chromosome scaffold fraction. *Chromosoma* **113**, 358–397.
3. Sone, T., Iwano, M., Kobayashi, S., Ishihara, T., Hori, N., Takata, H., et al. (2002) Changes in chromosomal surface structure by different isolation conditions. *Arch. Histol. Cytol.* **65**, 445–455.
4. Takata, H., Uchiyama, S., Nakamura, N., Nakashima, S., Kobayashi, S., Sone, T., et al. (2007) A comparative proteome analysis of human metaphase chromosomes isolated from two different cell lines reveals a set of conserved chromosome-associated proteins. *Genes Cells* **12**, 269–284.
5. Uchiyama, S., Kobayashi, S., Takata, H., Ishihara, T., Hori, N., Higashi, T., et al. (2005) Proteome analysis of human metaphase chromosomes. *J. Biol. Chem.* **289**, 16994–17004.
6. Uchiyama, S., Kobayashi, S., Takata, H., Ishihara, T., Sone, T., Matsunaga, S., et al. (2004) Protein composition of human metaphase chromosomes analyzed by two-dimensional electrophoreses. *Cytogenet. Genome Res.* **107**, 49–54.
7. Ohnuki, Y. (1968) Structure of chromosomes. I. Morphological studies of the spiral structure of human somatic chromosomes. *Chromosoma* **25,** 402–428.
8. Kuriki, H. and Haruo, T. (1997) Standardization of bivariate flow karyotypes of human chromosomes for clinical applications. *J. Clinic. Lab. Anal.* **11,** 169–174.

9. Spector, D. L., Goldman, R. D., and Leinwand, L. A. (1998) Chromosome isolation for biochemical and morphological analysis. In Janssen K. (ed), *Cells, A Laboratory Manual*, pp. 49.41–49.12. Cold Spring Harbor Laboratory Press, New York.

10. Gasser, S. and Laemmli, U. (1987) Improved methods for the isolation of individual and clustered mitotic chromosomes. *Exp. Cell Res.* **173**, 85–98.

11. Hardy, S. J., Kurland, C. G., Voynow, P., and Mora, G. (1969) The ribosomal proteins of *Escherichia coli*. I. Purification of the 30S ribosomal proteins. *Biochemistry* **8**, 2897–2905.

12. Izutsu, K., Wada, A., and Wada, C. (2001) Expression of ribosome modulation factor (RMF) in Escherichia coli requires ppGpp. *Gene Cells* **6**, 665–676.

13. Inoue, T. and Osatake, H. (1988) A new drying method of biological specimens for scanning electron microscopy: the *t*-butyl alcohol freeze-drying method. *Arch. Histol. Cytol.* **51**, 53–61.

11

Purification and Proteomic Analysis of Plant Plasma Membranes

Erik Alexandersson, Niklas Gustavsson, Katja Bernfur, Adine Karlsson, Per Kjellbom, and Christer Larsson

Summary

All techniques needed for proteomic analyses of plant plasma membranes are described in detail, from isolation of plasma membranes to protein identification by mass spectrometry (MS). Plasma membranes are isolated by aqueous two-phase partitioning yielding vesicles with a cytoplasmic side-in orientation and a purity of about 95%. These vesicles are turned inside-out by treatment with Brij 58, which removes soluble contaminating proteins enclosed in the vesicles as well as loosely attached proteins. The final plasma membrane preparation thus retains all integral proteins and many peripheral proteins. Proteins are separated by one-dimensional sodium dodecyl sulphate–polyacrylamide gel electrophoresis (SDS–PAGE), and protein bands are excised and digested with trypsin. Peptides in tryptic digests are separated by nanoflow liquid chromatography and either fed directly into an ESI-MS or spotted onto matrix-assisted laser desorption ionization (MALDI) plates for analysis with MALDI-MS. Finally, data processing and database searching are used for protein identification to define a plasma membrane proteome.

Key Words: Plant plasma membranes; two-phase partitioning; nanoflow liquid chromatography; mass spectrometry; ESI-MS; MALDI-MS; integral membrane proteins; plasma membrane proteome.

1. Introduction

The plant plasma membrane is the outermost membrane of the cell and constitutes the cell border across which nutrients are imported and metabolic products exported. It is the site for receptors recording the environment and

From: *Methods in Molecular Biology, vol. 432: Organelle Proteomics*
Edited by: D. Pflieger and J. Rossier © Humana Press, Totowa, NJ

for receptors responding to endogenous signals. It is also the site for synthesis of cellulose and other components of the cell wall. This makes the plasma membrane a key membrane for cell functioning and explains the great interest in its characterization. However, as the plasma membrane only constitutes a few percent of the total membranes of a plant cell, a strong demand is placed on the isolation procedure to obtain pure plasma membranes. The plasma membrane is probably the most diverse membrane of the cell, with a protein composition that varies with cell type, developmental stage, and environment, and it is likely to harbor more than a thousand different proteins, factors that make proteomic characterization of plasma membranes a very challenging task. Below, we describe how the plasma membrane is separated from all intracellular membranes resulting in plasma membrane preparations with high yield and

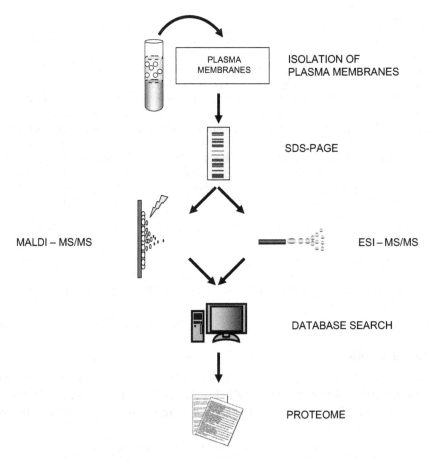

Fig. 1. Flow scheme—from isolation of plasma membranes to proteome analysis.

purity, retaining integral as well as peripheral proteins; and we describe how such preparations are used in proteomic studies based on either ESI-MS *(1)* or matrix-assisted laser desorption ionization-mass spectrometry (MALDI-MS) for protein identification (*see* **Fig. 1**).

2. Materials
2.1. Preparation of a Microsomal Fraction

1. Homogenization medium: 0.3 M sucrose, 50 mM 3-(*N*-morpholino)propane-sulfonic acid (MOPS)–KOH, 5 mM EDTA–KOH, pH 7.5, and 0.2% (w/v) casein (enzymatic hydrolysate; Sigma C7290, www.sigmaaldrich.com) (*see* **Note 1**).
2. Insoluble polyvinylpyrrolidone (PVPP) to bind free phenols.
3. Dithiothreitol (DTT), reducing agent (*see* **Note 2**).
4. Ascorbic acid, reducing agent (*see* **Note 2**).
5. 100 mM phenylmethylsulfonyl fluoride (PMSF) in isopropanol (*see* **Note 3**).
6. Protease inhibitor cocktail for plant cell and tissue extracts (Sigma P9599).
7. 50 mM ethylenediaminetetraacetic acid (EDTA)–KOH, pH 7.8.
8. Resuspension medium: 0.3 M sucrose, 5 mM potassium phosphate, pH 7.8 (*see* **Note 4**), 0.1 mM EDTA–KOH, pH 7.8 (use **item 7** above), and 1 mM DTT (*see* **Note 2**).
9. Kitchen blender.
10. Nylon cloth, with about 0.2-mm pore size.
11. Refrigerated centrifuge equipped with an angle rotor for 50-mL tubes.

2.2. Isolation of Plasma Membranes by Aqueous Two-Phase Partitioning

1. 20% (w/w) Dextran T 500 (Pharmacia, www.serva.de) in water (*see* **Note 5**).
2. 40% (w/w) polyethylene glycol 3350 (PEG) in water (Sigma P4338).
3. Sucrose.
4. 0.2 M potassium phosphate, pH 7.8 (*see* **Note 4**).
5. 2 M KCl.
6. 0.5 M DTT in 50 mM EDTA–KOH, pH 7.8 (*see* **Subheading 2.1, item 7**) (*see* **Note 2**).
7. Phase mixture: 7.32 g of 20% dextran solution, 3.66 g of 40% PEG solution, 1.85 g of sucrose, 0.450 mL of 0.2 M potassium phosphate, pH 7.8, 0.036 mL of 2 M KCl, and H^2O to a final weight of 18.00 g in a 50-mL centrifuge tube (*see* **Notes 6** and **7**).
8. Bulk phase system: 91.5 g of 20% dextran solution, 45.75 g of 40% PEG solution, 30.8 g of sucrose, 7.50 mL of 0.2 M potassium phosphate, pH 7.8, 0.45 mL of 2 M KCl, and H^2O to a final weight of 300 g (*see* **Notes 7** and **8**).
9. Centrifuge with swinging bucket rotor for 50-mL tubes and temperature control (4°C).

10. Ultracentrifuge with an angle rotor for 70-mL tubes.
11. Separatory funnel.

2.3. Protein Quantification

1. Bearden (2) reagent: 200 mg/L of Coomassie brilliant blue G-250, 17% (w/v) phosphoric acid (see **Note 9**).
2. 0.1% (w/v) Triton X-100 in water.
3. Spectrophotometer.

2.4. Brij 58 Treatment to Remove Soluble Proteins

1. 0.3 M sucrose and 5 mM potassium phosphate, pH 7.8 (see **Note 4**).
2. 2% (w/v) Brij 58 (polyethylene glycol hexadecyl ether; Sigma P5884) made fresh in 0.3 M sucrose and 5 mM potassium phosphate, pH 7.8.
3. 2 M KCl, 0.3 M sucrose, and 5 mM potassium phosphate, pH 7.8.
4. Ultracentrifuge with an angle rotor for 5- to 10-mL tubes.

2.5. Sodium Dodecyl Sulphate–Polyacrylamide Gel Electrophoresis

1. Standard equipment for sodium dodecyl sulphate–polyacrylamide gel electrophoresis (SDS–PAGE).
2. Staining solution: 2% (w/v) H_3PO_4, 1.29 M $(NH_4)_2SO_4$, and 34% (v/v) MeOH (see **Note 10**).
3. Coomassie brilliant blue G-250.
4. Rocking platform.

2.6. In Situ Enzymatic Proteolysis

1. Solution I: 50% (v/v) ethanol and 50 mM NH_4HCO_3-HCl, pH 7.8 (see **Note 11**).
2. Solution II: 10 mM DTT and 50 mM NH_4HCO_3-HCl, pH 7.8 (see **Notes 2** and **11**).
3. Solution III: 55 mM iodoacetamide in 50 mM NH_4HCO_3-HCl, pH 7.8 (see **Notes 11** and **12**).
4. Solution IV: trypsin at 10 ng/μL in 50 mM NH_4HCO_3-HCl, pH 7.8 (see **Note 13**).
5. Solution V: 0.5% (v/v) trifluoroacetic acid (TFA) in water (see **Note 14**).
6. 50 mM NH_4HCO_3-HCl, pH 7.8 (see **Note 11**).
7. Ethanol, p.a.
8. Ice bucket.
9. Speedvac.
10. Heating block or water bath pre-heated to 37°C.
11. Powder-free gloves.
12. 1.5-mL microfuge tubes for proteomic applications.

2.7. Nanoflow Liquid Chromatography MALDI-MS

1. Buffer A: acetonitrile (MeCN)/water/TFA, 1/98.9/0.1 (v/v/v).
2. Buffer B: MeCN/water/TFA, 90/9.9/0.1 (v/v/v).
3. Matrix solution: 5 mg/mL of α-cyano-4-hydroxycinnamic acid (CHCA) and 50 mM citric acid in MeCN/water/TFA, 50/49.5/0.5 (v/v/v) (*see* **Note 15**).
4. Nanoflow liquid chromatography system with a MALDI sample support fraction collector. We use the Agilent 1100 series fraction collector, but any system that can position the capillary delivering the mobile phase onto a MALDI sample support can be used.
5. MALDI-MS instrument with MS/MS capability, such as MALDI-TOF/TOF or MALDI-Q-TOF.

2.8. Nanoflow Liquid Chromatography ESI-MS

1. Buffer C: MeCN/water/formic acid (FA), 2.5/96.5/1 (v/v/v).
2. Buffer D: MeCN/water/FA, 90/9/1 (v/v/v).
3. Nanoflow liquid chromatography system.
4. ESI-MS instrument equipped with a nano-ESI source and with MS/MS capability, such as an ESI-Q-TOF or an ESI-ion trap.

3. Methods

3.1. Preparation of a Microsomal Fraction

1. Conduct all steps at 4°C using pre-cooled media and equipment.
2. Homogenize in a kitchen blender 100 g of plant tissue (*see* **Note 16**) in 150 mL of the following medium: 150 mL of homogenization medium with fresh additions of 0.9 g of PVPP, 0.12 g of DTT (final concentration is 5 mM), and 0.13 g of ascorbic acid (final concentration is 5 mM) (*see* **Note 2**). Homogenize with 4 bursts of 20 s each.
3. Filter the homogenate through a nylon cloth, gently squeezing out the remaining liquid, and immediately add 0.75 mL of 100 mM PMSF and 1.5 mL of the protease inhibitor cocktail to the filtrate.
4. Centrifuge the filtrate at 10,000 g for 10 min, discard the pellets.
5. Centrifuge the supernatants at 50,000 g for 30 min, save the pellets.
6. Resuspend the pellets in a total volume of 7 mL of resuspension medium + 70 μL of protease inhibitor cocktail to yield the microsomal fraction.

3.2. Isolation of Plasma Membranes by Aqueous Two-Phase Partitioning

1. Add 6.00 mL of the microsomal fraction (*see* **Subheading 3.1., item 6**) to the phase mixture (*see* **Fig. 2, step 1**, loading).

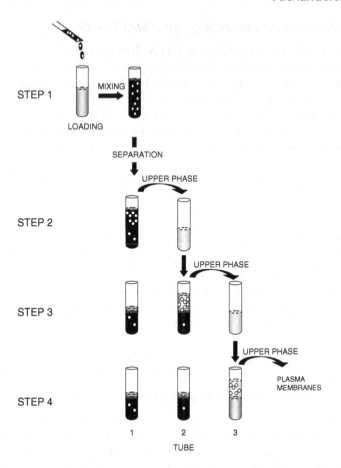

Fig. 2. Separation of plasma membranes (o) from intracellular membranes (dark matter) by aqueous two-phase partitioning. The microsomal fraction is loaded on a phase mixture (**step 1**). After mixing and separation, the upper phase is transferred to a fresh lower phase (**step 2**). The procedure is repeated (**steps 3** and **4**) to produce a final upper phase enriched in plasma membranes. The lower phase left in tube 1 is enriched in intracellular membranes and depleted in plasma membranes.

2. Mix the contents thoroughly by 20 inversions of the tube (*see* **Fig. 2, step 1**, mixing).
3. Centrifuge at 1500 *g* for ~5 min to get phase separation (*see* **Fig. 2, steps 1 and 2**, separation).
4. Transfer ~90% of the upper phase (use a disposable transfer pipette), without disturbing the interface, to a 50-mL centrifuge tube containing ~10 mL of lower phase obtained from the bulk phase system (*see* **Fig. 2, step 2**).
5. Mix and spin, that is, repeat **points 2** and **3** above.

6. Repeat **points 4** and **5** above (*see* **Fig. 2, step 3**).
7. Withdraw the final upper phase (*see* **Fig. 2, step 4**) and dilute it to ~70 mL (*see* **Note 17**) with resuspension medium (*see* **Subheading 2.1., item 8**) in a 70-mL ultracentrifuge tube. This fraction is highly enriched in plasma membranes, mainly as right-side out (cytoplasmic-side-in) vesicles.
8. Dilute the lower phase left in tube 1 (*see* **Fig. 2, step 4**) to ~100 mL with resuspension medium (*see* **Subheading 2.1., item 8**) and transfer 10 mL of this 100 mL to a 70-mL ultracentrifuge tube (*see* **Note 17**). Dilute with resuspension medium. This fraction is enriched in intracellular membranes and depleted in plasma membranes (*see* **Note 18**). Discard the remaining 90 mL.
9. Transfer ~1 mL of the microsomal fraction that was not added to the phase system to a 70 mL ultracentrifuge tube. Dilute with resuspension medium (*see* **Subheading 2.1., item 8**) (*see* **Note 18**).
10. Pellet all membranes by centrifugation at 100,000 *g* for 1 h.
11. Resuspend the pellets in 0.5 mL of resuspension medium (*see* **Subheading 2.1., item 8**) + 5 μL protease inhibitor cocktail (*see* **Subheading 2.1., item 6**), and store the samples in liquid nitrogen until use (*see* **Note 19**).

3.3. Protein Quantification

1. Add 100 μL of 0.1% (w/v) Triton X-100 to a 3-mL cuvette.
2. Add *x* μL of membrane preparation (*see* **Subheading 3.2., point 11**) corresponding to 5–15 μg of protein.
3. Add 900–*x* μL of water and mix well.
4. Add 1 mL of Bearden reagent, mix, and wait for 5 min.
6. Measure with the spectrophotometer at 595 and 465 nm. Take the difference between A_{595} and A_{465} and compare to a standard curve prepared with bovine serum albumin (BSA) (1 to 20 μg of protein) (*see* **Note 20**).

3.4. Brij 58 Treatment to Remove Soluble Proteins

1. Dilute plasma membranes with 0.3 M sucrose and 5 mM potassium phosphate, pH 7.8, to 3.0 mL in a 5-mL ultracentrifuge tube (in the cold room or on ice).
2. Add to the sample 2% (w/v) Brij 58, 0.3 M sucrose, and 5 mM potassium phosphate, pH 7.8, to give a detergent:protein ratio of 5:1 (w/w), mix and let stand for 5 min (*see* **Note 21**).
3. Add to the sample 0.3 M sucrose and 5 mM potassium phosphate, pH 7.8, to a final volume of 4.5 mL.
4. Add to the sample 0.5 mL of 2 M KCl, 0.3 M sucrose, and 5 mM potassium phosphate, pH 7.8, and mix.
5. Centrifuge (4°C) for 2 h at 100,000 *g* (*see* **Note 22**).
6. Resuspend in half the original volume (usually 250 μL) of resuspension medium (*see* **Subheading 2.1., item 8**) and 2.5 μL of protease inhibitor cocktail (*see* **Subheading 2.1., item 6**).

3.5. SDS–PAGE

1. SDS–PAGE is run according to standard procedures using either commercially available minigels or self-made large gels (12-cm separation gel, 12% acrylamide, or linear gradient from 12–20% acrylamide).
2. Load 10–20 μg of protein per lane with minigels (100–200 μg of protein per lane with large gels).
3. After electrophoresis is complete, place the minigel in a tray with 125 mL of staining solution (0.5 L of staining solution with a large gel) and keep it on a rocking platform for 1 h.
4. For minigels, add 0.25 g of Coomassie brilliant blue G-250 to 25 mL of staining solution (1.0 g Coomassie brilliant blue G-250 to 100 mL of staining solution for large gels). Stir the solution until the dye is evenly dispersed and add it to the tray containing the 125 mL (0.5 L with a large gel) of staining solution.
5. Keep the tray on the rocking platform for up to 5 days.
6. Destain the gel in water on the rocking platform (*see* **Note 23**).

3.6. In Situ Enzymatic Proteolysis

1. Excise gel pieces from the protein bands of interest using either a circular excision tool with a diameter of 1.5–2 mm or a scalpel (*see* **Note 24**).
2. Place each gel piece in a 1.5-mL microfuge tube (*see* **Note 25**).
3. Dispense 100 μL of solution I and incubate for 30 min. Remove the liquid. Repeat once.
4. Dispense 75 μL of ethanol. Incubate for 5 min. Remove the liquid.
5. Dispense 10 μL of solution II. Incubate for 30 min at 37°C to reduce disulfide bridges. Remove the liquid.
6. Dispense 75 μL of ethanol. Incubate for 5 min. Remove the liquid.
7. Dispense 10 μL of solution III. Incubate for 30 min in darkness to alkylate cystein residues. Remove the liquid.
8. Dispense 100 μL of solution I. Incubate for 5 min. Remove the liquid.
9. Dispense 75 μL of ethanol. Incubate for 10 min. Remove the liquid.
10. Make sure that all ethanol has evaporated before continuing (a few min in a Speedvac is enough).
11. Place the tubes with the gel pieces, the trypsin solution, and the buffer solution on ice before continuing.
12. Dispense solution IV: 4.5 μL for 1.5-mm diameter gel pieces and 6 μL for 2.0-mm diameter gel pieces.
13. Check that the gel pieces are covered with trypsin solution, and keep the tubes on ice for 30 min.
14. Dispense 50 mM NH_4HCO_3-HCl, pH 7.8: 12 μL for 1.5-mm gel pieces and 15 μL for 2.0-mm gel pieces.
15. Incubate overnight at 37°C.
16. Dispense solution V: 12 μL for 1.5-mm gel pieces and 15 μL for 2.0-mm gel pieces.

17. Incubate for at least 30 min at room temperature.
18. Collect the supernatants and store them at –80°C (or –20°C).

3.7. Nanoflow Liquid Chromatography MALDI-MS

1. Load a suitable amount (we routinely load 5 μL of a protein gel band tryptic digest corresponding to ∼20% of the total digest volume) of the peptide extracts onto a pre-column or directly onto the separation column (*see* **Note 26**) at a flow rate of 10–50 μL/min of buffer A.
2. Wash the peptides loaded on the pre-column for ∼10 min to remove salts before connecting the pre-column to the separation column and the binary pump delivering the nanoflow gradient by switching the appropriate valve of the chromatographic system.
3. Elute the peptides at a flow rate of 200–300 nL/min, with a gradient suitable for the complexity of the sample, for example, 10–36 min, 0–60% buffer B and 36–46 min, 60–100% buffer B. Collect fractions of desired volume and time intervals directly onto a MALDI sample support.
4. Add 0.5 μL of 5 mg/mL of CHCA and 50 mM citric acid in MeCN/water/TFA, 50/49.5/0.5 (v/v/v), to each of the fractions either by using an automatic liquid handler or by manual pipetting. Citric acid reduces background signals from the MALDI matrix.
5. Acquire mass spectra of all fractions. To achieve high-mass accuracy of the data, perform internal calibration by adding at least two peptides of known mass as standards to the matrix solution and use the corresponding signals for calibration of each collected mass spectrum. The mass accuracy of our TOF-MS data is typically improved from around 50 ppm to below 10 ppm by using internal calibration. Selection of precursor ions for MS/MS can either be performed automatically for each fraction (e.g., by selecting a maximum number of signals above a certain S/N threshold) or preferably for the complete MS data set afterwards (*see* **Note 27**). Acquire MS/MS data for the selected precursor ions.

3.8. Nanoflow Liquid Chromatography ESI-MS

1. Load a suitable amount of the peptide extracts onto a pre-column or directly onto the separation column (*see* **Note 28**) at a flow rate of 10–50 μL /min of buffer C.
2. Wash the peptides loaded on the column for ∼10 min to remove salts.
3. Elute peptides with a gradient suitable for the complexity of the sample, for example, 40 min, 0–100% buffer D with a flow rate of 50–200 nL/min.
4. Acquire tandem mass spectra (MS/MS) by data-dependent analysis. We routinely analyzed the three most abundant ions in each cycle; 0.3-s MS scan over the *m/z* range 400–2000 Da and maximum 4.8-s MS/MS scan over the *m/z* range 50–3000 Da, continuum mode, 60-s dynamic exclusion to avoid repeated acquisition of intense ions (*see* **Note 29**).

3.9. MS Data Processing and Database Searching

1. Query the database of choice with the acquired MS ion fragment data sets using search parameters appropriate to the experimental conditions and instrument performance. Typical settings using Mascot as search engine are enzyme missed cleavages, 1; precursor peptide mass tolerance, ±25 ppm; fragment ion mass tolerance, ±0.4 Da; fixed modifications, carbamidomethylation of cysteine residues; and variable modifications, oxidation of methionines. Peptide charge set to +1 for MALDI-MS and 2+ or 2+ and 3+ for ESI-MS. In our hands, a single gel band excised from a lane loaded with 100 µg of protein on a 12-cm-long separation gel resulted in 0–70 protein identifications. Usually, about 10–30 proteins per gel band were identified.

4. Notes

1. The casein hydrolysate is dissolved in a small volume of water and boiled for 10 min under stirring. The casein hydrolysate serves as competing substrate for proteases.
2. DTT and ascorbate are unstable in water (slowly oxidized) and are therefore added fresh at the time of usage.
3. PMSF is a protease inhibitor and is poisonous. It is unstable in water solution but stable in isopropanol at room temperature for several months.
4. Add 0.2 M dipotassium hydrogen phosphate to 0.2 M potassium dihydrogen phosphate (ratio about 10/1) to obtain a 0.2 M stock solution of potassium phosphate, pH 7.8.
5. We use Dextran T 500 from Pharmacia, Uppsala, Sweden. Since this supplier is no longer available, we suggest Serva, Germany. Dextran powder contains some percent of water. Therefore, the exact concentration of the solution has to be determined with a polarimeter. Layer 220 g of dextran on 780 g of water. Heat on a water bath with gentle stirring until all dextran is dissolved. Dilute about 5 g (note the exact amount and make duplicates) of the solution to 25 mL with water and measure the optical rotation with a polarimeter at 589 nm. The specific rotation is +199 degree mL/g/dm(the cuvette usually has an optical length of 1 dm), and thus the concentration (% w/w) is given by the following equation:

$$\frac{Optical\,rotation}{199} \cdot \frac{25}{Dextran\,(g)} \cdot 100$$

Adjust to 20% (w/w) concentration with water. If a polarimeter is not available, trust the percentage of water given on the package (usually around 5%) and make a large stock solution that can be used for many preparations to ensure reproducibility. Store solutions in the freezer. Store the dextran powder in a dry environment.

6. Phase partitioning is temperature dependent, therefore, all solutions should be equilibrated at 4°C before use, and all manipulations are done in the cold room (or on ice, if a cold room is not available).

7. The final polymer concentration of the described phase system is 6.1% (w/w) for both dextran T 500 and PEG 3350, the chloride concentration is 3 mM, and the phosphate concentration is 5 mM. This system is useful for many tissues, but the optimal concentrations for separation should be determined for each tissue studied. Usually, varying the polymer concentrations between 6.1 and 6.5% and the KCl concentration between 1 and 5 mM is sufficient to find the optimal phase composition for yield and purity (as long as the phosphate concentration is held at 5 mM). Start with some green material, such as spinach leaves, to get familiar with the system.

8. The bulk phase system is made in a separatory funnel, temperature-equilibrated over night at 4°C, thoroughly shaken, and left for at least 4 h for phase separation to occur. The upper and lower phases are withdrawn (discard the interface) and are stored at 4°C or frozen.

9. Dissolve 200 mg of Coomassie brilliant blue G-250 in 200 mL of 85% (w/v) phosphoric acid. Leave over night. Add 800 mL of water and mix. Filter (using filter paper) and store at room temperature in the dark.

10. Add 5.88 mL of 85% (w/v) H_3PO_4 to 150 mL of H_2O. Dissolve 42.5 g of $(NH_4)_2SO_4$ in the solution. Add 85 mL of MeOH and dissolve the salt again.

11. Prepare from a 1 M NH_4HCO_3-HCl, pH 7.8, stock solution (store at –20°C). The pH will be close to 7.8 and does not usually have to be adjusted with HCl. However, the pH will increase with time. Prepare new every second week.

12. Prepare immediately before use. Iodoacetamide is light sensitive.

13. Prepare immediately before use, keep on ice to minimize autoproteolysis.

14. For ESI-MS, use 5% (v/v) FA instead of TFA. TFA suppresses the MS signal in ESI-MS, and FA is therefore often preferred even if it gives lower chromatographic resolution than TFA. See Garcia *(3)* for a discussion on advantages and disadvantages of different mobile phase additives used prior to ESI-MS.

15. Optionally, peptide standards can be included in the matrix solution at concentrations of 5–10 fmol/µL for internal calibration of MALDI-TOF mass spectra.

16. If smaller amounts of plant material are used, the whole procedure can be scaled down, including the size of the phase system. However, it is not wise to use phase systems below 4 g, as a high accuracy in weighing up the system is needed.

17. The polymers make the phases viscous, particularly the dextran in the lower phase. Therefore, the phases should be well diluted (about three times for the upper phase, at least 10 times for the lower phase) to ascertain that most membrane vesicles are sedimented in the subsequent ultracentrifugation step.

18. The parts of the microsomal fraction and the lower phase from tube 1, which are saved, may be used as reference fractions *(4)*. The lower phase will, similarly to the microsomal fraction, contain all membranes present in the cell but will be depleted in plasma membranes.

19. The recovery of protein in the plasma membrane fraction varies with material. About 5 mg of protein per 100 g of leaves can be expected with Arabidopsis.

20. In the method described by Bearden *(2)*, the absorbance of the unbound Coomassie G-250 (peak at 465 nm) and the absorbance of the protein-bound Coomassie G-250 (peak at 595 nm) are subtracted and then compared to a BSA standard curve determined in the same way. Other standard procedures for protein quantification may also be used. Test the one you are most familiar with first.

21. Example: starting with 4 mg of plasma membrane protein, 1 mL of Brij 58 solution should be added. The plasma membrane vesicles, which are mainly cytoplasmic-side-in vesicles when freshly prepared and still more than 50% cytoplasmic-side-in after freezing and thawing, are all turned cytoplasmic-side-out by the Brij treatment *(5)*. This removes soluble proteins enclosed in the vesicles, and the 0.2 M KCl helps remove proteins only loosely attached to the membrane. True peripheral membrane proteins should largely remain associated with the membrane. This represents an additional purification step.

22. The vesicles produced by the Brij treatment are smaller, therefore the centrifugation time is increased to 2 h to increase recovery, which is 50–60%.

23. Background staining is usually low and destaining is rapid. This is the so-called collodial Coomassie staining *(6)*, which is more compatible with MS. However, a standard protocol for Coomassie staining may also be used.

24. The protocol is adjusted for gels with a thickness of 1.0 mm. If gels of different thickness are used, the volumes of reagent solutions in each step of the protocol should be adjusted accordingly. If complete protein gel bands are excised, each band should be divided into smaller pieces and the volumes of reagent solutions in each step of the protocol should be adjusted according to the total volume of the gel band. Use powder-free gloves. Samples are very sensitive to contamination, especially proteins such as keratin.

25. To minimize peptide losses, we recommend 1.5-mL microfuge tubes especially made for proteomic applications.

26. We routinely use a 5-mm-long C18 pre-column with an inner diameter of 300 μm, particle size of 5 μm, and pore size of 300 Å and a 15-cm-long analytical C18 column with an inner diameter of 75 μm, particle size of 3.5 μm, and pore size of 300 Å, such as Zorbax SB300-C18.

27. One major difference between using an online ESI interface and an offline MALDI-MS interface with the nanoflow liquid chromatographic system lies in the procedure for MS/MS precursor ion selection. Using online ESI, selection of precursor ions is limited to the time frame during which the peptides elute from the column. Using an offline MALDI interface, the peptides are collected onto the MALDI sample plate, and only a small part of each fraction is consumed for MS data acquisition. This allows the user to evaluate the complete MS data set before MS/MS precursor ions are selected from fractions with high signal intensity and minimal overlap of peptides. If the selection of precursors for MS/MS is instead based only on the individual spots in which the signals occur,

the instrument might select a precursor from a spot where there is a considerable overlap with other signals.

28. We routinely use a 5-mm-long C18 pre-column with an inner diameter of 300 µm, particle size of 5 µm, and pore size of 100 Å and a 15-cm-long analytical C18 column with an inner diameter of 75 µm, particle size of 3 µm, and pore size of 100 Å.

29. In our case, the criteria were purely time dependent. What time you chose for MS/MS acquisition is dependent on the concentration and the complexity of your sample. If your sample only contains a few peptides yielding relatively strong signals, identification with a faster MS/MS acquisition time is possible. On the other hand, if the peptide signals are weaker, longer acquisition time is needed. This, however, prolongs the sampling time and/or reduces the number of acquired spectra.

Acknowledgments

We thank Gerhard Saalbach for his help and comments on the ESI-MS section. Work in the authors' laboratories was supported by grants from the Swedish Foundation for Strategic Research (C.L.), the Swedish Research Council (C.L.), the Knut and Alice Wallenberg Foundation (C.L.), and Formas (P.K.).

References

1. Alexandersson, E., Saalbach, G., Larsson, C., and Kjellbom, P. (2004) *Arabidopsis* plasma membrane proteomics identifies components of transport, signal transduction and membrane trafficking. *Plant Cell Physiol.* **45**, 1543–1556.

2. Bearden, J. C., Jr (1978) Quantification of submicrogram quantities of protein by an improved protein-dye binding assay. *Biochim. Biophys. Acta*, **533**, 525–529.

3. Garcia, M. C. (2005) The effect of the mobile phase additives on sensitivity in the analysis of peptides and proteins by high-performance liquid chromatography–electrospray mass spectrometry. *J. Chromatogr. B. Analyt. Technol. Biomed. Life Sci.* **825**, 111–123.

4. Nelson, C. J., Hegeman, A. D., Harms, A. C., and Sussman, M. R. (in press) A quantitative analysis of *Arabidopsis* plasma membrane using trypsin-catalyzed ^{18}O labeling. *Mol. Cell. Proteomics* **5**, 1382–1395.

5. Johansson, F., Olbe, M., Sommarin, M., and Larsson, C. (1995) Brij 58, a polyoxyethylene ether, creates membrane vesicles of uniform sidedness. A new tool to obtain inside-out (cytoplasmic side-out) plasma membrane vesicles. *Plant J.* **7**, 165–173.

6. Neuhoff, V., Arold, N., Taube, D., and Ehrhardt, W. (1988) Improved staining of proteins in polyacrylamide gels including isoelectric focusing gels with clear background at nanogram sensitivity using Coomassie Brilliant Blue G-250 or R-250. *Electrophoresis* **9**, 255–262.

12

Protocol to Enrich and Analyze Plasma Membrane Proteins from Frozen Tissues

Jacek R. Wiśniewski

Summary

This chapter presents procedures for preparation and analysis of fractions enriched in plasma membranes from frozen tissue. It consists of a method for extraction and fractionation of membranes and a method for enzymatic digestion of membrane proteins without use of detergents. The method for isolation of membranes comprises a stepwise depletion of non-integral membrane proteins from entire tissue homogenate by high-salt, carbonate, and urea washes followed by treatment of the membranes with sublytic concentrations of digitonin and enrichment of the plasma membranes by a density gradient fractionation. Reduction, carboxymethylation, and digestion with endoproteinase Lys-C are carried out on non-solubilized membranes. The entire procedure allows processing and preparation of samples of minute amounts as 10–20 mg tissue and therefore can be extremely helpful for proteomic profiling of small pieces of tissue and clinical material.

Key Words: Frozen tissue; membrane proteins; plasma membrane; mouse brain.

1. Introduction

Identification and characterization of integral and associated with plasma membrane proteins is pivotal in discovery of novel disease markers and drug targets. The plasma membrane proteins are receptors for signaling pathways and sensory mechanisms, channels, and transporters exchanging membrane-impermeable substances. Adhesion molecules constitute another important group of plasma membrane proteins that mediate intracellular interactions and modulate tissue plasticity. For studying these proteins in clinical samples, methods are required that allow in-depth analysis of proteins and their relative

From: *Methods in Molecular Biology, vol. 432: Organelle Proteomics*
Edited by: D. Pflieger and J. Rossier © Humana Press, Totowa, NJ

quantification using low, biopsy-size amounts of tissue. For practical reasons as shipment, storage, and the possibility to repeat experiments after longer periods of time, the vast majority of the biological material is available only in a frozen state. Unfortunately, the classical subcellular fractionation techniques, including those for purification of membrane proteins, were developed and work well with fresh tissue. While frozen material is used, often DNA released from broken nuclei causes aggregation of other organelles that can be detrimental to isolation of plasma membranes. Therefore, the demand for methods allowing preparation of plasma membranes from limited amount of frozen tissues is obvious, in particular, for proteomic studies of small organs consisting of different types of tissues and cells.

Another, more general, challenge in studying membrane proteins is their limited solubility in aqueous solutions, resulting in poor resolution in the commonly used two-dimensional gel electrophoresis technique, as well as weak recovery from the gels *(1,2)*. Even though the use of strong detergents allows quantitative solubilization of most membrane proteins, their use for in-solution sample treatments cannot circumvent these problems: the efficient detergent removal, which is a prerequisite for successful mass spectrometric analysis, may be difficult or at least requires additional sample cleaning steps.

The protocols described in this chapter attempt to address these issues. Firstly, I describe a method for isolation of membranes consisting of a stepwise depletion of non-integral membrane molecules from entire tissue homogenate by high-salt, carbonate, and urea washes followed by treatment of the membranes with sublytic concentrations of digitonin and an enrichment of the plasma membranes by a density gradient fractionation. Secondly, assays for measuring the extent of plasma membrane enrichment are described. Finally, I describe how the integral membrane proteins can be digested without solubilization of the membrane, thus avoiding the use of detergents. In this method, proteins are directly digested on membranes under mild denaturing conditions *(3)*. The combination of both methods followed by off-line reverse-phase chromatographic separation of the released peptides and their online LC-MS/MS analysis allowed identification of 1670 proteins from 15 mg of frozen mouse hippocampus *(4)*. Recently, the protocol to enrich and analyze plasma membrane proteins from frozen tissues was applied to quantitative mapping of membrane proteins in distinct areas of mouse brain *(5,6)*. Using the label-free approach, relative abundance of 967 proteins in fore- and hindbrain was measured *(5)*. Eighty-one percent of them were known membrane proteins and 38% of the protein sequences were predicted to contain transmembrane helices *(5)*. In the second study, which employed isotope-coded affinity tagging reagent

HysTag *(3)*, 555 proteins including 197 known plasma membrane proteins were quantified between cortex, hippocampus, and cerebellum *(6)*.

2. Materials

2.1. Membrane Extraction

1. High salt buffer: 2 M NaCl, 10 mM HEPES–NaOH, pH 7.4, and 1 mM ethylene-diaminetetraacetic acid (EDTA). Store at 4°C.
2. IKA Ultra Turbax blender.
3. Ultracentrifuge Sorval S150AT or Beckman MLA 130.
4. Carbonate buffer: 0.1 M Na_2CO_3 and 1 mM EDTA, pH 11.3. Store at 4°C.
5. Wash buffer: 4 M urea, 100 mM NaCl, 10 mM HEPES–NaOH, pH 7.4, and 1 mM EDTA. Prepare fresh and store up to 1 day at 4°C.
6. Protease inhibitors: protease inhibitor cocktail tablets from Roche Diagnostics (Hvidovre, Denmark). Each of the above buffers should be supplemented with the recommended amount of inhibitors.

2.2. Density Gradient Centrifugation

1. Sucrose stock solution: 2 M sucrose in water. Store at 4°C.
2. Gradient buffer: 0.25 M sucrose, 100 mM NaCl, 10 mM HEPES–NaOH, pH 7.4, and 1 mM EDTA.
3. Percoll from Amersham Biosciences (Piscataway, NJ).
4. Digitonin stock solution: 40 mg/mL digitonin (high purity) from Calbiochem (La Jolla, CA).

2.3. γ–Glutamyl Transpeptidase Assay

1. Substrate stock: 2 mg/mL γ-L-glutamic acid 7-amido-4-methylcoumarin from Glycosynth (Warrington, UK) in ethanol. Prepare fresh.
2. Assay buffer: 10 mM $MgCl_2$, 10 mM glycyl glycine, and 100 mM Tris–HCl, pH 9.0 (prepare fresh).
3. Standard solution: 1 mg/mL 4-methyl umbelliferone (Sigma) in ethanol.
4. Spectrofluorometer and standard 1 cm × 1 cm cuvettes.

2.4. Protein Assay

1. 100 mM sodium dodecyl sulphate (SDS).
2. 0.1 M -mercaptoethanol.
3. 8 M urea in water.
4. Standard solution: 2 µg/mL L-tryptophanamide hydrochloride in gradient buffer.

2.5. Thiol–Alkylation and Digestion of Membrane Proteins

1. Reduction buffer: 0.1 M Na_2CO_3 and 10 mM dithiothreitol (DTT), pH 11.3.
2. Reduction buffer 2: 0.2 M NaBr, 0.2 M KCl, 50 mM Tris–HCl, pH 8.0, and 10 mM DTT.
3. Alkylation solution: 4 M urea, 100 mM iodoacetamide, and 100 mM Tris–HCl, pH 8.
4. Carbonate buffer: 0.1 M Na_2CO_3, pH 11.3.
5. Digestion buffer: 4 M urea, 100 mM NaCl, and 100 mM Tris–HCl, pH 8.0.
6. Endoproteinase Lys-C from Wako Bioproducts (Richmond, VA).

3. Methods

The following protocol describes the isolation of membranes from frozen mouse brain. This method comprises three extraction steps and a density gradient purification step. In the first extraction step, high ionic strength allows impairing of ionic bonds. The selected salt concentration is high enough to break very tight bonds between DNA and the core histones (7). In the second step, the high pH of the carbonate buffer introduces discontinuities into membrane vesicles (8). The discontinuities allow efficient removal of soluble proteins from both artificial vesicles of microsomes assembled during homogenization and from intact organelles such as mitochondria and lysosomes. Finally, the membranes are treated with 4 M urea, a concentration that is high enough for globular domains of the vast majority of proteins to melt (9,10) but that does not affect the integrity of the transmembrane helices in the lipid bilayer. In this step, non-integral membrane proteins that were unaffected by the carbonate treatment can be washed out.

Fractionation of the total membranes is achieved by using a self-generating density gradient of Percoll. The differences in the buoyant densities are affected by addition of sublytic amounts of the non-ionic detergent digitonin. This technique was initially developed for enrichment of microsomal membranes from rat liver in a sucrose gradient (11). In that study, digitonin caused a selective increase of the density of microsomes, whereas organelles remained little affected. Treatment of purified membranes from brain tissues with the detergent results in a shift of all membranes, but the shift of membranes originating from the plasma membrane is less prominent, and therefore the membrane fractions with a lower buoyant density are enriched in plasma membrane (4). In the described method, the selection of fractions with the highest enrichment in plasma membranes is based on the activity measurement of γ-glutamyl transpeptidase, a plasma membrane marker enzyme, and on the determination of total protein concentration. As the method is designed for analysis of small amounts of biological sample, sensitive fluorometric assays are selected.

3.1. Total Membrane Extraction from Frozen Tissue

In this method, all steps should be carried out in a cold room at 4°C and/or on ice.

1. Thaw 20–50 mg of tissue and add 1 mL of high salt buffer.
2. Blend the tissue using an IKA Ultra Turbax blender at maximum speed (~25,000 rpm) for 30 s.
3. Ultracentrifuge the suspension in a Sorval S150AT or Beckman MLA 130 at 900,000 g for 10 min. The tubes should be balanced to a difference less than 50 mg.
4. Discard the supernatant and homogenize the pellet in 1 mL of carbonate buffer as in **step 2**.
5. Incubate for 30 min (*see* **Note 1**).
6. Collect the non-soluble material by centrifugation as in **step 3** (**steps 4–6** can be repeated two to three times, *see* **Note 2**).
7. Discard the supernatant and resuspend the pellet in wash buffer as described in **step 2**.
8. Collect the crude membranes by centrifugation as in **step 3**.

3.2. Enrichment of Plasma Membranes by Density Gradient Centrifugation

1. The following protocol describes the fractionation of membranes using a Sorvall S100AT6 (or Beckman TLA110) rotor accommodating 4.1 mL (or 4.7 mL) tubes.
2. Resuspend the crude membranes in 1 mL of gradient buffer using an IKA Ultra Turbax blender at maximum speed (~25,000 rpm) for 30 s.
3. Transfer the suspension into a centrifugal vial and add

   a. 1.44 mL (or 1.65 mL in TLA 110) of Percoll,
   b. 0.205 mL (0.27 mL) of 2 M sucrose,
   c. 1 mL (1.31 mL) of gradient buffer, and
   d. 0.41 mL (0.47 mL) of digitonin solution (*see* **Note 3**).

4. Close the tubes and centrifuge at 150,000 g for 30 min. Decelerate the rotor over 5–10 min. To balance the tube, another one is prepared with the same composition of solutions of **step 3a–d** and 1 mL of gradient buffer instead of the sample suspension.
5. Remove the sample tube from the rotor and assemble the fractionation device (*see* **Note 4** and **Fig. 1**). Collect fractions by the bottom gradient displacement with 2 M sucrose using a peristaltic pump at a flow rate of 0.5 mL/min. Collect at least 10 0.4-mL fractions. To remove Percoll, dilute the fractions with 0.6 mL of gradient buffer and centrifuge in a Sorval S150AT or Beckman MLA 130 at 900,000 g for 15 min. The membranes accumulate on the surface of a transparent Percoll pellet (*see* **Note 5**).

6. Resuspend each pellet in 200–300 μL of gradient buffer. Aliquots of 40–50 μL are subjected to measurement of the plasma membrane marker activity and to total protein amount determination.

3.3. Biochemical Analysis of Plasma Membranes

3.3.1. γ–Glutamyl Transpeptidase

1. The enzymatic activity is measured at 365 and 460 nm for excitation and emission of the liberated 4-methylumbelliferol, respectively, at a constant temperature of 37°C (see **Note 6**).
2. Mix 50 mL of assay buffer with 0.5 mL of substrate stock solution.
3. Add 3 mL of this diluted substrate to the 1 cm × 1 cm standard spectrofluorometer cuvettes and preincubate for a few minutes to reach the assay temperature of 37°C.
4. Add sample (usually 2–20 μL of the analyzed subcellular fraction obtained in **Subheading 3.2, step 6**) and mix thoroughly.
5. Start recording the intensity changes at 460 nm.
6. Measure changes for 5–10 min. The slope of the recorded reaction should be straight linear because it is a zero-order reaction. If the velocity of the reaction decreases during the recording time, reduce the amount of sample for the measurement (**step 4**).
7. Repeat each measurement at least once.
8. Calculate for each fraction the relative specific activity and the yield of plasma membrane.
9. For absolute measurement of the released 4-methylumbelliferol, measure the intensity of fluorescence of defined amounts of the standard solution in the range of the measured activity.

3.3.2. Protein Assay

1. Mix 30 μL of the sample obtained in **Subheading 3.2, step 6**, with 7 μL of 100 mM SDS (final concentration of 20 mM) and 1 μL of 0.1 M -mercaptoethanol. For measurements of protein concentrations in the initial homogenate or crude membrane fractions, dilute the samples appropriately.
2. Incubate at 96°C for 5 min and centrifuge at ∼15,000 g for 10 min.
3. Set the spectrofluorometer for measurements at 295 and 350 nm for excitation and emission, respectively (see **Note 7**).
4. Prepare the standard curve by mixing 2 mL of 8 M urea with different amounts of tryptophanamide ranging from 1–20 μL of the stock solution and by measuring the emission intensity. The added amounts of tryptophanamide correspond to 2–40 ng.
5. Introduce 10 μL of the analyzed fraction fraction prepared in **step 1** in 2 mL of 8 M urea and measure emission intensity. The protein amounts can be calculated assuming tryptophan accounts for 1.3% of their total weight in average.

3.4. Digestion of Membrane Proteins

1. Combine fractions with the highest plasma membrane marker activities and collect the membranes by centrifugation at 900,000 g for 15 min. Remove the supernatant.
2. Resuspend the membrane pellet in 1 mL of reduction buffer and incubate on ice for 30 min.
3. Ultracentrifuge the suspension and as in **step 1**. Discard the supernatant.
4. Resuspend the pellet in 1 mL of reduction buffer 2 and incubate on ice for 30 min. Centrifuge as in **step 1**. Discard the supernatant (*see* **Note 8**).
5. Resuspend the pellet in 1 mL of alkylation solution and incubate at 25°C for 2 h. Cysteine residues become acetylated.
6. Centrifuge as in **step 1**. Discard the supernatant.
7. Resuspend the pellet in 1 mL of carbonate buffer and incubate on ice for 30 min. In this step, the iodoacetamide that is trapped in membrane vesicles is removed.
8. Centrifuge as in **step 1**. Discard the supernatant.
9. Resuspend the pellet in 100–200 μL of digestion buffer, add 1 μg of endoproteinase Lys-C per 100 μg of protein, and incubate overnight at 20°C.
10. Centrifuge as in **step 1** and collect the supernatant containing peptides.
11. The peptides can be desalted using a solid phase extraction cartridge or directly fractionated on a reverse-phase column (*11*).

4. Notes

1. Incubations should be carried out in a thermomixer (Eppendorf) with gentle mixing of the sample.
2. The introduction of discontinuities in the membrane vesicles requires a high excess of sodium carbonate. Usually, at least a 50-fold (v/v) excess of the solution over membranes is required (*8*). If the content of integral membrane proteins in the purified membranes (final product) is not at least 30–40% of total identified proteins, the sodium carbonate extraction can be repeated two or three times.
3. The concentration of digitonin is critical in the procedure. The concentration of 4 mg/mL was found to be optimal for purification of mouse brain plasma membranes. At a digitonin concentration of 8 mg/mL, lysis of membranes was observed. When working with other tissues, the amount of digitonin required for the density shift should be determined experimentally.
4. Fractionation of the density gradient is an important step; it should therefore be prepared in advance and performed thoroughly. Fractions can be collected manually by piercing the bottom of the tube and collecting drops or by aspirating with a pipette from the top of the gradient. In terms of reproducibility and resolution, the best results can be achieved using a fractionation device in which the gradient is displaced upward by pumping dense sucrose from the bottom, and the fractions are collected from the top (*see* **Fig. 1**).
5. Percoll can be removed by high-speed centrifugation. After centrifugation, membranes form a thin pellet over a firm and transparent pellet of the density

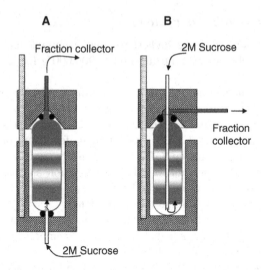

Fig. 1. Two types of "bottom displacement" gradient fractionation devices. (**A**) The bottom of the centrifugal tube is pierced by a needle and a dense solution (2 M sucrose) is pumped from the bottom. (**B**) The dense solution is pumped through a long capillary introduced from the top to the bottom. In both devices, gradient elutes from the top of the tube.

medium. The collection of the membrane pellet is achieved by resuspending it in a small volume of buffer.

6. It is important to keep the temperature constant during the assay because both the reaction velocity and the fluorescence intensity are temperature dependent.

7. Assaying protein concentration by measurement of tryptophan florescence is a sensitive and simple method. The sensitivity of the assay depends on the type of fluorometer and on the selected setting. It typically allows reliable measurements of protein amounts as low as 10 ng.

8. The reduction buffer 2 contains 0.2 M NaBr. Bromide is a moderate chaotropic reagent that facilitates removal of proteins that are only weakly associated with the membranes, and it can increase accessibility of disulfide bridges to the reducing agent. Two reduction steps employed in our procedure are combined with membrane washes with carbonate and chaotropic agent/salt, respectively. Our experience is that both steps are required to achieve best results.

Acknowledgments

The author thanks Peter Aa. Nielsen (Proxeon A/S, Odense, Denmark) for excellent technical assistance and contribution to development of the described methods.

References

1. Lubec, G., Krapfenbauer, K., and Fountoulakis, M. (2003) Proteomics in brain research: potentials and limitations. *Prog. Neurobiol.* **69**, 193–211.
2. Vercauteren, F. G. G., Bergeron, J. J. M., Vandesande, F., Arckens, L., and Quirion, R. (2004) Proteomic approaches in brain research and neuropharmacology. *Eur. J. Pharmacol.* **500**, 385–398.
3. Olsen, J. V., Andersen, J. R., Nielsen, P. Aa., Nielsen, M. L., Figeys, D., Mann, M., et al. (2004) HysTag–a novel proteomic quantification tool applied to differential display analysis of membrane proteins from distinct areas of mouse brain. *Mol. Cell. Proteomics* **3**, 82–92.
4. Nielsen, P. Aa., Olsen, J. V., Podtelejnikov, A. V., Andersen, J. R., Mann, M., and Wiśniewski, J. R (2005) Proteomic mapping of brain plasma membrane proteins. *Mol. Cell. Proteomics* **4**, 402–408.
5. Le Bihan, T., Goh, T., Stewart, I. I., Salter, A.-M., Bukhman, Y., Dharsee, M., et al. (2006) Differential analysis of membrane proteins in mouse fore- and hindbrain using a label-free approach. *J. Proteome Res.* **5**, 2701–2710.
6. Olsen, J. V., Nielsen, P. Aa., Andersen, J. R., Mann, M., and Wiśniewski, J. R (2007) Quantitative proteomic profiling of membrane proteins from the mouse brain cortex, hippocampus, and cerebellum using the HysTag reagent: Mapping of neurotransmitter receptors and ion channels. *Brain Res.,* **1134**, 95–106.
7. Burton, D. R., Butler, M. J., Hyde, J. E., Phillips, D., Skidmore, C. J., and Walker, I. O. (1978) The interaction of core histones with DNA: equilibrium binding studies. *Nucleic Acids Res.* **5**, 3643–3663.
8. Fujiki, Y., Hubbard, A. L., Fowler, S., and Lazarow, P. B. (1982) Isolation of intracellular membranes by means of sodium carbonate treatment: application to endoplasmic reticulum. *J. Cell Biol.* **93**, 97–102.
9. Wiśniewski, J. R., Heßler, K., Claus, P., and Zechel, K. (1997) Structural and functional consequences of mutations within the hydrophobic cores of the HMG1-box domain of the *Chironomus* high-mobility-group protein 1a. *Eur. J. Biochem.* **243**, 151–159.
10. Pace, C. N., Shirley, B. A., and Thompson, J. A. (1989) Measuring the conformational stability of a protein. In *Protein Structure – Practical Approach* (Creighton, T. E., ed.), pp. 311–330, IRL Press, Oxford, New York, and Tokyo.
11. Amar-Costesec, A., Beaufay, H., Wibo, M., Thines-Sempoux, D., Feytmans, E., Robbi, M., et al. (1974) Analytical study of microsomes and isolated subcellular membranes from rat liver. II. Preparation and composition of the microsomal fraction. *J. Cell Biol.* **61**, 201–212.
12. Kristensen, D. B, Børnd, J. C., Nielsen, P. Aa, Andersen, J. R., Sørensen, O. T., Jørgensen, V., et al. (2004) Experimental peptide identification repository (EPIR): an integrated peptide-centric platform for validation and mining of tandem mass spectrometry data. *Mol. Cell. Proteomics* **3**, 1023–1038.

13

Proteomic Characterization of Membrane Protein Topology

Adele R. Blackler and Christine C. Wu

Summary

Integral membrane proteins are represented by 20–30% of the eukaryotic genome and crucial for cellular functions including cell signaling, nutrient influx, toxin efflux, and maintenance of osmotic balance. Importantly, over 70% of all drugs are targeted at membrane proteins. Because of their hydrophobicity, however, methods used to characterize the structure of soluble proteins, such as NMR and X-ray crystallography, are generally not suitable to the study of membrane proteins *(1)*. The methods described in this chapter facilitate the identification and mapping of both extracellular and cytoplasmic-soluble domains of integral plasma membrane proteins using mass spectrometry. By combining a classical protease protection approach with recently developed proteomic methods, protease-accessible peptides (PAPs) are digested from proteins embedded in their native lipid environment and identified to characterize the topologies of integral membrane proteins.

Key Words: Integral membrane protein; topology; MudPIT; proteomics.

1. Introduction

The methods detailed in this chapter facilitate the high-throughput profiling of membrane proteins in an enriched membrane fraction, as well as the determination of the sidedness of exposed, protease-accessible peptides (PAPs) of membrane proteins. A soluble, non-sequence-specific protease (proteinase K) is used to digest PAPs from integral membrane proteins while embedded in their native lipid bilayer environment. Under physiological conditions, proteinase K is a robust protease that cleaves proteins into di- and tri-peptides unsuitable for

From: *Methods in Molecular Biology, vol. 432: Organelle Proteomics*
Edited by: D. Pflieger and J. Rossier © Humana Press, Totowa, NJ

proteomic analysis by mass spectrometry. By exposing the enriched membrane fraction to a high pH with agitation, sealed membrane structures open to form stable membrane sheets *(2)*, allowing protease access to both sides of the membrane *(3)*. High pH also attenuates the activity of proteinase K, resulting in peptides with average lengths of 4–25 amino acids, which are ideal for analysis by tandem mass spectrometry. However, the resultant peptide mixture is extremely complex and requires a specialized separation strategy *(3)*.

Multidimensional protein identification technology (MudPIT) couples multi-dimensional high-pressure liquid chromatography separation with tandem mass spectrometry and automated peptide identification using cross-correlation software *(4)*. This strategy facilitates the separation and identification of the overlapping peptides that map to the exposed domains of the lipid-embedded integral membrane proteins.

The methods described in this chapter detail (1) the enrichment of plasma membranes from HeLa cells, (2) sample preparation resulting in PAPs for proteomic analysis, (3) MudPIT, and (4) data analysis and generation of membrane protein topological maps.

2. Materials

2.1. Plasma Membrane Enrichment

1. Rubber policeman for cell harvest.
2. Phosphate buffer: 400 mM KH_2PO_4/K_2HPO_4, pH 6.7. Make 1 L of 400 mM dibasic potassium buffer (K_2HPO_4) and 1 L of 400 mM monobasic potassium buffer (KH_2PO_4). Add the monobasic slowly to the dibasic while monitoring the pH until the pH is 6.7. Store at 4°C.
3. 100 mM phosphate buffer: 100 mM KH_2PO_4/K_2HPO_4, pH 6.7. Dilute 25 mL of the 400 mM potassium phosphate buffer with 75 mL of ultrapure water. This buffer can be stored at room temperature. Cool to 4°C before use.
4. Magnesium chloride ($MgCl_2$): dissolve at 1.0 M concentration in ultrapure water. Stable at room temperature.
5. Protease inhibitor cocktail for mammalian cell culture in DMSO (Sigma, St Louis, MO).
6. Sucrose: dissolve at 2.0 M concentration in ultrapure water. Apply low heat (40°C) to help sucrose dissolve initially. Store at 4°C.
7. Buffers for sucrose gradient (keep on ice or at 4°C): prepare according to instructions in **Table 1**.
8. Glass homogenizer and teflon pestle.
9. Refrigerated tabletop centrifuge.
10. Swinging bucket rotor capable of handling 135,000 *g*.
11. Narrow-bore transfer pipettes.
12. 1cc insulin syringes.

Table 1
Sucrose Gradient Solutions

Reagents	Sucrose solutions				Final concentration
	0.25 M	0.5 M	0.86 M	1.3 M	
400 mM KH_2PO_4/ K_2HPO_4 buffer	12.5 mL	12.5 mL	12.5 mL	12.5 mL	100 mM
1.0 M $MgCl_2$	0.25 mL	0.25 mL	0.25 mL	0.25 mL	5 mM
2.0 M sucrose	6.25 mL	12.5 mL	21.5 mL	32.5 mL	Variable
Ultrapure water	31 mL	24.75 mL	15.75 mL	4.75 mL	–
Protease inhibitor cocktail	50 μL	50 μL	50 μL	50 μL	–
Final volume	50 mL	50 mL	50 mL	50 mL	–

13. Sodium carbonate buffer: dissolve Na_2CO_3 at 0.2 M concentration in ultrapure water and adjust pH to 11.0 with concentrated HCl solution. Stable at room temperature.

2.2. Sample Digestion

1. Solid urea (ultrapure molecular grade).
2. Dithiotheitol (DTT): dissolve at 0.5 M concentration in ultrapure water and aliquot into single-use aliquots (recommended: 200 μL). Store aliquots at –20°C (stable for 1 year). Do not freeze-thaw aliquots.
3. Iodoacetamide (IAA): dissolve at 0.5 M concentration in ultrapure water and aliquot into single-use aliquots (recommended: 200 μL). Store aliquots at –20°C (stable for 1 year). Do not freeze-thaw aliquots.
4. Recombinant proteinase K (Roche, Indianapolis, IN): resuspend the proteinase K at a concentration of 1 mg/mL in ultrapure water, aliquot (200-μL aliquots) and store at –20°C (stable for 1 year). Do not freeze-thaw aliquots.

2.3. Liquid Chromatography and Mass Spectrometry

1. 90% formic acid (J.T. Baker) (used for samples and chromatography buffers).
2. Buffers for reverse-phase chromatography—mass spectrometry. Buffer A: acetonitrile/formic acid/water, 5/0.1/95 (v/v/v). Use mass spectrometry grade acetonitrile and water (J.T. Baker, Phillipsburg, NJ). Buffer B: water/formic acid/acetonitrile, 5/0.1/95 (v/v/v). Approximately 100 mL of each buffer will be needed for one MudPIT analysis.
3. Ammonium acetate (CH_3COONH_4) salt solutions for salt pulses. Start with a 5 M salt concentration in buffer A and dilute with buffer A to make the necessary

concentrations. There are 10 concentrations needed for salt pulses: 200 mM, 300 mM, 400 mM, 500 mM, 600 mM, 700 mM, 800 mM, 900 mM, 1 M, and 5 M. 2 mL of each salt concentration will last for ~15 MudPIT analyses. Stable at room temperature.

4. Mass spectrometry grade methanol.
5. Aqua C18 reverse-phase material (Phenomenex, Torrance, CA). Add a small aliquot (~10 mg) to 0.5 mL of mass spectrometry grade methanol in a micro-centrifuge tube just before use and spin briefly so the solid phase forms a loose pellet at the bottom of the tube. Do not store long-term in methanol.
6. Partisphere strong cation exchange (SCX) solid phase (Whatman Inc., Florham Park, NJ). Add a small aliquot (~10 mg) to 0.5 mL of mass spectrometry grade methanol in a microcentrifuge tube just before use and spin briefly so that the solid phase forms a loose pellet at the bottom of the tube. Do not store long-term in methanol.
7. 100 µm internal diameter fused silica tubing (Polymicro Technologies, Phoenix, AZ) for use as chromatography columns.
8. 250 µm silica tubing (Polymicro Technologies) for use as a loading column.
9. In-line filter assembly (Upchurch, Oak Harbar, WA) for loading column.
10. Ceramic scribes.
11. Laser puller (Sutter, Novato, CA) for pulling chromatography column tips.
12. High-pressure bomb (built in-house, plans available upon request) for packing columns capable of handling 2000 psi.
13. Helium for use with high-pressure bomb
14. LTQ Linear Ion Trap mass spectrometer (ThermoFinnigan, San Jose, CA).
15. Autosampler and HPLC (Agilent, Santa Clara, CA).
16. Microspray source (built in-house, plans available upon request).

3. Methods

3.1. Plasma Membrane Enrichment

This step takes ~4–5 h.

1. Harvest two 500-cm^2 plates of HeLa cells at 80% confluence using a rubber policeman. Do not use trypsin to harvest cells (*see* **Note 1**). After harvest, pellet cells with centrifugation (700 *g*, 5 min, 4°C) and remove any liquid (*see* **Note 2**) (*see* **Fig. 1**).
2. Resuspend cell pellet in 5 mL of cold 0.5 M sucrose buffer and homogenize with a cold teflon pestle and glass homogenizer slowly for 25 strokes (*see* **Notes 3–6**).
3. After homogenization, pellet nuclei and unbroken cells by low-speed centrifugation (700 *g*, 5 min, 4°C) and collect the post-nuclear supernatant (PNS).
4. Gently layer the PNS onto a four-step discontinuous sucrose gradient with the following sucrose concentrations (in order from bottom to top) in a thin-walled

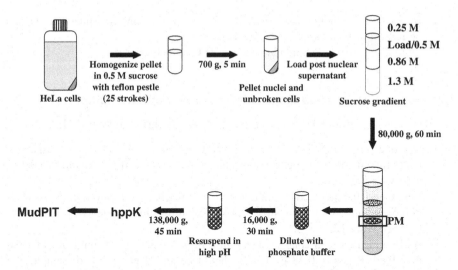

Fig. 1. Workflow for proteomic analysis of membrane proteins. An enriched-membrane fraction is isolated using a discontinuous sucrose gradient. Enriched plasma membranes are collected at the interface between 1.3M and 0.86M sucrose. After dilution of the sucrose with phosphate buffer, the membranes are enriched by high pH extraction. The enriched membrane sample is then processed for mass spectrometry using the hppK method, and the resulting peptides are analyzed using multidimensional protein identification technology (MudPIT).

12.5-mL centrifugation tube: 3 mL of 1.3 M sucrose, 3.5 mL of 0.86 M sucrose, 5 mL of 0.5 M sucrose containing the PNS, and 2.5 mL of 0.25 M sucrose (*see* **Note 7**).

5. Centrifuge in a swinging bucket rotor (80,000 *g*, 4°C, 60 min) (*see* **Note 8**). The plasma membrane-enriched fraction will settle to the SIII layer between the 0.86 M and 1.3 M sucrose layers.

6. Collect the plasma membrane-enriched fraction from the top of the gradient to the bottom using transfer pipets in 0.5-mL aliquots in 1.5-mL microcentrifuge tubes.

7. Immediately after collection, add 1 volume (if fractions were collected in 0.5-mL aliquots, then add 0.5 mL) of dilute, cold 100 mM phosphate buffer to the plasma membrane fraction (*see* **Note 9**).

8. Spin the diluted samples in a tabletop microfuge to pellet the fractions (30 min, 16,000 *g*, 4°C) and discard the supernatants (*see* **Note 10**).

9. Resuspend the membrane pellets in 0.5 mL of Na_2CO_3 buffer and agitate on ice using 5 strokes of an insulin syringe every 15 min for 1 h (*see* **Note 11**).

10. Transfer each sample to a thick-walled 2.5-mL centrifuge tube and pellet the membranes by centrifugation in an ultracentrifuge (138,000 *g*, 45 min, 4°C).

11. Resuspend the membrane pellets in 0.5 mL of 100 mM phosphate buffer and perform a protein assay (*see* **Note 12**).
12. Store the resuspended membranes at –20°C until further use.

3.2. Preparation of Total Protease-Accessible Peptides Sample

This step—the high pH/proteinase K (hppK) method—takes ~3 h.

1. Remove a volume corresponding to 150 µg of protein from the suspended membrane fraction and pellet using a tabletop microcentrifuge (30 min, 16,000 g, 4°C) (*see* **Fig. 2**).
2. Resuspend the pelleted membranes at a 1 mg/mL concentration in Na_2CO_3 buffer and agitate on ice using 5 strokes of an insulin syringe every 15 min for 1 h (*see* **Note 13**).
3. Add solid urea to a final concentration of 8 M and vortex until all urea has dissolved. Roughly 0.48 g of urea should be added per 1 mL of sample (*see* **Note 14**).

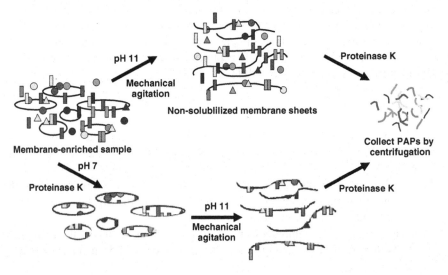

Fig. 2. Workflow for proteomic analysis of protease-accessible domains (PAPs). Top: a membrane-enriched sample is agitated at high pH. This results in the formation of membrane "sheets" and allows proteinase K to digest proteins from both membrane surfaces and generates PAPs from both extracellular and cytoplasmic domains. Bottom: a membrane-enriched sample is digested with proteinase K at neutral pH. This results in the removal of all exposed (extracellular) protein domains. This membrane sample is reisolated and agitated at high pH. This results in the formation of membrane "sheets" and allows proteinase K to digest proteins from the lumenal surface and generates PAPs from only the cytoplasmic domain.

4. Add 500 mM DTT to a final concentration of 5 mM and incubate for 20 min (60°C, with gentle shaking) (*see* **Note 15**).

5. After incubation, place the sample briefly on ice to cool to room temperature and add 500 mM IAA to a final concentration of 12 mM and incubate for 20 min (22°C, in the dark) (*see* **Note 16**).

6. Add proteinase K to a 1:50 (w/w) enzyme to substrate ratio and incubate for 5–15 h (37°C, with shaking).

7. After incubation, double the volume of the sample by adding 1 volume of mass spectrometry buffer A.

8. Adjust the sample to a final concentration of 5% (v/v) formic acid.

9. Centrifuge in a tabletop centrifuge to pellet insoluble debris (16,000 *g*, 30 min, 4°C).

10. Collect the supernatant and store at –20°C until MudPIT analysis.

This analysis will yield both extracellular and cytoplasmic PAPs of integral membrane proteins.

3.3. Preparation of Cytoplasmic Protease-Accessible Peptides Sample

This step—protease protection followed by the high hppK method—takes ~2 h.

1. Obtain an enriched membrane fraction of sample by following the steps under **Subheading 3.1**.

2. Remove a volume corresponding to 150 μg of protein from the suspended membrane fraction and pellet using a tabletop microcentrifuge (30 min, 16,000 *g*, 4°C).

3. Resuspend the membrane pellet at a 1 mg/mL concentration in cold 0.25 M sucrose buffer (*see* **Fig. 2**).

4. Add proteinase K (1:100 enzyme : substrate ratio) and incubate for 30 min with rotation at 37°C.

5. Layer the sample onto a sucrose cushion (250 μL of 0.5 M sucrose) in a thick-walled 2.5-mL centrifuge tube. Reisolate the membranes by centrifugation in an ultracentrifuge (80,000 *g*, 45 min, 4°C) and discard the supernatant. The supernatant contains proteinase K and the peptides cleaved from the extracellular side of the membrane (*see* **Note 17**).

6. Resuspend the pellet at 1 mg/mL concentration in Na$_2$CO$_3$ buffer and proceed with proteinase K digestion, **Subheading 3.2**, starting at **step 2**.
This analysis will yield only cytoplasmic PAPs of integral membrane proteins.

3.4. Peptide Analysis by MudPIT

This step takes ~30 h.

1. Pull a 100-μm inner diameter fused silica column to a 5-μm tip using a laser puller.

2. Place the microcentrifuge tube containing a slurry of methanol and Aqua C18 reverse-phase resin in the high-pressure bomb and seal the lid.

3. Put the chromatography column in the top of the bomb with the tip of the column facing upward. Make sure the bottom of the column is not touching the bottom of the microcentrifuge tube containing the reverse-phase material. Apply 600 psi of pressure to the bomb (*see* **Fig. 3**).

4. Verify that methanol is coming out of the tip of the column. It may be necessary to gently open up the column with a ceramic scribe.

5. While holding onto the column, gently loosen the ferrule holding the column by a quarter of a turn. This allows for the column to be manipulated without venting the bomb. Gently tap the column to the bottom of the microcentrifuge tube a few times. Packing the column a few millimeters at a time allows for more control and reproducible packing dimensions. When the column is not touching the base of the microcentrifuge tube, retighten the ferrule so that the column will stay in the bomb. The resin can be viewed traveling up the column (*see* **Note 18**).

6. Pack the column to 8 cm of reverse-phase material.

7. After packing the first phase on the column, pack the second phase consisting of 4 cm of SCX resin, prepared as a slurry in methanol. Pack by applying 600 psi using a high-pressure bomb. Use the method described for packing the reverse-phase material (*see* **Note 19**).

8. Equilibrate the column on a HPLC pump for a minimum of 30 min by running buffer A through it at a rate of 0.1 mL/min.

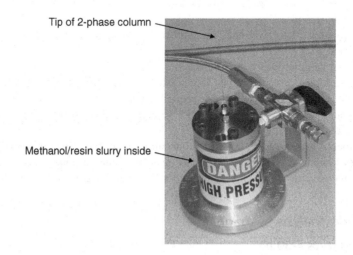

Fig. 3. Custom-built high-pressure bomb used for packing and loading microcapillary columns.

9. Cut two 8-cm sections of 250-μm silica tubing, and place both sections in opposite ends of an in-line filter assembly. This is used as a loading column (*see* **Note 20**).

10. Put the microcentrifuge tube of the reverse-phase slurry in the bomb and seal the lid.

11. Put the 250-μm column in the in-line filter assembly in the top of the bomb so that the column going into the filter is in the bomb and will be loaded with reverse-phase material and the waste line is out of the bomb.

12. Pack the loading column with 2 cm of Aqua C18 resin in methanol and equilibrate for 15 min in the same manner as the column was packed and equilibrated.

13. Thaw the sample and pressure load a 150-μg aliquot at 800 psi onto the loading column after equilibration.

14. After the sample is loaded, desalt it by washing for 10 min with buffer A (0.1 mL/min).

15. Remove the waste line from the loading column filter assembly and put the two-phase column in its place, so the two columns are now in-line with one another. Place both on a microspray source in-line with an ion trap mass spectrometer and align so the tip of the column is in-line with the heated source of the mass spectrometer.

16. Start running buffer A through the columns at 0.1 mL/min. Adjust the waste line on the mass spectrometer to keep the pressure to 25–35 bar.

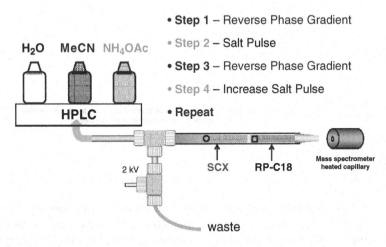

Fig. 4. Workflow for in-line multidimensional chromatographic separation. A two-phase microcapillary column is placed in-line with the mass spectrometer. Salt (NH_4OAc) displaces subsets of peptides based on charge from the strong cation exchange (SCX) material onto the reverse-phase (RP) C18 material where the peptides are separated based on hydrophobicity using an acetonitrile (MeCN) gradient mobile phase.

17. Analyze the sample using an automated MudPIT containing 12 2-h steps (*see* **Fig. 4**). After the first step, the peptide sample that was loaded onto the loading column should now be loaded onto the SCX material in the two-phase column.

18. Remove the loading column and continue with **steps 2–12** of the MudPIT with just the two-phase column in-line with the mass spectrometer.

19. The acetonitrile gradient for each step of the MudPIT starts at 100% buffer A for 5 min and gradually decreases to 35% buffer A, 65% buffer B over 80 min.

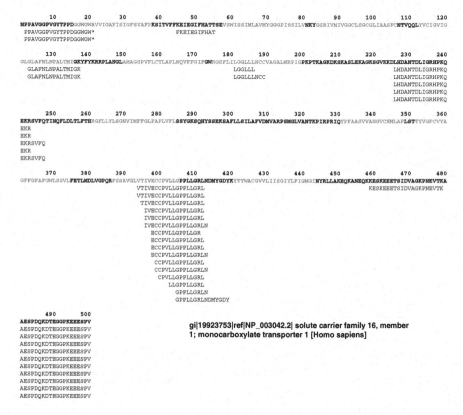

Fig. 5. Characterization of membrane protein topology by mapping identified peptides to protein sequence. The amino acid sequence of solute carrier family 16, member 1 (monocarboxylate transporter 1) is shown in bolded lettering. This protein is predicted to have 11 transmembrane domains (TMDs, bolded gray) and was identified with 39 peptides (listed beneath the protein sequence) using the hppK method. The majority of the peptides identified overlapped with domains predicted to be protease accessible (bolded black). Overlap with TMDs 4, 5, and 10 suggest that these predicted TMDs are protease accessible and not embedded in lipid. *Duplicate sequences indicate tandem mass spectra acquired from both +2 and +3 parent ions.

The gradient then returns to 100% buffer A over 10 min and remains at 100% buffer A for the rest of the 120 min.

20. At the beginning of **steps 2–12** of the MudPIT, a salt injection is delivered through the autosampler. The salt pulses for **steps 2–12** have the following volumes and concentrations of ammonium acetate: 50 µL of 200 mM, 50 µL of 300 mM, 50 µL of 400 mM, 50 µL of 500 mM, 50 µL of 600 mM, 50 µL of 700 mM, 50 µL of 800 mM, 50 µL of 900 mM, 50 µL of 1 M, 75 µL of 1 M, and 100 µL of 5 M. Each salt injection displaces peptides from the SCX onto the reverse-phase material.

21. Analyze the resulting tandem spectra.

3.5. Protein and Peptide Identification Using Sequest

1. Convert the RAW files into. ms2 files containing the tandem mass spectra using the Windows-based software program makeMS2 *(5)*. Search the. ms2 files against a human sequence database (*see* **Note 21**) concatenated to a randomized database using SEQUEST-Norm (*see* **Note 22**) *(6,7)*.

2. Assemble the identified peptides into proteins and filter to minimize false positives using the program DTASelect 1.9 *(8)*. General starting parameters for filtering are normalized Xcorr = 0.15, DeltCN = 0.15, % ion = 0.2, and minimum peptide length = 6.

3. Calculate the false discovery rate (FDR) by dividing the total number of hits to the randomized database by the total number of hits to the real, non-randomized database *(9)*. Adjust the DTASelect parameters to maintain an FDR less than 5%.

4. The identified peptides for the protein of interest map to the soluble protease-accessible domains of the protein (*see* **Fig. 5**).

4. Notes

1. Trypsin may digest off exposed protein domains from the cell surface.

2. Make sure that all liquid is removed from surface of cell pellet. The presence of extra buffer will dilute the sucrose for the sucrose gradient. Cell pellets may be stored at –80°C until use.

3. Homogenization conditions reported here are optimized for HeLa cells. Homogenization conditions need to be optimized separately for each cell type or tissue.

4. It is important during cell fractionation to keep everything on ice to reduce proteolytic digestion.

5. Over-homogenization of cells and tissue results in DNA contamination from broken nuclei. This will result in poor separation of membrane fractions.

6. Analysis of membrane proteins that do not localize to the plasma membrane requires enrichment strategies optimized for that organelle.

7. Make sure that all interfaces are crisp and not diffuse. Handle the sucrose gradients gently to avoid disturbing the interfaces.

8. It is necessary to stop the centrifuge slowly (or leave the brake off when the centrifuge is stopping) to avoid disturbing the sucrose interfaces.

9. This step dilutes the sucrose enough so that the plasma membranes will pellet to the bottom instead of remaining suspended in the sucrose gradient.

10. At this point, the membrane pellets can be stored at –20°C if necessary.

11. This high pH treatment opens membrane structures and releases the soluble luminal proteins. This extraction method serves as an enrichment step for membranes.

12. It is ideal for the subsequent digestion to have the initial protein concentration at 1 mg/mL. However, if very little sample is present, the protein assay can be skipped.

13. High pH extraction with mechanical agitation causes membrane vesicles to open and form stable membrane sheets with free edges. This step is important to allow the protease access to both sides of the membrane.

14. Do not worry about the subsequent volume increase from the addition of urea. This step helps denature the soluble domains of integral membrane proteins without removing them from the lipid bilayer.

15. The addition of DTT reduces disulfide bonds.

16. IAA is light sensitive, so thaw the aliquot in the dark. IAA alkylates sulfhydryl groups to inhibit the reformation of disulfide bonds.

17. These peptides are not suited for analysis by tandem mass spectrometry because of their short length of less than 3 amino acids.

18. A black backboard should be placed on the wall next to the bomb to facilitate viewing of the resin in the column.

19. If no micro-columns are available for mass spectrometry, analyses can be carried out with commercially available, single-phase columns. The sample can be fractionated off-line using the salt pulses from **subheading 3.4, step 20**, and the fractions can be collected and loaded separately onto a reverse-phase column.

20. Because of the presence of urea in the sample, a loading column is often needed to load the sample, as loading it directly onto a 100-μm column may clog the column. The section of 250-μm tubing not going into the filter is called the waste line.

21. If animal tissue or non-human cell lines are being used, search against the protein sequence database corresponding to the appropriate species.

22. The program Bioworks (ThermoFinnigan) or Mascot (Matrix Science Ltd. London, UK) can also be used to analyze the raw data to identify peptides.

References

1. Wu, C. C., and Yates, J. R., III (2003) The application of mass spectrometry to membrane proteomics. *Nat. Biotechnol.* **21**, 262–267.

2. Howell, K. E. and Palade, G. E. (1982) Hepatic Golgi fractions resolved into membrane and content subfractions. *J. Cell Biol.* **92**, 822–832.

3. Wu, C. C., MacCoss, M. J., Howell, K. E., and Yates, J. R., 3rd. (2003) A method for the comprehensive proteomic analysis of membrane proteins. *Nat. Biotechnol.* **21**, 532–538.

4. Washburn, M. P., Wolters, D., and Yates, J. R. (2001) Large-scale analysis of the yeast proteome by multidimensional protein identification technology. *Nat. Biotechnol.* **19**, 242–247.

5. McDonald, W. H., Tabb, D. L., Sadygov, R. G., MacCoss, M. J., Venable, J., Graumann, J., et al. (2004) MS1, MS2, and SQT-three unified, compact, and easily parsed file formats for the storage of shotgun proteomic spectra and identifications. *Rapid Commun. Mass Spectrom.* **18**, 2162–2168.

6. Eng, J. K., McCormack, A. L., Yates J. R., III (1994) An approach to correlate tandem mass spectra of modified peptides to amino acid sequences in a protein database. *J. Am. Soc. Mass Spectrom.* **5**, 976–989.

7. MacCoss, M. J., Wu, C. C., and Yates, J. R., III (2002) Probability-based validation of protein identifications using a modified SEQUEST algorithm. *Anal. Chem.* **74**, 5593–5599.

8. Tabb, D. L., McDonald, W. H., and Yates, J. R., III (2002) DTASelect and Contrast: tools for assembling and comparing protein identifications from shotgun proteomics. *J. Proteome Res.* **1**, 21–26.

9. Blackler, A. R., Klammer, A. A., MacCoss, M. J., and Wu, C. C. (2005) Quantitative comparison of proteomic data quality between a 2D and 3D quadrupole ion trap. *Anal. Chem.* **78**, 1337–1344.

14

Free Flow Isoelectric Focusing

A Method for the Separation of Both Hydrophilic and Hydrophobic Proteins of Rat Liver Peroxisomes

Markus Islinger and Gerhard Weber

Summary

Peroxisomes take part in various metabolic pathways related to the regulation of lipid homeostasis. Although detailed information on the enzymes involved in the peroxisomal lipid metabolism was acquired in the past, the mechanisms of metabolic exchange between peroxisomes and the cytosol or other organelles still remain an enigma. Therefore, a detailed analysis of the peroxisomal membrane proteome could help identify potential metabolite transporters. However, because of their highly hydrophobic character, membrane proteins tend to precipitate in aqueous media, making their fractionation still a challenging task. To overcome these obstacles, we have elaborated a protocol for the separation of both hydrophilic as well as hydrophobic proteins using free flow isoelectric focusing (FF-IEF). Similar to traditional gel-based isoelectric focusing, a denaturing electrophoresis buffer containing a mixture of urea, thiourea and detergents is applied to keep highly hydrophobic proteins in solution. Electrophoresis is conducted on a BD Free Flow Electrophoresis System with a linear pH gradient from 3 to 10 and sampled into 96 fractions. As a second dimension, sodium dodecyl sulphate–polyacrylamide gel electrophoresis (SDS–PAGE) is used to further separate and visualize the protein pattern of the peroxisomal subfractions of matrix, peripheral and integral membrane proteins. The identification of the known peroxisomal membrane proteins PMP22, PMP70 as well as mGST in the subsequent matrix-assisted laser desorption ionization time of flight mass spectrometry (MALDI-TOF MS) analysis of the 100 most prominent protein bands has documented the suitability of this new technique for the analysis of hydrophobic proteins.

Key Words: Peroxisomes; free flow electrophoresis; membrane proteins; organelle separation.

From: *Methods in Molecular Biology, vol. 432: Organelle Proteomics*
Edited by: D. Pflieger and J. Rossier © Humana Press, Totowa, NJ

1. Introduction

Peroxisomes are single-membrane-enclosed organelles often referred to as 'multipurpose organelles' as they catalyze a broad variety of catabolic as well as anabolic reactions *(1–3)*. Main functions attributed to peroxisomes are the beta-oxidation of very long-chain fatty acids, unsaturated long-chain fatty acids and prostaglandins but also the biosynthesis of cholesterol, bile acids and ether phospholipids, indicating that peroxisomes are of considerable importance for lipid homeostasis *(1,4)*. Furthermore, peroxisomes are also involved in the breakdown of glyoxylate, purines, pipecolate and phytenate *(2,5,6)*. In contrast to most other organelles, peroxisomes do not fulfil oxidative reactions by dehydrogenases but rather by oxidases, thereby generating H_2O_2, broken down by the classical peroxisomal marker enzyme catalase.

Whereas, more than 50 peroxisomal matrix enzymes have been identified and enzymatically characterized in the past *(3,7,8)*, data on membrane proteins are still scarce and there is considerable lack of information concerning the transmission of metabolites across the peroxisomal membrane *(9)*. However, the characterization of multipass membrane proteins is commonly hampered by their high hydrophobicity, making their investigation in an aqueous solution a challenging task. Classical methods of protein separation, such as 2D isoelectric focusing (IEF)/sodium dodecyl sulphate–polyacrylamide gel electrophoresis (SDS–PAGE), generally fail, as a consequence of protein precipitation when the membrane proteins are focused at their isoelectric point, minimizing their already low polarity and thus preventing their detection in MS *(10)*. In contrast, shotgun mass spectrometry allows the detection of even multipass membrane proteins *(11,12)* yet is limited by the resolution, scanning speed and dynamic range of the mass spectrometer. Prefractionation methods used on- or offline to MS, such as 1D SDS–PAGE or HPLC-based separations, offer the possibility to increase the peak capacity at the protein or peptide level but certainly do not offer the enormous capacity of 2D separation approaches. In this context, free flow IEF (FF-IEF) provides a valuable alternative, combining the benefits of gel-free electrophoresis with the potential of isoelectric focusing *(13,14)*. As proteins do not have to cross surface barriers in FF-IEF, precipitation is drastically reduced compared to IPG 2D electrophoresis *(15)*. Furthermore, the fluid material can be continuously sampled and directly subjected to downstream separation steps.

2. Materials

2.1. Preparation of Optiprep Density Gradients

1. Refractometer for the preparation of the density gradients.
2. Optiprep: 60% (w/v) iodixanol in water (Axis Shield, Rodel∅kka, Sweden).
3. 40-mL centrifugation tube.

2.2. Isolation of Peroxisomes

1. Motor-driven Potter–Elvehjem tissue grinder with loose-fitting pestle (clearance 0.1 – 0.15 mm).
2. Homogenization buffer (HB): 250 mM sucrose, 5 mM MOPS, 1 mM ethylenediaminetetraacetic acid (EDTA), 0.1% ethanol, 2 mM phenylmethylsulphonylfluoride (PMSF), 1 mM dithiothreitol (DTT), 1 mM ε-aminocaproic acid; adjust pH to 7.4 with KOH.
3. Gradient buffer (GB): 5 mM MOPS, 1 mM EDTA, 0.1% (w/v) ethanol, 2 mM PMSF, 1 mM DTT and 1 mM ε-aminocaproic acid, pH 7.4.
4. Refrigerated centrifuge with a fixed angle rotor, e.g. Beckman Avanti J-25 and JA-20 rotor.
5. Ultracentrifuge, rotor and centrifugation tubes: Beckman Optima LE-80K with a Beckman VTi 50 vertical-type rotor and Quick Seal polyallomer tubes 25 × 89 mm.
6. Animals: rats of 220–225 g body weight (~8 week old or mice of the same age) (*see* **Note 1**).
7. Peroxisome lysis buffer: 1 mM $NaHCO_3$, 1 mM EDTA, 0.1% (w/v) ethanol and 0.01% (w/v) Triton X-100, pH adjusted to 7.6 with HCl.

2.3. Subfractionation into Membrane and Matrix Compartments

1. Stripping buffer A: 250 mM KCl in GB.
2. Stripping buffer B: 100 mM Na_2CO_3 in water.
3. Membrane solubilization buffer: 7 M urea, 2 M thiourea, 2% (w/v) CHAPS, 2% (w/v) ASB-14 (Calbiochem, Darmstadt, FRG), 2 mM tributylphosphine and 0.2% (w/v) carrier ampholytes (Bio-Rad, Munich, FRG).
4. Buffer exchange and sample concentration: Viva Spin 5000 MWCO PES Columns (Viva Science, Sartorius, Goettingen, FRG).
5. Protein quantification in membrane fraction: Plus One 2D Quant Kit (GE Healthcare, Munich, FRG).

2.4. Free Flow Isoelectric Focusing (see Note 2)

1. Anodic stabilization medium: included in the FF-IEF Kit (BD Diagnostics, Munich, Germany). The medium is based on a 100 mM H_2SO_4, 7 M urea and 2 M thiourea solution, pH 2.42 ± 0.10 supplemented with a mixture of selected buffers under proprietary rights of BD.
2. Separation medium: 250 mM mannitol, 17.5% (w/w) ProLyte 4–9 (BD Diagnostics), 7 M urea, 2 M thiourea (*see* **Note 3**) and 1% (w/w) ASB-14; pH 6.91 ± 0.10, cond. 323 ± 15 μS/cm.
3. Cathodic stabilization medium: included in the FF-IEF Kit (BD Diagnostics). Based on 150 mM NaOH, 7 M urea and 2 M thiourea, pH 10.75, supplemented with a mixture of selected buffers under proprietary rights of BD.
4. Counterflow medium: 12.5% (w/w) glycerol, 7 M urea and 2 M thiourea.

5. Electrode anode circuit: 100 mM H_2SO_4.
6. Electrode cathode circuit: 100 mM NaOH.
7. Pump tubes (Tygon standard; Ismatec, Wertheim-Mondfeld FRG): media, seven tubes of 0.64 mm ID; counterflow, one tube of 0.78 mm ID, one tube of 1.4 mm ID, one tube of 0.78 mm ID and sample: one tube of 0.51 mm ID.
8. SPADNS solution: for the stripe test sulfanilic acid, azochromotrope (SPADNS) is diluted 1:100 in water ($\sim$50 mL of SPADNS solution is required for each test).
9. pI-Marker: for performance test dilute BD™ Free Flow Electrophoresis pI-Marker 1:10 (v/v) with separation medium.
10. Parameters given are suitable for a BD™ Free Flow Electrophoresis System. However, with slight modifications, the procedure should be adaptable to the OCTOPUS or TECAN FFE instruments.

2.5. SDS–PAGE

1. NuPAGE Novex Bis–Tris 4–12% polyacrylamide gels (Invitrogen, Karlsruhe, Germany), 20× NuPAGE MES SDS running buffer and 4× NuPAGE LDS sample buffer.
2. Silver staining fixative 1: methanol/acetic acid/water, 50/10/40 (v/v/v), containing 500 μL/L of a 37% formaldehyde solution.
3. Silver staining fixative 2: ethanol/water, 50/50 (v/v).
4. Sensitizer solution: 2 mg/L of sodium thiosulfate in water.
5. Staining solution: 0.2% (w/v) $AgNO_3$ and 750 μL/L of a 37% formaldehyde solution in water.
6. Developing solution: 50 μg/L of sodium thiosulfate, 60 g/L of Na_2CO_3 and 250 μL/L of a 37% formaldehyde solution in water.
7. Stop solution: methanol/acetic acid/water, 50/10/40 (v/v/v).

3. Methods

Peroxisomes are fragile, leaky organelles. Thus, all fractions obtained during separation must be kept on ice and processed rapidly until the final centrifugation step is completed. After density gradient centrifugation in Optiprep, peroxisomes usually band at two distinct buoyant densities of 1.185 and 1.203 g/cm³, designated low- (lPos) and high-density peroxisomes (hPos). Both differ in their degree of mitochondrial contamination with the latter showing a quite high purity (>95%). The isolated peroxisomes can be stored resuspended in lysis buffer at –80°C until further subfractionation. As a quality control, SDS–PAGE of the isolated fractions is recommended. Hepatic peroxisomes of rodents should show their most prominent protein bands at $\sim$60 and $\sim$35 kDa, respectively – the molecular weights of the two marker enzymes, catalase and uricase (primates do not possess a functional uricase gene). In contrast, a thick band occurring at the height of 170 kDa, comprising carbamoyl phosphatase, points to mitochondrial contamination (*see* **Fig 1**).

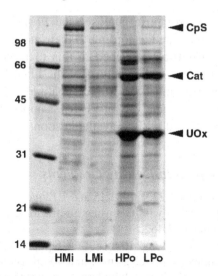

Fig. 1. Polypeptide pattern of peroxisomal fractions after Coomassie staining (10 μg of protein was loaded per lane). HMi, heavy mitochondrial fraction; LMi, light mitochondrial fraction – enriched in peroxisomes; HPos, high-density peroxisomes and LPos, low-density peroxisomes. Protein bands of catalase (Cat, peroxisomes), urate oxidase (UOx, peroxisomes) and carbamoylphosphate synthase (CpS, mitochondria) are marked with arrowheads.

Because of the protein ratio of matrix versus membrane proteins (usually more than 95% of total proteins correspond to matrix constituents), a thoroughly conducted fractionation into matrix and membrane compartments is a prerequisite for the detection of low-abundance integral membrane proteins. As already reported by Alexson and coworkers (16), peroxisomal matrix proteins exert a large range of 'leakiness', which means that they can be more or less easily extracted from the isolated organelles. Whereas easily to intermediately soluble proteins like catalase or acyl-CoA oxidase 1 can be nearly totally extracted by sonication in 0.01% Triton X-100, other peroxisomal matrix proteins like the L- and the D-bifunctional enzymes remain partially assembled with the peroxisomal membrane. These proteins can be released by a treatment with a buffer of high ionic strength, such as the 100 mM KCl stripping buffer used in this protocol. This stickiness was attributed to a fibrillar network associated with the peroxisomal membrane and core. As a consequence, the peripheral membrane fraction also contains peroxisomal matrix enzymes with such a hydrophobic character. Carbonate stripping as published by Fujiki and coworkers (17) is able to completely fragment peroxisomes into membrane sheets and dissolve the crystalloid core mainly comprising urate oxidase.

Before their mass spectrometric analysis, the FF-IEF fractionated protein samples can be further separated according to their molecular weight by SDS–PAGE as second dimension as described in this text. Theoretically, there is also the possibility for a direct MS analysis of the FF-IEF samples. However, detergents used in FF-IEF must be MS compatible or must be efficiently removed. Protocols for direct analysis of FF-IEF samples by MS are currently under development at BD and will be available soon.

3.1. Preparation of Optiprep Density Gradients

1. Prepare dilutions of 1.26, 1.22, 1.19, 1.15 and 1.12 g/cm³ from the 60% Optiprep (1.32 g/cm³) stock solution with GB using a refractometer and the formula: $\rho =$ 3.350× refractive index − 3.462.
2. Layer sequentially 4, 3, 6, 7 and 10 mL of the Optiprep dilutions, by decreasing order of concentration (1.26–1.12 g/cm³), in a 40-mL centrifugation tube.
3. Freeze the discontinous gradient in liquid nitrogen and store at −80°C (*see* **Note 4**).
4. Thaw the gradient at room temperature immediately before use in a metallic stand.

3.2. Isolation of Peroxisomes

1. All steps should be performed at a temperature of 4°C.
2. Anaesthetize rats (*see* **Note 5**), open the body cavity and rapidly excise the liver (*see* **Note 6**). After weighing, the organ is cut into pieces of ~2.5 mm³ and suspended in ice-cold HB at a ratio of 3 mL/g tissue.
3. Homogenize the liver pieces using a motor-driven Potter–Elvehjem tissue grinder and a loose-fitting pestle with exactly 1 stroke of 1000 rpm in 2 min.
4. Centrifuge the homogenate at 600 g_{av} for 10 min at 4°C to sediment nuclei and cellular debris.
5. Store the supernatant on ice and resuspend the pellet in at least 10 mL of HB per liver. Homogenize the pellet for a second time using a stroke of 1000 rpm in 1 min. Centrifuge again at 600 g_{av}.
6. Combine both supernatants and centrifuge at 1950 g_{av} for 10 min to collect most of the mitochondria (heavy mitochondrial fraction).
7. Decant the supernatant and save it on ice. Resuspend the heavy mitochondrial pellet manually with a glass rod and centrifuge at 1950 g for a second time.
8. Combine the supernatant with the first one and centrifuge at 25,300 g_{av} for 20 min. This prefraction contains a mixture of mainly light mitochondria, microsomes and peroxisomes.
9. Aspirate the supernatant including the reddish fluffy layer on top of the pellet mainly comprising microsomes.
10. Resuspend the pellet with a glass rod, add HB drop-wise until a homogenous suspension is gained, then adjust with HB to a volume of ~2 mL/g. Centrifuge at 25,300 g for 15 min.

11. Carefully, resuspend the final pellet comprising a crude peroxisomal fraction (light mitochondrial fraction) as described in **step 9** in a HB volume of 1–2 mL/g.
12. Layer 5 mL of the crude peroxisomal fraction on top of the Optiprep gradient. Overlay with GB and seal the tubes (*see* **Note 7**).
13. Centrifuge in a vertical-type rotor (e.g. VTi 50) at an integrated force of 1256 $\times 10^6 \times g \times$ min (g_{max} = 33,000) with slow acceleration/deceleration at 4°C.
14. Puncture the tubes with a syringe and remove the peroxisomes banding at 1.21 and 1.23 g/cm^3 (*see* **Fig. 2**).
15. Remove the Optiprep by diluting the fraction 1:4 with HB, and pellet the organelles at 25,300 g. Resuspend the pellet with a glass rod by adding drop-wise Lysis Buffer. Proceed to subfractionation protocol or store at –80°C.

3.3. Subfractionation of Peroxisomes

1. Thaw peroxisomes on ice and sonicate the suspension for 2 × 15 s with 100 W.
2. Centrifuge for 30 min at 100,000 g at 4°C. The resulting supernatant 1 comprises the bulk of peroxisomal matrix proteins. To increase purity of the pellet, optionally wash it with lysis buffer and centrifuge again.
3. Thoroughly stir up the pellet by using a small glass rod, add drop-wise stripping buffer A, keep for 5 min on ice.

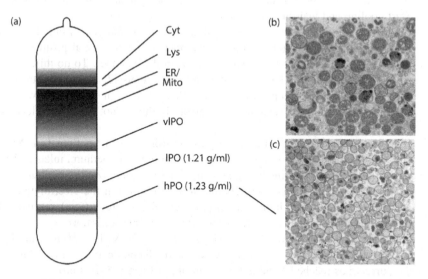

Fig. 2. (**a**) Schematic drawing of an Optiprep gradient after centrifugation. Electron micrographs of the isolated peroxisomal hPo-fraction (**c**) documenting their purity compared to the light mitochondrial pellet (**b**). Cyt, cytosol; Lys, lysosomes; ER, endoplasmic reticulum; Mito, mitochondria; vlPo, peroxisomes of very low density; lPo, peroxisomes of low density and hPo, peroxisomes of high density.

4. Centrifuge for 30 min at 100,000 g at 4°C. The resulting supernatant 2 contains proteins peripherally attached to the peroxisomal membrane. You can repeat washing in stripping buffer A to increase the purity of your membrane sample. The remaining pellet now consists of enriched integral membrane proteins plus the peroxisomal crystalline core.

5. For a subsequent mass spectrometric analysis, it is indispensable to remove the core, consisting of more than 95% of uricase; otherwise the high amounts of this enzyme will severely disturb protein separation by FFE. Therefore, resuspend the pellet in stripping buffer B and incubate on ice for 30 min.

6. Centrifuge for 30 min at 100,000 g at 4°C. Discard the supernatant and optionally wash the pellet once in GB. Stir up the now extremely hydrophobic pellet with a small glass rod adding drop-wise membrane solubilization buffer. Incubate at room temperature until all material has dissolved properly; gently agitate the suspension from time to time.

7. Dialyze both supernatants against water and lyophylize. Solubilize the pellet in solubilization buffer.

8. Quantify protein concentration of your samples using the Plus One 2D Quant Kit according to the protocol provided by the manufacturer.

3.4. Free Flow IEF of Peroxisomal Proteins

1. Assemble the chamber as described in the operating manual of your FFE-system. Start cooling and wait until the chamber has reached 10°C.

2. Fill the separation chamber with water and remove any trapped air bubbles, check for leakages and lower the separation chamber to the horizontal position. Carry out a stripe test to evaluate the laminar flow in the chamber. To do this, prepare a 1:100 dilution of SPADNS in water and place inlets I2, I4 and I6 into it (*see* **Fig. 3**). Run media pump with 40 rpm. Check for even, parallel flow of the coloured stripes in the separation chamber. If this is not the case, clean and reassemble your system.

3. Change to separation media according to the following instructions (*see* **Note 8** and **Fig. 3**): hang inlets 1 and 2 into anodic stabilization medium, inlets 3–5 into separation medium and inlets 6 and 7 into cathodic stabilization medium. Place counterflow tubings C 1–3 into counterflow medium. Fill the electrolyte anode and cathode circuit with liquid circuit (+ve) and liquid circuit (–ve), respectively. Equilibrate the chamber for at least two transition times (15 min).

4. After medium change, switch on the high voltage (520 V; I_{max}, 50 mA and P_{max}, 60 W) to establish the pH gradient. Switch on media pump at 45 mL/h. Wait until the current has reached a stable minimum of $\sim$18 mA (15–20 min).

5. To ensure that your chamber has reached a separation capacity of high quality, carry out a performance test with a pI marker. To do this, dilute the BD Free Flow Electrophoresis System pI-Marker 1:10 in separation medium and apply the mixture via the middle sample inlet at a flow rate of 1 mL/h. When coloured drops are visible at the collecting tube outlets, wait for another 10 min. Then, position a

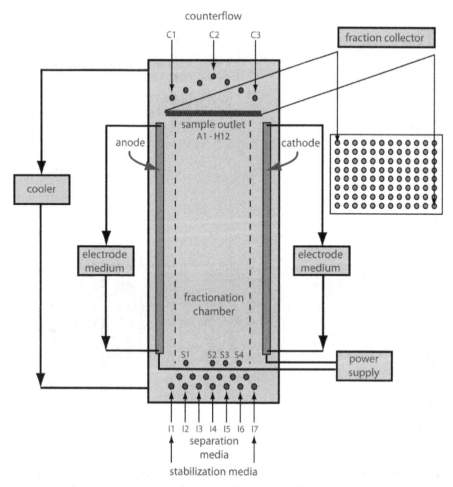

Fig. 3. Schematic overview of the free flow electrophoresis system. The abbreviations used for the various buffer inlets correspond to the terms used in the text section. In brief, counterflow inlets C1–3 are operated with counterflow medium, media inlets I1, I2 and I6, I7 with anodic and cathodic stabilization buffer, respectively, and inlets I3–I5 with separation medium.

microtiter plate on the collector's drawer, tap fractionation plate and slide in the drawer under the fractionation tubes. Switch off the sample pump and wait until the microtiter plate is filled three-fourth. The pI marker should be separated into five discrete components focused each in 1–3 wells of the microtiter plate.

6. Dilute sample 1:2 in separation medium. To visualize your sample during separation, FFE SPADNS can be optionally added. The amount of 1% SPADNS added in μL is calculated by taking five times the volume of the sample in mL divided by the sample pump speed in mL/h.

(a)

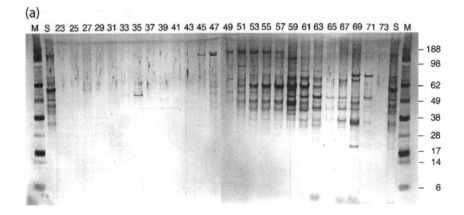

(b)

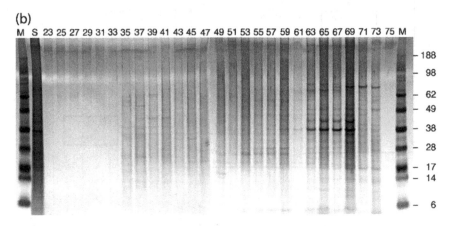

Fig. 4. Sodium dodecyl sulphate–polyacrylamide gel electrophoresis (SDS–PAGE) separation of peroxisomal subfractions obtained by free flow isoelectric focusing (FF-IEF). The FF-IEF was conducted with a prolyte mixture spanning pH 4–9. As shown exemplarily for the separation of the membrane fraction, the pH profile extends in a largely linear range from pH 3.5 to 10.5 (**d**). After IEF aliquots of the peroxisome fractions were loaded onto NuPAGE 4–12% Bis–Tris gels, the resulting protein pattern was visualized by silver staining as described in the method part. Note the characteristically different protein pattern for each subcompartment (**a**, matrix; **b**, integral membrane fraction; **c**, peripheral membrane fraction). Especially, the integral membrane fraction shows a highly complex pattern of low abundant proteins, which cannot be seen in direct 1D SDS–PAGE of that fraction. Because of the basic character of the bulk of peroxisomal proteins, as calculated by their theoretical pI, most proteins migrated to the cathodic side of the FFE chamber. Therefore, a flattened pH gradient over pH 6–12 should allow optimizing resolution for peroxisomal protein fractions.

(c)

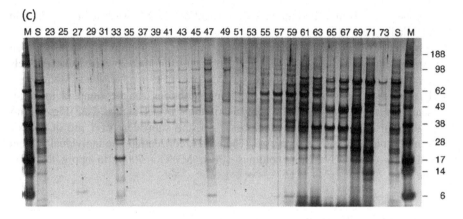

(d)

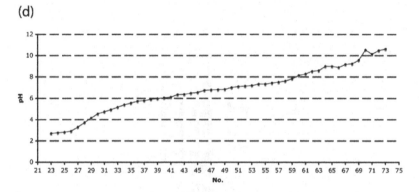

Fig. 4. *(Continued)*

7. Apply the sample to the separation chamber through the middle sample inlet at a flow rate of 0.5–2 mL/h, which should correspond to a protein throughput of 1–4 mg/h. Residence time in the separation chamber under these conditions is ∼23 min.

3.5. Analysis of the Separation by SDS–PAGE

1. To visualize the protein separation, run an aliquot of ∼10 μL of every second fraction of your microtiter plate on NuPAGE 4–12% acrylamide gels in a Xcell SureLock Mini Cell at a constant voltage of 150 V using a MES buffer system (*see* **Note 9** and **Fig. 4**).
2. After SDS–PAGE, fix the gels in silver staining fixative 1 for 2 × 15 min and in silver staining fixative 2 for 3 × 10 min.
3. Impregnate the gels for 1 min with sensitizer solution and wash for 2 × 30 s with distilled water.

4. Incubate for 20 min in staining solution and afterwards wash again for 30 s in distilled water.

5. Visualize protein bands in developing solution to an appropriate intensity, then rapidly change to the stop solution; incubate for another 20 min. Store the gels in ethanol/glycerol/water, 20/2/78 (v/v/v).

6. If results of the separation are satisfying, fractions can be sampled in larger scale for the subsequent MS analysis of their in-gel digests (*see* **Fig. 5 and Table 1**). For a preparative Coomassie gel, the equivalent of 1 mg of peroxisomal protein should be separated. Thereafter, concentrate the individual protein solutions using Viva Spin columns (PES; MWCO 5000) to apply them on the gel.

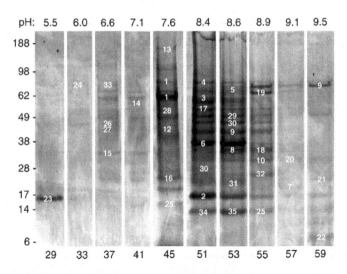

Fig. 5. Bands of proteins from the integral membrane fraction of isolated rat liver peroxisomes after separation by free flow ITP identified by matrix-assisted laser desorption ionization time of flight mass spectrometry (MALDI-TOF MS) analysis of their in-gel digests (mass fingerprinting). Protein bands are numbered according to Table 1. It appears that not all polypeptides reach their theoretical pI during focusing. Nevertheless, most of the proteins migrate according to their pI, indicating that free flow isoelectric focusing (FF-IEF) is a suitable pH-dependent electrophoresis comparable to NEPHGE. Calculated pI and molecular weight for each protein can be found in Table 1. MALDI-TOF MS data were interpreted by the MASCOT search algorithm (version 1.7) using the SwissProt/Trembl database. They allowed detection of typical peroxisomal transmembrane proteins, such as PMP22 and PMP70, or the microsomal GST1, which peroxisomes share with mitochondria and the endoplasmic reticulum (*22*).

Table 1

Mass spectrometric analysis of the 100 most prominent bands observed after free flow isoelectric focusing (FF-IEF) of the integral peroxisomal membrane fraction followed by sodium dodecyl sulphate–polyacrylamide gel electrophoresis (SDS–PAGE) as shown in Fig. 5

No	Protein name	Accession number	pI	GRAVY	MW	Local	Score	Coverage	TMH
1	Long-chain fatty acid acyl-CoA synthase	P18163	6.6	-0.080	78.2	Po	263	52	1
2	PMP22	Q07066	10.2	0.403	22.4	Po	202	32	4
3	PMP70	P16970	9.3	-0.152	75.2	Po	182	32	3
4	Estradiol-17β dehydrogenase 4	P97852	8.8	-0.104	79.3	Po	241	43	0
5	AOX2	P97562	7.6	-0.243	76.8	Po	225	43	0
6	Uricase	P09118	8.2	-0.458	34.8	Po	142	54	0
7	AOX1	P07872	8.6	-0.248	74.7	Po	66	21	0
8	Phytanoyl-CoA dioxygenase	P57093	8.8	-0.459	38.6	Po	86	24	0
9	Peroxisomal bifunctional protein	P07896	9.3	-0.109	78.5	Po	364	45	0
10	PEX14	Q9R0A0	5.0	-0.785	41.3	Po	109	28	1
11	PEX11	O70597	9.8	-0.008	27.9	Po	*	1	
12	Hydroxyacid oxidase 1	Q9WU19	7.6	-0.084	41.0	Po	65	38	0
13	Carbamoylphosphate synthase	P07756	6.3	-0.117	164.6	Mito	157	25	0
14	Sulfite oxidase	Q07116	5.8	-0.408	54.3	Mito	56	27	0
15	Prohibitin	P67778	5.6	0.009	29.8	Mito	351	65	0
16	ATP-synthase D-Chain	P31399	6.2	-0.718	18.6	Mito	120	70	0
17	ATP-synthase alpha-Chain	P15999	9.2	-0.138	58.8	Mito	72	29	0
18	Voltage-dependent anion selective channel protein 1	Q60932	8.6	-0.334	32.3	Mito	86	48	0

Table 1
(Continued)

No	Protein name	Accession number	pI	GRAVY	MW	Local	Score	Coverage	TMH
19	Calcium-binding mito-chondrial carrier protein	Q9QXX4	8.8	0.022	74.5	Mito	143	38	0
20	3,2-Trans-enoyl-CoA isomerase	P23965	9.6	−0.156	32.2	Mito	69	30	0
21	ATP synthase B chain	Q9CQQ7	9.1	−0.149	28.9	Mito	39	30	0
22	ATP synthase e chain	Q06185	9.4	−0.529	8.1	Mito	130	71	0
23	Cytochrome b5	P00173	4.9	−0.598	15.2	ER	143	54	1
24	78 kDa glucose-regulated protein	P06761	5.1	−0.481	72.3	ER	229	42	0
25	mGST-1	P08011	9.6	0.182	17.3	ER	65	29	3
26	Beta-actin	P48975	5.2	0.200	41.7	Cyto	277	37	0
27	Fructose-1,6-bisphosphatase	P19112	5.5	−0.132	39.5	Cyto	227	57	0
28	Myosin heavy chain (beta)	P02564	5.6	−0.801	223.1	Cyto	82	18	0
29	Myosin heavy chain (alpha)	Q02566	5.6	−0.822	223.6	Cyto	81	13	0
30	Betaine-homocysteine-S-methyltransferase	O09171	8.0	−0.355	45.0	Cyto	153	50	0
31	Peroxoredoxin	Q63716	8.3	−0.222	22.1	Cyto	141	56	0
32	Glutathione-S-transferase Yc-1	P04904	8.8	−0.334	25.2	Cyto	165	43	0
33	Serum albumin precursor	P02770	6.1	−0.389	68.7	Secret.	164	31	0
34	Haemoglobin beta chain	P11517	8.0	−0.055	15.9	Blood	141	53	0
35	Haemoglobin alpha1 and alpha2	P01946	7.9	−0.130	15.2	Blood	169	39	0

MW, molecular weight; local, cellular localization according to the SwissProt annotation; score, $-10 \times \log(P)$, where P is the probability that the observed match is a random result; protein scores greater than 54 are significant ($p < 0.05$); coverage, protein sequence coverage (%); TMH, number of transmembrane helices predicted by TMHMM; GRAVY, grand average index of hydrophathy (23).

[a]This protein was identified by immunoblotting.

4. Notes

1. Usually females are preferred, as their hepatic peroxisomes are less fragile compared to those of males. The liver of a 250-g rat weighs around 5–7 g, the liver weight of an adult mice is ∼1 g. Thus, use six mice to obtain the same amount of peroxisomes as from one rat. All animals should be starved overnight to minimize glycogen concentration, which otherwise interferes with the subcellular fractionation.

2. It is suggested to prepare all media for FFE freshly every day as decreased performance might be encountered due to degradation of the urea-containing media.

3. Take care to use urea with a low conductance, e.g. from SERVA Urea analytical grade PIN 292410009000. The media can be prepared in a lukewarm water bath to prevent the cooling effect due to urea dissolution.

4. Even when Optiprep gradients are used right after their preparation, the freezing/thawing process cannot be left out, because it is a prerequisite for a functional density gradient. During the thawing process, the still frozen, solid core floats upwards and consequently transforms the step gradient into a gradient of slightly sigmoidal shape.

5. Anaesthetization of the animal, e.g. by ip injection of Nembutal or chloralhydrate, must be carried out by a properly trained operator with the appropriate license.

6. The isolation procedure described was optimized for peroxisome isolation out of rat and mouse liver *(18)*. Nevertheless, it was further applied to other mammalian species *(19)* as well as to the isolation of renal peroxisomes *(20)* and peroxisomes from the digestive gland of molluscs *(21)*. However, for tissue-bearing microperoxisomes, e.g. brain, testis and lung, centrifugation conditions have to be newly adapted.

7. To ensure a proper separation of a crude peroxisomal fraction, do not apply more than the equivalent of 10-g rat liver tissue on top of the density gradient (maximum two rats per gradient).

8. The media temperature should not exceed the working temperature by more than 15°C. To avoid degassing, the media should not be below the chamber temperature. It is recommended to use media with a temperature similar to room temperature.

9. The MES running buffer can be used for up to three runs as long as the voltage is kept constant for each successive run.

Acknowledgments

The authors thank Professor Alfred Voelkl for the scientific advice and encouragement as well as for carefully reading this manuscript. Further on, we thank Heribert Mohr, Christian Obermayer and Ute Sukopp for their excellent technical assistance.

References

1. Reddy, J. K. and Mannaerts, G. P. (1994) Peroxisomal lipid metabolism. *Annu. Rev. Nutr.* **14**, 343–370.
2. Mannaerts, G. P. and van Veldhoven, P. P. (1993) Metabolic pathways in mammalian peroxisomes. *Biochemie* **75**, 147–158.
3. van den Bosch, H., Shutgens, R. B., Wanders, R. J., and Tager, J. M. (1992) Biochemistry of peroxisomes. *Annu. Rev. Biochem.* **61**, 157–197.
4. Osmundsen, H., Bremer, J., and Pedersen, J. I. (1991) Metabolic aspects of peroxisomal beta-oxidation. *Biochim. Biophys. Acta* **1085**, 141–158.
5. Singh, I. (1997) Biochemistry of peroxisomes in health and disease. *Mol. Cell. Biochem.* **167**, 1–29.
6. Lazarow, P. B. (1995) Peroxisome structure, function, and biogenesis – human patients and yeast mutants show strikingly similar defects in peroxisome biogenesis. *J. Neuropathol. Exp. Neurol.* **54**, 720–725.
7. Alexson, S. E., Fujiki, Y., Shio, H., and Lazarow, P. B. (1985) Partial disassembly of peroxisomes. *J. Cell Biology* **101**, 294–304.
8. Hashimoto, T. (2000) Peroxisomal beta-oxidation enzymes. *Cell. Biochem. Biophys.* **32**, 63–72.
9. Wanders, R. J. A. (2004) Peroxisomes, lipid metabolism and peroxisomal disorders. *Mol. Genet. Metab.* **83**, 16–27.
10. Klein, C., Garcia-Rizo, C., Bisle, B., Scheffer, B., Zischka, H., Pfeiffer, F., et al. (2005) The membrane proteome of Halobacterium salinarium. *Proteomics* **5**, 180–197.
11. Blonder, J., Conrads, T. P., and Veenstra, T. D. (2004) Characterization and quantitation of membrane proteomes using multidimensional MS-based proteomic technologies. *Expert Rev. Proteomics* **1**, 153–163.
12. Zhang, N., Li, N., and Li, L. (2004) Liquid chromatography MALDI MS/MS for membrane proteome analysis. *J. Proteome Res.* 3, 719–727.
13. Wang, Y., Hancock, W. S., Weber, G., Eckerskorn, C., and Palmer-Toy, D. (2004) Free flow electrophoresis coupled with liquid chromatography-mass spectrometry for a proteomic study of the human cell line (K562/CR3). *J. Chromatogr. A* **1053**, 269–278.
14. Obermaier, C., Jankowski, V., Schmutzler, C., Bauer, J., Wildgruber, R., Infanger, M., et al. (2005) Free-flow isoelectric focusing of proteins remaining in cell fragments following sonication of thyroid carcinoma cells. *Electrophoresis* **26**, 2109–2116.
15. Weber, G., Islinger, M., Weber, P., Eckerskorn, C., and Voelkl, A. (2004) Efficient separation and analysis of peroxisomal membrane proteins using free-flow isoelectric focusing. *Electrophoresis* **25**, 1735–1747.
16. Alexson, E. H., Fujiki, Y., Shio, H., and Lazarow, P. B. (1985) Partial disassembly of peroxisomes. *J. Cell Biol.* **101**, 294–305.
17. Fujiki, Y., Fowler, S., Shio, H., Hubbard, A. L., and Lazarow, P. B. (1982) Polypeptide and phospholipids composition of the membrane of rat liver peroxisomes: comparison with endoplasmic reticulum and mitochondrial membranes. *J. Cell Biol.* **93**, 103–110.

18. Voelkl, A. and Fahimi, H. D. (1985) Isolation and characterization of peroxisomes from the liver of normal untreated rats. *Eur. J. Biochem.* **149**, 257–265.
19. Fahimi, H. D., Baumgart, E., Beier, K., Pill, J., Hartig, F., and Voelkl, A. (1993) Ultrastructural and biochemical aspects of peroxisome proliferation and biogenesis in different mammalian species. In: *Peroxisomes: Biology and Importance in Toxicology and Medicine* (ed. G. G. Gibson and B. Lake), pp. 395–424. Taylor and Francis Ltd., London.
20. Zaar, K. (1992) Structure and function of peroxisomes in the mammalian kidney. *Eur. J. Cell Biol.* **59**, 233–254.
21. Cajaraville, M. P., Voelkl, A., and Fahimi, H. D. (1992) Peroxisomes in digestive gland cells of the mussel Mytilus galloprovincialis Lmk. Biochemical, ultrastructural and immunocytochemical characterization. *Eur. J. Cell Biol.* **59**, 255–264.
22. Islinger, M., Lüers, G. H., Zischka, H., Ueffing, M., and Völkl, A. (2006) Insights into the membrane proteome of rat liver peroxisomes: Microsomal glutathione-S-transferase is shared by both subcellular compartments. *Proteomics* **6**, 804–816.
23. Kyte, J. and Doolittle, R. F. (1982) A simple method for displaying the hydrophobic character of a protein. *J. Mol. Biol.* **157**, 105–132.

15

Use of Gas-Phase Fractionation to Increase Protein Identifications
Application to the Peroxisome

Jacob Kennedy and Eugene C. Yi

Summary

Gas-phase fractionation (GPF), defined as iterative mass spectrometric interrogations of a sample over multiple smaller mass-to-charge (m/z) ranges, enables the ions selected for collision-induced dissociation to come from a greater number of unique peptides compared to the ions selected from the wide mass range scan in automated liquid chromatography-tandem mass spectrometry (LC-MS/MS) analysis. GPF is described as a means to achieve higher proteome coverage than multiple LC-MS/MS analyses of unfractionated complex peptide mixtures. It is applied to organellar proteomics through analysis of yeast peroxisomal proteins obtained from a discontinuous Nycodenz gradient fraction known to be enriched with yeast peroxisomal membrane proteins.

Key Words: Liquid chromatography-tandem mass spectrometry; mass spectrometry; gas-phase fractionation; yeast; peroxisome; organelles; data-dependent duty cycle; subcellular; *Nycodenz* gradient fraction.

1. Introduction

Automated liquid chromatography-tandem mass spectrometry (LC-MS/MS) is being widely applied for the analysis of large-scale protein expression in cells or tissues. Although LC-MS/MS permits the consolidated analysis of proteome, comprehensive global proteome analysis is limited because of issues associated with sample complexity and dynamic range of protein mixtures in which individual components differ in abundance by six or more orders

From: *Methods in Molecular Biology, vol. 432: Organelle Proteomics*
Edited by: D. Pflieger and J. Rossier © Humana Press, Totowa, NJ

of magnitude. These limitations have been addressed on a biochemical level through considerable efforts on sample fractionation and simplification at cell organelles, large protein complexes, or protein and peptide levels. These limitations have also been addressed from the instrument perspective by exploring modified data acquisition mode in conjunction with automated data-dependent tandem MS. In a typical data-dependent tandem MS analysis, the most intense peptide ion surveyed in the selected mass-to-charge (m/z) scan range is selected for fragmentation (*see* **Fig. 1**). After fragmentation, the precursor ion m/z value is written to a dynamic list for exclusion from further selection for fragmentation. This process is carried out iteratively as the next most intense ion in the precursor survey scan mass spectrum is chosen for fragmentation. Typically, precursor ion survey scans are carried out across a wide m/z range (e.g., 400–2000 Th). However, the use of the wide m/z range for ion selection from complex mixtures of peptides results in a seemingly poor reproducibility of peptide ion selection between LC-MS/MS replicates mainly because of two reasons. First, as shown in **Fig. 1**, the most abundant ions can be associated

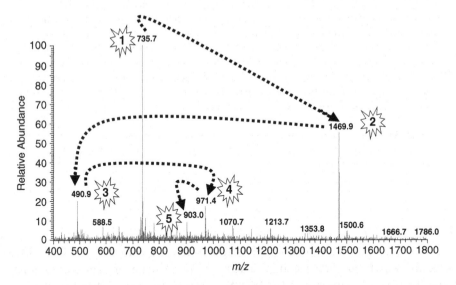

Fig. 1. Typical data-dependent ion selection in tandem mass spectrometry analysis. Large numbers in bold indicate the order of ion selection that the data-dependent software controlling ion selection will follow, beginning with the base peak (*1*). As each ion is fragmented, the parent ion value is written to a list and dynamically excluded from further selection for CID. Note that most ions will not be selected for CID, resulting in loss of information and low reproducibility of protein identifications between replicate liquid chromatography-tandem mass spectrometry (LC-MS/MS) analyses. Reproduced with permission from reference (*3*).

with the same peptide. This multiple charging phenomenon of electrospray ionization results in redundant fragmentation of the same peptide. Second, the duty cycle associated with ion surveying, selection and fragmentation is not fast enough to select all of the peptides as they elute from the reversed-phase column. This data-dependent duty-cycle limitation allows only a small portion of peptide/protein identification, predominantly the most abundant proteins, in the given complex sample mixtures. To improve reproducibility and efficiency of duty-cycle time for data-dependent peptide ion selection, and thus increase peptide/protein identifications, Spahr et al. first employed a direct iterative approach using gas-phase fractionation (GPF) by MS in the *m/z* dimension ($GPF_{m/z}$) in automated LC-MS/MS *(1,2)*. In $GPF_{m/z}$ analysis, a sample is repeatedly analyzed by automated data-dependent tandem MS using many narrow *m/z* ranges (e.g., 400–500 or 500–600 Th) as survey scans from which to select ions for fragmentation rather than a single full range (e.g., 400–2000 Th). In addition to the *m/z* dimension, GPF in the relative peptide ion intensity dimension, denoted as GPF_{RI}, was explored for increasing the number of protein identifications *(3)*. In GPF_{RI}, peptide ion selection for fragmentation begins not by the normal top–down process (e.g., first or second most intense ion), which starts ion selection with the most intense peak, but with an ion of lower intensity (e.g., fourth or fifth most intense ion) in the survey scan.

Protein quantity limitations can prevent fractionation prior to LC-MS/MS analysis. In these cases, multiple LC-MS/MS analyses can be performed to increase the number of unique peptide identifications. The unfractionated mixture of peptides generated by proteolysis of a yeast whole cell lysate was studied by triplicate LC-MS/MS analysis, by data-dependent ion selection over a survey scan range 400–1800 Th, and by $GPF_{m/z}$ over 400–800, 800–1200, and 1200–1800 Th. Comparison of protein identifications produced the following results *(3)*: more proteins (~1.5 times) can be identified by $GPF_{m/z}$ for an equal amount of effort (i.e., three LC-ESI-MS/MS analyses) and proteins identified by $GPF_{m/z}$ have a lower average codon bias value (CBV) *(3)*. Use of GPF_{RI} identifies more proteins per *m/z* unit scanned than $GPF_{m/z}$ or triplicate analysis over a wide *m/z* range.

We further applied the utility of GPF to the study of yeast peroxisome. After tryptic digestion of all the proteins contained in a discontinuous Nycodenz gradient fraction known to be enriched with yeast peroxisomal membrane proteins, we detected 94% (46/49) of known and predicted peroxisomal proteins using $GPF_{m/z}$ (400–800, 800–1200, and 1200–1800 Th), but only 77% using a standard wide *m/z* range survey scan (400–1800 Th) (*see* **Table 1**) *(3)*.

This chapter provides description of instrumental control methods for GPF and application of the GPF methods to increase protein identifications of a yeast subcellular organelle, peroxisome, in which protein quantity limits the use of

Table 1
Peroxisomal proteins identified by a wide *m/z* range (e.g., 400–2000 Th) liquid chromatography-tandem mass spectrometry (LC-MS/MS) analysis and by GFP$_{m/z}$ (400–800, 790–1200, and 1190–2000 Th)

ORF	Name	Peroxisomal localization	Cellular localization	Full range LC-MS/MS	GPF LC-MS/MS
YBR041W	FAT1	Integral [P]	Lipid particles [E], Perox [E]	X (1)	X (1)
YBR222C	FAT2	Soluble [E], Peripheral [E]	Peroxisomal [E]	X (62)	X (49)
YCR005C	CIT2	Soluble [E]	Peroxisomal [E]	X (5)	X (2)
YDL078C	MDH3	Soluble [E]	Peroxisomal [E]	X (28)	X (26)
YDR142C	PEX7	Soluble [E]	Peroxisomal [E]	X (4)	X (1)
YDR234W	LYS4		Mitoch [E], Perox [E]	X (3)	
YDR244W	PEX5	Soluble [E]	Cyto[E], Perox[E]	X (3)	X (1)
YDR256C	CTA1		Lysosome Vacuole [E], Perox [E]	X (10)	X (11)
YDR329C	PEX3	Integral [E]	Peroxisomal [E]	X (7)	X (3)
YER015W	FAA2	Peripheral [E]	Peroxisomal [E]	X (31)	X (32)
YGL153W	PEX14	Peripheral [E]	Peroxisomal [E]	X (7)	X (4)
YGL184C	STR3	Soluble [P]	Peroxisomal [E]	X (2)	X (5)
YGL205W	POX1		Peroxisomal [E]	X (51)	X (43)
YGLO67W	NPY1		Peroxisomal [E]	X (3)	X (1)

YGR077C	PEX8	Peripheral [E]	Peroxisomal [E]	X (2)	X(2)
YGR133W	PEX4	Peripheral [E]	Peroxisomal [E]		X (1)
YGR154C	YGR154C		Peroxisomal [E]	X (7)	X (2)
YIL160C	POT1		Peroxisomal [E]		X (7)
YIR031C	*DAL7*		*Cyto [E], Perox [E]*		
YIR034C	LYS1	Integral [E]	Peroxisomal [E]	X (6)	X (8)
YJL210W	PEX2		Peroxisomal [E]	X (1)	X (2)
YJR019C	TES1	Integral [P]	Peroxisomal [E]	X (21)	X (12)
YKL188C	PXA2		Peroxisomal [E]		X (6)
YKR009C	FOX2		Peroxisomal [E]	X (1)	X (1)
YLR027C	*AAT2*	*Soluble [E]*	*Cyto [E], Perox [E]*		X (3)
YLR151C	PCD1		Peroxisomal [E]		X(2)
YLR191W	PEX13	Integral [E]	Peroxisomal [E]	X (2)	X (2)
YLR284C	ECI1	Soluble [P]	Peroxisomal [E]	X (7)	X (5)
YML042W	*CAT2*	*Soluble [E]*	*Mitoch [E], Perox [E]*	X (7)	X (7)
YMR026C	PEX12	Integral [P]	Peroxisomal [E]		
YNL009W	IDP3	Soluble [E]	Peroxisomal [E]	X (5)	X(3)
YNL117W	MLS1		Peroxisomal [E]	X (2)	X (8)
YNL202W	SPS19		Peroxisomal [E]	X (10)	X (10)
YNL214W	PEX17	Peripheral [E]	Peroxisomal [E]		X (1)
YNL329C	PEX6	Integral [P]	Peroxisomal [E]	X (7)	X(1)
YOL044W	PEX15	Integral [E]	Peroxisomal [E]		X(1)

(Continued)

Table 1
(Continued)

ORF	Name	Peroxisomal localization	Cellular localization	Full range LC-MS/MS	GPF LC-MS/MS
YOL126C	*MDH2*	*Soluble [P]*	*Cyto [E], Perox [E]*	*X (4)*	*X (3)*
YOL147C	PEX11	Peripheral [E]	Peroxisomal [E]	X (13)	X (8)
YOR180C	DCI1		Peroxisomal [E]	X (2)	X (3)
YPL147W	PXA1	Integral [E]	Peroxisomal [E]		X (4)
YPR128C	ANT1	Integral [E]	Peroxisomal [E]		X (2)
Number of identifications	30	38			
Total peroxisomal [E]	41	41			
overage	73%	93%			

Proteins with dual cellular localization according to information in yeast proteome database (www.incyte.proteome.org) are indicated in italics. A blank box indicates no information or that the protein was not identified. Soluble and peripheral refer to protein localization in the membrane. Perox, cyto, and mito refer to cellular location of a protein. X, protein identified by the method in the corresponding column header. #, number of peptides identified from protein in row header; E, experimentally shown and P, predicted.

Reproduced with permission from reference (3).

multidimensional chromatographic fractionation. Although the main focus of this chapter is intended to describe the modified data acquisition scheme using GPF approach to improve protein identifications, **Headings 2** and **3** will be extended to the isolation of peroxisomes as described previously *(3)*.

2. Materials

2.1. Cell Culture and Lysis

1. Wild-type diploid BY4743 yeast strains (Research genetics, Huntsville, AL).
2. YPD medium: 2% yeast extract, 1% peptone, and 2% glucose.
3. SCIM medium: 0.5% yeast extract, 0.1% peptone, 0.79 g/L complete synthetic medium mix (Q-biogene, Carlbad, CA), 0.5% ammonium sulphate, 1.7 g/L yeast nitrogen base, 0.1% Tween 40, 0.1% glucose, and 0.15% Oleic acid.
4. Zymolyase.
5. LB buffer: 0.65 M sorbitol, 5 mM 2-MES-HCl, pH 5.5, 1 mM EDTA, and 1 mM potassium chloride, containing protease inhibitors (PINS: 0.2 mM PMSF, 2 µg/mL leupeptin, 2 µg/mL aprotinin, and 0.4 µg/mL pepstatin A).
6. 8 g/mL Nycodenz solution.
7. Beckman NVT65 rotor (Beckman-Coulter, Fullerton, CA) or equivalents.
8. Peroxisomal markers [anti-multifunctional enzyme type 2 (MFE2) and anti-Ser-Lys-Leu-COOH tripeptide (SKL)].
9. Trichloroacetic acid.
10. Methanol.
11. Ti8 buffer: 10 mM Tris–HCl, pH 8.0, 5 mM ethylenediaminetetraacetic acid (EDTA), and PINS.
12. Sodium dodecyl sulphate–polyacrylamide gel electrophoresis (SDS–PAGE) separation kit.
13. Western blot analysis kit.

2.2. Microcapillary Electrospray Device and LC-MS System

These procedures were performed on an LC-MS system capable of peptide tandem MS at the attomole level on an ion trap mass spectrometer with automated routine operation. Sample introduction was performed by an integrated microcapillary electrospray ionization emitter device that consists of a microcapillary pre-column and a microcapillary analytical column. The preparation of this microcapillary electrospray ionization emitter device has been described previously *(3)*, and it is integrated with a LC-MS setup consisting of a binary pump, an autosampler, and an ion trap mass spectrometer. While instrumental components of the LC-MS setup are the choice of investigators, the instruments used for the GPF experiment described in this chapter are listed here.

1. HP1100 binary pump (Agilent Technologies, Wilmington, DE).
2. Famos Micro-autosampler (Dionex, San Francisco, CA).

3. LCQ DECA ion trap mass spectrometer (Thermo Finnigan, San Jose, CA).
4. Polyimide-coated fused-silica capillary tube (75 μm ID × 360 μm OD) (Polymicro Technologies, Phoenix, AZ).
5. Magic C18 resins (particles of 5 μm diameter, with a porosity of 100 and 200 Å) (Michrom BioResources, Auburn, CA).
6. HPLC-grade formic acid, water, and acetonitrile.
7. OASIS® MCX (mixed-mode cation-exchange reversed-phase, Waters, Milford, MA).
8. Modified trypsin (Promega, Madison, WI).
9. Speedvac evaporator (Savant).

3. Methods

The method used to isolate peroxisomes is that of Yi et al. *(3)*, and further details can be found elsewhere *(4,5)*.

3.1. Subcellular Fractionation and Isolation of Peroxisomes

1. Grow wild-type diploid BY4743 yeast strains to mid-log phase in 25 mL of YPD medium, harvest by centrifugation, and seed into 1 L of SCIM medium overnight at 30°C for the induction of peroxisome proliferation.
2. Harvest cells by centrifugation, wash with SCIM medium, and convert into spheroplasts with 1 mg of Zymolyase 100T per gram of cells for 1 h at 30°C and lyse spheroplasts by Potter homogenization in LB buffer.
3. Centrifuge the homogenate for 10 min at 2000 g to generate a postnuclear supernatant.

All subsequent steps from this point should be carried out at 4°C and in the presence of PINS.

4. Centrifuge the postnuclear supernatant for 30 min at 20,000 g_{max} to yield a supernatant and a pellet enriched in peroxisomes, mitochondria, and other organelles.
5. Resuspend the pellet in 1 mL of LB buffer and determine the protein concentration by Bradford assay.
6. Layer 1 mg of the resuspended material onto a 12-mL four-step gradient consisting of 17, 25, 35, and 50% Nycodenz in LB buffer and centrifuge at 34,500 g for 2 h in a Beckman NVT65 rotor to isolate peroxisomes at the 35–50% Nycodenz boundary.
7. Collect 1-mL fractions from the bottom of these gradients and analyze an aliquot of each by SDS–PAGE and western blotting. Identify fractions containing isolated peroxisomes by Western blotting analysis using antibodies directed to known peroxisomal markers (MFE2 and SKL).

3.2. Extraction of Peroxisome Matrix Proteins and Peroxisomal Membrane Fraction Isolation

1. Lyse the isolated fractions containing peroxisomes by addition of 10 volumes of Ti8 buffer and leave on ice for 1 h.

2. Centrifuge the suspension at 200,000 g_{max} for 1 h at 4°C in a Beckman TLA 100.2 rotor to isolate peroxisomal membranes and collect the resulting supernatant (STi8) containing matrix proteins.

3. Resuspend the membrane pellet (PTi8) in Ti8 buffer and save a portion of the suspension for both SDS–PAGE and immunoblotting analyses. Repellet proteins by centrifugation at 15,000 g for 15 min.

4. Precipitate proteins in STi8 by addition of TCA to a 12% final concentration for 10 min, followed by centrifugation at 15,000 g for 15 min.

5. Wash the precipitate of STi8 and the PTi8 fractions with 90% methanol, followed by centrifugation at 15,000 g for 15 min.

6. Store the pellets at –20°C until further use.

3.3. Sample Digestion, Purification, and Mass Spectrometry Analysis

1. Dissolve peroxisomal protein pellets obtained in **Subheading 3.2, step 6**, in 50 mM NH_4HCO_3 containing 0.5% SDS by heating at 60°C for 30 min, with occasional vortexing.

2. Dilute peroxisomal fractions with 50 mM NH_4HCO_3 until SDS is at a final concentration of 0.05% (*see* **Note 1**).

3. Add sequencing-grade trypsin at a 1:100 (w/w) enzyme/protein ratio and incubate overnight at 37°C. Quench the reaction by adding 6 N acetic acid to reach a pH of 3.0.

4. Purify the resulting peptide mixtures by OASIS® MCX cartridges, following the manufacturer's protocol, prior to LC-MS analysis (*see* **Note 2**).

5. Dry the purified peptide mixture completely in a Speedvac evaporator.

6. Add LC-MS loading buffer [0.1% (v/v) formic acid or 0.4% acetic acid in water] to the dried peptides. Vortex, centrifuge, carefully transfer to a microcentrifuge tube, and store at –80°C until ready for μLC-MS/MS.

7. Inject peptide mixtures onto a home-built trap (25) for clean up using an autosampler and then pass onto a 10 cm length × 75 μm ID microcapillary HPLC (μLC) column by initiating linear gradient flow of acetonitrile from the binary pump. For sample analyses, a binary solvent composition gradient with HPLC buffer A [0.1% (v/v) formic acid in water] and buffer B [0.1% (v/v) formic acid in acetonitrile] is developed from 5 to 35% B in 90 min, followed by 80% B for 5 min and 5% B for 20 min (*see* **Note 3**).

3.4. Mass Spectrometry Data Collection

The LCQ mass spectrometer parameters for both GPF experiments are as follows: nanospray voltage (2.4 kV) and heated capillary temperature (200°C). The data-dependent MS/MS parameters are as follows: three MS/MS scans after every MS scan, thee microscans averaged for every MS/MS scan, 3 min dynamic exclusion period, 3-Th (±1.5 Th) isolation width of ions selected for MS/MS, and 35% collision energy for MS/MS.

3.4.1. GPF$_{m/z}$

The utility of GFP has been adopted as a means to increase proteome coverage as well as enhance confidence level of protein identifications as a consequence of getting more peptide identifications from unfractionated peptide mixtures, in which protein quantity hampers the use of multidimensional chromatographic fractionation. In GPF$_{m/z}$, the choice of the number of m/z ranges depends on the available amount of sample and instrument time. We adopt the GFP approach when the sample amount is greater than 50 μg and inject ~2–3 μg per LC-MS/MS analysis. The following example describes two sets of experimental conditions for GPF covering an m/z range of 400–2000 Th but differing in the number and width of m/z ranges used. The first set of experimental conditions employs three narrow m/z ranges covering 400–800, 790–1200, and 1190–2000 Th and the second set employs 16 very narrow m/z ranges covering 400–510, 490–610, 590–710, 690–810, 790–910, 890–1010, 990–1110, 1090–1210, 1290–1310, 1390–1410, 1490–1610, 1590–1710, 1690–1810, 1790–1910, and 1890–2000. Adjoining survey scans should overlap by at least 10 Th to maximize ion selection. Compared to the data from triplicate analysis over a wide m/z range (e.g., 400–2000 Th), the three narrow GPF$_{m/z}$ produced 25% additional good-quality MS/MS spectra, which in turn resulted in 26% more protein identifications. Thus, low copy number proteins are more easily detected. It is worth noting that the GPF$_{m/z}$ method decreases the possibility of redundant peptide analysis that occurs when different charge states of the same peptide are selected for MS/MS.

We evaluated the data from the two different ion selection approaches with CBV as a means to imply that a given technique is capable of detecting less abundant species. **Figure 2** plots the CBV of proteins identified in both experiments and shows that lower CBV proteins were identified from the three narrow GPF$_{m/z}$ strategy than from triplicate analysis over a single, wide m/z range. More exhaustive GPF$_{m/z}$ strategy can be applied using many narrow m/z ranges (e.g., 400–500 or 500–600 Th) as survey scans to lead to more protein identifications as predicted by the ability to detect less abundant species as well as to identify more proteins per unit of m/z scanned.

3.4.2. GPF$_{RI}$

In addition to the GPF$_{m/z}$ strategy, the GPF$_{RI}$ strategy can be explored for increasing the number of protein identifications *(3)*. In GPF$_{RI}$, peptide ions are selected for fragmentation based on their relative intensity. In this approach the ion selection routine for automated data-dependent tandem MS is conducted from less intense precursor ions (e.g., fourth or fifth most intense ion) to avoid selecting precursor ions derived from high abundance species. With the GPF$_{RI}$

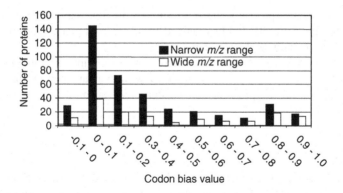

Fig. 2. Codon bias values (CBV) determined by GPF and standard wide *m/z* survey scan. Proteins identified by liquid chromatography-tandem mass spectrometry (LC-MS/MS) analysis from triplicate analysis over the wide *m/z* range 400–1800 Th or from GPF over the narrow ranges 400–800, 800–1200 and 1200–1800 Th. The sample was a mixture of all peptides produced by proteolytic action of tyrpsin on all proteins in yeast whole-cell lysate. Reproduced with permission from reference *(3)*.

strategy, more proteins can be identified than with the $GPF_{m/z}$ strategy *(3)*. The following example describes four sets of experimental conditions covering an *m/z* range of 400–2000 Th. Ions at the following relative intensities were selected for an MS/MS scan following the survey scan over 400–2000 Th: (1) 4th–6th, (2) 7th–9th, (3) 10th–12th, and (4) 13th–15th. Although the majority of proteins can be identified from the 4th–6th and 7th–9th GPF_{RI} ion selections, noticeably low CBV proteins can still be detected while extending the ion selection down to the 15th most abundant.

4. Notes

Below are some technical notes to bear in mind for LC-MS sample preparation and data acquisitions. Technical considerations associated with the isolation method of peroxisomes can be found in the **Notes** section of the Chapter 24 by Marelli et al.

1. High percentage of SDS [e.g., 1–2% (w/v) SDS] and agitation are often required to solubilize the protein sample isolated from peroxisomal fractionation. Prior to trypsin digestion, it is important to dilute the SDS concentration with the digestion buffer below 0.05% to ensure efficient trypsin activity for the digestion.

2. To avoid column clogging during LC-MS sample run, it is necessary to clean up the trypsin-digested sample mixtures by eliminating buffers, salts, and detergents. The MCX cartridge (mixed-mode cation-exchange reversed-phase) from Waters

(Milford, MA) is probably the most suitable for this sample clean-up process, as SDS cannot be efficiently removed by a reversed-phase desalting cartridge because of its long aliphatic chain.

3. Prior to LC-MS sample analysis, microcapillary column performance and MS parameters should be checked and tuned with standard peptide mixtures. The length of the HPLC gradient can be optimized according to the amount and complexity of the sample.

References

1. Spahr, C. S., Davis, M. T., McGinley, M. D., Robinson, J. H., Bures, E. J., Beierle, J., et al. (2001) Towards defining the urinary proteome using liquid chromatography-tandem mass spectrometry I. Profiling an unfractionated tryptic digest. *Proteomics* **1**, 93–107.

2. Davis, M. T., Spahr, C. S., McGinley, M. D., Robinson, J. H., Bures, E. J., Beierle, J., et al. (2001) Towards defining the urinary proteome using liquid chromatography-tandem mass spectrometry II. Limitations of complex mixture analyses. *Proteomics* **1**, 108–117.

3. Yi., E. C., Marelli, M., Lee, H., Purvine, S. O., Aebersold, R., Aitchison, J. D., et al. (2002) Approaching complete peroxisome characterization by gas-phase fractionation. *Electrophoresis* **23**, 3205–3216.

4. Marelli, M., Smith, J. J., Jung, S., Yi, E., Nesvizhskii, A. I., Christmas, R. H., et al. Quantitative mass spectrometry reveals a role for the GTPase Rho1p in actin organization on the peroxisome membrane (2004) *J. Cell Biol.* **167**, 1099–1112.

5. Bonifacino, J. S., Dasso, M., Harford, J. B., Lippincott-Schwartz, J., Yamada, K. M. (2001) *Current Protocols in Cell Biology*, Vol 1.

16

Purification and Proteomic Analysis of Lysosomal Integral Membrane Proteins

Huiwen Zhang, Xiaolian Fan, Rick Bagshaw, Don J. Mahuran, and John W. Callahan

Summary

Lysosomes are essential for normal function of cells. This is best illustrated by the occurrence of greater than 40 lysosomal storage diseases. While the enzymes of the luminal compartment have been widely studied usually in the context of these diseases, the composition of the enveloping membrane has received scant attention. Advances in mass spectrometry and proteomics have laid the necessary groundwork to facilitate investigation of membranes such as those of lysosomes, mitochondria, and other organelles to find novel proteins and novel functions. Pure lysosomes are a prerequisite, and we have successfully identified an abundance of membrane proteins from lysosomes of rat liver. Here, we describe two comparable and easy methods to isolate lysosomes from mouse or rat liver in sufficient quantities for proteomics studies. Also included is a comparison of the soluble, luminal proteins obtained from each of the two preparations separated by 2D immobilized pH gradient (IPG) sodium dodecyl sulphate–polyacrylamide gel electrophoresis (SDS–PAGE).

Key Words: Tritosomes; Tyloxapol; Percoll; lysosomes; proteomics.

1. Introduction

Lysosomes occur in most animal cells and tissues. They vary in size from less than one to several microns in diameter, are bounded by a single membrane, and contain a wide variety of acid hydrolases that have the capacity to degrade all of the macromolecules and intracellular organelles of cells and tissues (*1*). Recent evidence has shown that lysosomes are essential for repair of plasma membranes, neurite expansions, and MHC antigen presentation

From: *Methods in Molecular Biology, vol. 432: Organelle Proteomics*
Edited by: D. Pflieger and J. Rossier © Humana Press, Totowa, NJ

(2,3). Chediak–Higashi and Hermansky–Pudlak syndromes are rare autosomal recessive disorders of humans, characterized by severe immunologic defects, partial albinism, a mild bleeding tendency and recurrent severe infections *(4–6).* These lysosomal disorders are distinguished by the occurrence of enlarged or "giant" lysosomes where the underlying genetic bases involve proteins localized to the cytosol. Histochemical and immunofluorescence studies have shown that giant dense organelles were positive for marker proteins of late endosomes and lysosomes, for example, cathepsin D, lysosome-associated membrane protein (Lamp) 1, Lamp2, and a 120-kDa lysosome glycoprotein, while negative for mannose-6-phosphate receptor *(7,8).* The underlying basis for the occurrence of "giant lysosomes" is unknown even though mutations in the causative genes have been described *(2,6,9–11).*

Mass spectrometry-based proteomics has been successfully used to identify organelle-localized proteins such as in mitochondria, phagolysosomes, and melanosomes/lysosomes *(7,12–14).* Similarly, we recently reported a proteomics study of the lysosomal membrane in the expectation that the results would guide further study into the biogenesis and mechanisms for control of lysosome function *(7).*

Lysosomes are routinely isolated from liver, spleen, brain, kidney, cultured cells, and other sources, traditionally with the use of differential centrifugation techniques usually employing isotonic sucrose solutions. One of the earliest methods developed to purify lysosomes that remains in use employed injection of a neutral detergent, Triton WR 1339 (Tyloxapol) *(15).* This substance is rapidly endocytosed by liver Kupffer cells and liver hepatocytes into the lysosome compartment, and if sufficient amounts are taken up, it alters their buoyant density such that they can be separated by centrifugation from their usual contaminants, peroxisomes, and mitochondria. We have successfully used this method to isolate lysosomes (tritosomes) from rat livers and developed a proteomic approach to study the composition of the lysosome membrane *(7).* This method is also applicable to the isolation of lysosomes from mouse liver. More recently, Percoll has been used to subfractionate organelles, and the highest-purity preparation of lysosomes reported was obtained by this method *(8).*

2. Materials

2.1. Preparation of Tritosomes

1. Obtain fresh livers from two mice, about 3–4 g (mice are 25 g each; 8–10 weeks old).
2. 10% Tyloxapol (Sigma, St. Louis, MO, USA) solution: 2 g of the liquid is weighed in a weighing boat and dissolved completely in 10 mM phosphate-150 mM saline, pH 7.4 (phosphate-buffered saline, PBS) to a final volume of 20 mL.
3. Homogenization solution: 0.25 M sucrose, prepared on the day of use and supplemented with proteinase inhibitors (Sigma)—1 mg/mL leupeptin, 1 mg/mL

pepstatin A, and 0.1 mM phenylmethylsulfonyl fluoride (PMSF). The stock solution of proteinase inhibitors is made 1000-fold higher than working solution and stored at $-20°C$. Leupeptin, pepstatin A and PMSF are dissolved in deionized water, methanol, and isopropanol, respectively.

4. Thomas glass homogenizer with loose pestle driven by a Cole-Parmer Constant Torque control instrument, model 4420.

5. Sucrose solutions at 45, 34.5, and 14.3 (w/v) are prepared by introducing 45, 34.5, and 14.3 g of sucrose in water to reach a final volume of 100 mL.

6. SW 41 Ti rotor, ultracentrifuge, and 11-mL centrifuge tubes (Beckman, Mississauga, ON, Canada).

7. 0.25 M cold sucrose.

8. For measurement of total β-hexosaminidase activity: 1.6 mM 4-methylumbelliferyl β-N-acetylglucosaminide (Sigma) in 0.1 M citric acid, adjusted to pH 4.2 by adding 0.2 M sodium phosphate.

9. 0.1 M 2-methyl-2-amino-1-propanol, adjusted to pH 10.0 with 0.1 M HCl.

10. 4-Methylumbelliferone.

11. LS-50B Perkin-Elmer spectrofluorometer.

2.2. Lysosome Isolation by Percoll

1. Obtain fresh livers from two mice, about 3–4 g (mice are 25 g each; 8–10 weeks old).

2. Homogenization solution: 0.25 M sucrose prepared on the day of use and supplemented with proteinase inhibitors (Sigma): 1 mg/mL leupeptin, 1 mg/mL pepstatin A, and 0.1 mM PMSF.

3. Thomas glass homogenizer with loose pestle driven by a Cole-Parmer Constant Torque control instrument, model 4420.

4. 100 mM $CaCl_2$.

5. 0.25 M cold sucrose.

6. 40% (v/v) Percoll: 20 mL of Percoll reagent (GE Healthcare, Bail d' urfe, PQ, Canada) is diluted with 30 mL of homogenization solution. The pH is adjusted to 7.4 with a few drops of 5 M HCl (before adjustment the pH of the solution is 8.5–9.1).

7. 50% (w/v) sucrose solution: dissolve 50 g of sucrose in water to a final volume of 100 mL.

8. Quick-seal Beckman tube.

9. Beckman TL 100 ultracentrifuge with 100.3 rotor.

Unless otherwise specified, all solutions are kept at 4°C, and all centrifuge steps are performed at 4°C.

2.3. Lysosomal Subcellular Fractionation

1. Soluble fraction buffer: 50 mM Tris–HCl, pH 8.0, 0.2 M NaCl, and 1 mM ethylenediaminetetraacetic acid (EDTA) containing protease inhibitors (5 mg/mL leupeptin, 5 mg/mL pepstatin A, and 0.5 mM PMSF, made fresh).

2. Membrane-associated protein fraction buffer: 0.1 M Na_2CO_3, adjusted with a few drops of 1 M NaOH to pH 11 with protease inhibitors (5 mg/mL leupeptin, 5 mg/mL pepstatin A, and 0.5 mM PMSF, made fresh).
3. Membrane fraction buffer: 6 M urea, 1% (w/v) octyl-beta-glucopyranoside, and 50 mM Tris–HCl, pH 8.5, 0.1% (w/v) SDS with the protease inhibitors as above.

2.4. 2D Immobilized pH Gradient IEF/SDS–PAGE

1. Rehydration solution: 7 M urea, 2 M thiourea, 0.5% (v/v) Triton X-100, 0.5% (v/v) immobilized pH gradient (IPG) buffer (Amersham Biosciences, GE Healthcare, Bail d' urfe, PQ, Canada), 2% (w/v) ASB-14 zwitterionic detergent (Calbiochem, EMO Bio sciences, Son Diega, CA, USA), and 100 mM dithiothreitol (DTT).
2. IPG 18 cm 3–10NL strips and IPGPhor system (Amersham Biosciences).
3. Equilibration DTT solution: 50 mM Tris–HCl, pH 8.0, 6 M urea, 30% (v/v) glycerol, 2% (w/v) SDS, and a speck of bromophenol blue. DTT is added to 125 mM final concentration on the day of use.
4. Equilibration iodoacetamide solution: 50 mM Tris–HCl, pH 8.0, 6 M urea, 30% (v/v) glycerol, 2% (w/v) SDS, and a speck of bromophenol blue. Iodoacetamide is added to 0.45% (w/v) on the day of use.
5. Laemmli standard SDS buffer (2×): 25 mL of 0.5 M Tris–HCl, pH 6.8, 10 mL of glycerol, 4 g of sodium dodecylsulfate, 3.1 g of DTT, and 1 mg of bromophenol blue diluted to a final volume of 100 mL with water.
6. In tricine gels, the Tris–HCl in the Laemmli buffer system is substituted by the same concentration of tricine (*N*-(tri(hydroxymethyl)methyl)glycine).

2.5. CM-Sepharose Ion Exchange Chromatography

1. CM-Sepharose FF columns (weak cation exchanger, 0.09–0.13 mmol/mL) from GE Healthcare (Amersham Biosciences, GE Healthcare, Bail d' urfe, PQ, Canada).
2. Dissolving solution: 6 M guanidine-HCl and 40 mM DTT.
3. Iodoacetamide.
4. Column buffer: 50 mM formic acid, pH 3.6, 8 M urea, and 1% (v/v) Elugent™ detergent (Calbiochem).
5. Eluting solutions are made from column buffer to which salt is added to give final concentrations of 20 mM, 40 mM, 80 mM, and 250 mM NaCl. These solutions are prepared on the day of use.

2.6. Trypsin Digestion

1. PCR tubes (GeneAmp 2400 PCR system; Perkin-Elmer, Shelton, CT, USA).
2. Trypsin solution for digestion of dried gel fragments: dissolve modified sequencing grade trypsin in 50 mM ammonium bicarbonate, and 1 mM $CaCl_2$ (0.6 µg trypsin/30–50µL). Additional 50 mM ammonium bicarbonate may be added if the gel absorbs all the protease-containing liquid.
3. Acetonitrile/50 mM ammonium bicarbonate/trifluoroacetic acid, 50/49/1 (v/v/v).

4. Formic acid/2-propanol/acetonitrile/water, 20/15/25/40 (v/v/v/v).
5. Acetonitrile/water, 80/20 (v/v).
6. Sample dissolving solution: 0.8 M guanidine-HCl and 2.5% (v/v) trifluoroacetic acid in water.
7. C18 ZipTips (Millipore, Billerica, MA, USA).
8. Acetonitrile/water/acetic acid, 50/50/0.1 (v/v/v).

2.7. Mass Spectrometry

1. Matrix-assisted laser desorption ionization time of flight mass spectrometry (MALDI-TOF MS) instrument (Applied Biosystems Voyageur DE STR, Concord, ON, Canada).
2. QqTOF: Q-Star (MDS-Sciex, Concord, ON, Canada) equipped with a MALDI ionization source.
3. Matrix solution: 20 mg/mL of 2,5-dihydroxybenzoic acid in acetonitrile.
4. For LC-MS/MS analyses: Waters CapLC system equipped with a PicoFrit C18 column, online coupled to a Q-TOF mass spectrometer (Waters-Micromass, Milford, MA, USA).

3. Methods

3.1. Isolation of Triton-Filled Lysosomes from Mouse Liver

1. Mice (usually two) are given an intraperitoneal injection of Tyloxapol (85 mg/100 g of animal weight) 2 or 3 days before killing in a CO_2 chamber (*see* **Note 1**).
2. Mouse livers are minced with scissors and gently homogenized in homogenization solution (1:4, w/v) (5–6 strokes, 60 rpm). The homogenate is then centrifuged at 3000 *g* for 10 min to yield a postnuclear supernatant (PNS) and a pellet containing nuclei and intact cells.
3. The PNS is centrifuged at 34,000 *g* for 15 min to yield a crude organelle pellet and a supernatant containing primarily cytosolic proteins, ER, and Golgi. The crude organelles are resuspended in 10 mL of 45% (w/v) sucrose.
4. Two 11-mL centrifuge tubes (Beckman) are prepared. The crude organelles (5 mL for each tube) are layered beneath a discontinuous gradient of 4 mL of 34.5% (w/v) sucrose and 2 mL of 14.3% (w/v) sucrose in each centrifuge tube.
5. The tubes are centrifuged at 77,000 *g* for 2 h with a SW 41 Ti rotor.
6. Tyloxapol-filled lysosomes are obtained at the 14.3–34.5% interface.
7. The sample is diluted with 0.25 M sucrose (1:3, w/v) and pelleted at 28,000 *g* for 30 min.
8. Total β-hexosaminidase activity, used as the lysosome purification marker, is assayed using the synthetic substrate, 1.6 mM 4-methylumbelliferyl β-*N*-acetylglucosaminide in 0.1 M citric acid and 0.2 M sodium phosphate buffer, pH 4.2, in a final volume of 0.2 mL at 37°C for 30 min. The enzyme solution is diluted 1/1000 into the same buffer containing bovine serum albumin (1 mg/mL) for assay. Reactions are stopped with 2.0 mL of 0.1 M 2-methyl-2-amino-1-propanol, pH 10.

Fluorescence is compared to a 4-methylumbelliferone standard curve and measured in a LS-50B Perkin-Elmer spectrofluorometer with a sipper attachment set at an excitation wavelength of 365 nm and an emission wavelength of 450 nm. Protein amount is measured by the BCA method *(16)*.

9. The final lysosome sample is enriched greater than 50-fold (*see* **Table 1** and **Fig. 1**), and the usual yield from 4 g of liver of normal black mice is about 1 mg of total lysosome protein

3.2. Lysosomal Isolation by Percoll

1. Mouse livers are minced with scissors and gently homogenized in homogenization solution (1:4, w/v) as above with 6–8 strokes at 100 rpm.
2. The homogenate is centrifuged at 1000 *g* for 10 min.
3. The supernatant is collected and the pellet, containing mostly nuclei and unbroken cells, is rehomogenized in the original volume (5 strokes, 100 rpm) to increase yield.
4. The rehomogenate is centrifuged at 1000 *g* for 10 min, and the supernatant is collected. The two supernatants are combined and regarded as the PNS.
5. A 100 mM $CaCl_2$ solution is added to the combined PNS to a final concentration of 1 mM $CaCl_2$ (*see* **Note 2**).
6. The combined PNS is incubated at 37°C for 5 min and then centrifuged at 15,000 *g* for 15 min.
7. The crude organelles are overlaid with 6 mL of 0.25 M cold sucrose and resuspended by hand using 8 strokes of a loosely fitting Thomas homogenizer.
8. An 8-mL aliquot of 40% (v/v) Percoll is injected into the quick-seal Beckman centrifuge tube and overlayed with 1.5 mL of crude organelle (prepare four tubes).
9. Centrifuge at 60,000 *g* for 45 min with a 50 Ti Beckman rotor.

Table 1
Tritosome Isolation from Mouse Liver

Fraction	Specific activity (μM/h/mg)	Yield (%)	Purification factor
Homogenate	1.79	100	1
PNS	1.5	60.1	1
Crude organelle	3.34	25	2
Lysosome	94.9	15.1	53

A typical preparation is presented based on 2 g of mouse liver. Activity of hexosaminidase, used as the lysosome marker, and protein amount were measured as described. PNS refers to the postnuclear supernatant. The purification factor is calculated as the ratio of specific activities (micromoles/h/mg of protein) at each step in the procedure where the specific activity in the PNS is set to 1.0.

(a)

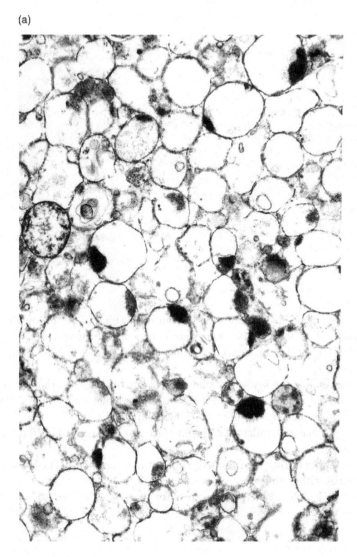

Fig. 1. Electron microscopy examination of lysosome preparations. Panel **A** shows lysosomes from normal mice isolated by Triton WR-1339 (21,000×). In panel **B**, lysosomes purified by the Percoll method were examined by conventional low magnification EM (4000×): nearly all organelles were recognizable as lysosomes. Panel **C**, the high magnification (40,000×) showed characteristic lysosomal structures.

(b)   (c)

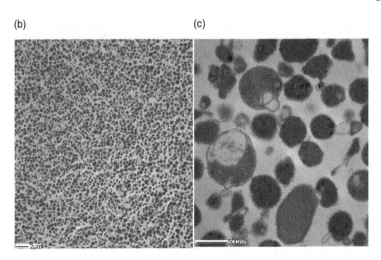

Fig. 1. (*Continued*)

10. Fractions are collected from the bottom of the tube in the following volumes: fractions 1–5 (0.5 mL; lysosomes are enriched in fractions 2–4), 6–7 (1.0 mL, containing the ER elements and mitochondria), and 8 (1.5 mL, largely Golgi elements).

11. To concentrate the lysosomes and remove Percoll, each fraction is underloaded with 0.2 mL of 50% (w/v) sucrose cushion and centrifuged at 300,000 g for 1 h in a Beckman TL 100 ultracentrifuge with 100.3 rotor.

12. The lysosomes are collected at the interface and resuspended in 10 mL of 0.25 M sucrose.

13. These are centrifuged at 25,000 g for 20 min, and the packed lysosomes are stored at –20°C until used.

14. The usual yield from two normal black mice is 0.5–1 mg of total lysosome protein with a purification factor of 50–100-fold (*see* **Table 2** and **Fig. 1**).

3.3. Lysosomal Protein Subfractionation

1. Frozen pelleted lysosomes are suspended in 0.2–0.3 mL of soluble fraction buffer (50 mM Tris–HCl, pH 8.0, 0.2 M NaCl, and 1 mM EDTA containing protease inhibitors) and pooled. The pooled fraction is frozen in dry ice/ethanol and thawed at 37°C. This process is repeated five times to efficiently lyse lysosomes.

2. The extract is centrifuged for 30 min at 355,000 g at 4°C. The resulting supernatant containing soluble lysosomal proteins is collected, analyzed for total protein amount (about 50% of the total lysosomal protein is soluble protein), and stored at –20°C until processed further.

Table 2
Lysosome Isolation by Percoll Gradient from Mouse Liver

Fraction	Specific activity (μM/h/mg)	Yield (%)	Purification factor
Homogenate	1.76	100	1
PNS	1.44	63.5	1
Crude organelle	3.34	38.1	1.9
Fraction 2	148.48	6.1	103
Fraction 3	135.2	4.8	93.9

A typical preparation is presented based on 2 g of mouse liver. Activity of hexosaminidase, used as the lysosome marker, and protein amount were measured as described. PNS refers to the postnuclear supernatant. The purification factor is calculated as the ratio of specific activities (micromoles/h/mg of protein) at each step in the procedure where the specific activity in the PNS is set to 1.0.

3. The resulting pellet is then suspended in a minimum volume (0.1 mL) of cold 0.1 M Na_2CO_3, pH 11.0, vortexed to disperse the proteins, and incubated on ice for 30 min.
4. This is then centrifuged at 355,000 *g* for 30 min to generate a supernatant containing the membrane-associated lysosomal proteins (about 25% of the total lysosomal protein) and a pellet containing the integral membrane proteins (about 25% of the total lysosomal protein).
5. The membrane fraction is dissolved in 6 M guanidine-HCl and 40 mM DTT to give a 1 mg/mL solution for CM-Sepharose chromatography or in the membrane fraction buffer for direct analysis.
6. These samples are analyzed for total protein amount and stored at –20°C until processed further.

3.4. Fractionation of Soluble and Membrane-Associated Lysosomal Proteins by 2D Immobilized pH Gradient IEF/SDS–PAGE

1. The rehydration solution is introduced into the samples by microdialysis overnight at room temperature.
2. Isoelectric focusing (IEF) is performed for 70,000 Vh using IPG 18 cm 3–10NL strips and the IPGPhor system. Typically, 0.2–0.3 mg of protein is used.
3. Strips are incubated in the equilibration DTT solution for 15 min at room temperature.
4. The strip is transferred to a fresh tube with equilibration-iodoacetamide solution for an additional 15 min in the dark.
5. The second dimension sodium dodecylsulfate-polyacrylamide gel electrophoresis (SDS–PAGE) separation is performed on a 10% tricine gel.

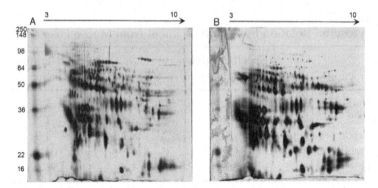

Fig. 2. Two-dimensional gel electrophoresis of soluble lysosomal proteins. Sample A (0.2 mg) was from the lysosomal-soluble protein prepared by the tritosome method while B (0.2 mg) was prepared by the Percoll method. Strips of non-linear pH ranging from 3 to 10 were used. The arrow above panel **A** and **B** indicates the direction from cathode to anode. Molecular weight markers (kDa) are shown on the left side of panel A. Proteins were revealed with silver nitrate. The main spots from the two methods are comparable.

6. Gels are stained by a modification of the technique reported by Mortz et al. *(17)*. Typical patterns for the soluble proteins obtained from Tritosomes and from Percoll gradients are shown in **Fig. 2**. Gel pieces are then recovered and processed for in-gel digestion with trypsin.

3.5. Fractionation of Lysosomal Integral Membrane Proteins on CM-Sepharose

The conditions for use of this adsorbent were chosen so that the abundant, acidic lysosome-associated membrane proteins (such as Lamp 1) would be collected in the unbound fraction. This was expected to allow enrichment of the lower abundance membrane proteins.

1. The membrane fraction dissolved in 6 M guanidine-HCl and 40 mM DTT to give a 1 mg/mL solution is incubated at 45°C for 30 min to ensure complete reduction of disulfide bonds.
2. Iodoacetamide is added to give a final concentration of 150 mM, and the solution is incubated for 1 h in the dark at room temperature.
3. The protein solution is then twice dialyzed overnight at room temperature against column buffer.
4. Conductivity measurements of the dialysis sac contents and the dialysate are made to ensure complete removal of guanidine cation which would affect the binding capacity of the resin.

5. The protein solution is clarified by centrifugation at 10,000 *g* for 15 min at 18°C.
6. The Sepharose CM-cation exchange column is pre-equilibrated in column buffer.
7. The sample is then loaded on Sepharose CM-cation exchange column (nominal capacity: 50 mg/mL resin).
8. The unbound fraction containing the Lamps and other unbound proteins is collected.
9. The bound proteins are eluted stepwise with separate solutions of 20 mM, 40 mM, 80 mM, and 250 mM NaCl dissolved in column buffer. Each fraction is handled separately.
10. To each of the protein fractions, 5 volumes of cold (–20°C) acetone is added to precipitate the proteins. Sample tubes are allowed to stand overnight at –20°C to ensure complete protein precipitation. After centrifugation and removal of acetone, the protein pellets are allowed to air dry. The pellets are dissolved in Laemmli standard SDS buffer.
11. The proteins are separated by standard one-dimensional SDS-polyacrylamide gel electrophoresis in 12% tube gels (5 mm × 12 cm).
12. The gels are recovered and proteins are fixed for 2 h in methanol/acetic acid/water, 50/10/40 (v/v/v).
13. Gels are then placed in a gel slicer and cut into 2–4 mm slices.
14. Slices are then processed for in-gel digestion with trypsin.

3.6. In-Gel Digestion and Isolation of Tryptic Peptides

1. Spots of interest are excised from the 2D gels by a scalpel blade and placed in a silanized (trimethylchlorosilane; Supelco, Bakville, ON, Canada) 1.5-mL microfuge tube. Gel slices are destained with several changes of 30 mM ammonium bicarbonate/acetonitrile, 60/40 (v/v), and gentle shaking on a rotary platform.
2. After aspirating the final wash liquid, the gel pieces are crushed by using a disposable pipet tip and dried in a SpeedVac. The dried gel pieces are transferred to a small PCR tube, rehydrated with the trypsin solution, and incubated at 37°C for 18 h.
3. After digestion, the liquid above the gel is removed and placed in a siliconized microfuge tube. The gel is then washed with 50 mM ammonium bicarbonate/acetonitrile/trifluoroacetic acid, 49/50/1 (v/v/v), then with formic acid/2-propanol/acetonitrile/water, 20/15/25/40 (v/v/v/v), and then with acetonitrile/water, 80/20 (v/v). In each instance the tube is vortexed briefly (20 s) and microfuged. The supernatant fluids are combined in the siliconized microfuge tube and completely dried in the SpeedVac concentrator.
4. The dried samples are dissolved in the sample dissolving solution and applied to C18 ZipTips. Peptides are eluted with acetonitrile/water/acetic acid, 50/50/0.1 (v/v/v) and analyzed.
5. For MALDI-TOF MS analysis, 1 μL of extracted peptides is overlaid with the same volume of matrix solution on the MALDI target and allowed to dry in air. MALDI-TOF MS/MS analysis is performed on the Q-Star with the same sample preparation.

6. For ESI-MS/MS analysis, recovered peptides are subjected to C18 nano-LC separation online coupled to a Q-TOF mass spectrometer.

3.7. Protein Identification

1. Although the steps before in-gel digestion and isolation of tryptic peptides for the soluble and hydrophobic proteins are different, from this point all samples are handled in the same way.
2. The MS/MS data obtained are analyzed using either MS/MS ions search (Mascot, http://www.matrixscience.com), sequence tags (PeptideSearch, EMBLBioana-lytical Research Group), or de novo sequencing (Pepseq; Waters-Micromass, Milford, MA, USA).
3. All proteins that score as significant database "hits" are verified by manual inspection of MS/MS data for the presence of interpretable y-ion and b-ion sequences. Proteins identified by one unique peptide must have MS/MS spectra that contain an interpretable y-ion series suitable for a peptide sequence tag search and "identity" Mascot peptides scores.
4. Search criteria include mass accuracy of maximum 50 ppm in MS mode, an MS/MS mass accuracy of 0.3 Da, and one possible missed trypsin cleavage site. Occasionally, the N-terminal peptide from proteins is found to be N-acetylated.
5. A non-redundant list of identified proteins is produced by removing any duplication through BLAST searching each protein versus a database of all the identified proteins.

4. Notes

1. When using the tritosome method to isolate lysosomes from mouse liver, the procedure is the same as for isolation of tritosomes from rat liver. One difference between the procedures is to remember to use a smaller tube for the discontinuous gradient as less original material is used. It is essential to maintain the visual separation between the different layers so that small adjustments in the volumes of the sucrose layers may be made.
2. In the Percoll method, the addition of 1 mM $CaCl_2$ to the PNS serves to swell light mitochondria, allowing their separation from lysosomes.

References

1. De Duve, C. (1973) The lysosome in retrospect. In *Lysosomes in Biology and Pathology* (Dingle, J. T. and Fell, H. B., eds.), pp. 3–40, North-Holland, Amsterdam.
2. Huynh, C., Roth, D., Ward, D. M., Kaplan, J. and Andrews, N. W. (2004) Defective lysosomal exocytosis and plasma membrane repair in Chediak-Higashi/beige cells. *Proc. Natl. Acad. Sci. U.S.A.* **101**, 16795–16800.

3. Reddy, A., Caler, E. V. and Andrews, N. W. (2001) Plasma membrane repair is mediated by Ca(2+)-regulated exocytosis of lysosomes. *Cell* **106**, 157–169.

4. Huizing, M., Anikster, Y. and Gahl, W. A. (2001) Hermansky-Pudlak syndrome and Chediak-Higashi syndrome: disorders of vesicle formation and trafficking. *Thromb. Haemost.* **86**, 233–245.

5. Perou, C. M., Leslie, C. M., Green, W., Li, L., McVey-Ward, D. and Kaplan, J. (1997) The Beige/Chediak-Higashi syndrome gene encodes a widely expressed cytosolic protein. *J. Biol. Chem.* **272**, 29790–29794.

6. Introne, W., Boissy, R. E. and Gahl, W. A. (1999) Clinical, molecular, and cell biological aspects of Chediak-Higashi syndrome. *Mol. Genet. Metab.* **68**, 283–303.

7. Bagshaw, R. D., Mahuran, D. J. and Callahan, J. W. (2005) A proteomics analysis of lysosomal integral membrane proteins reveals the diverse composition of the organelle. *Mol. Cell. Proteomics* **4**, 133–143.

8. Yamada, H., Hayashi, H. and Natori, Y. (1984) A simple procedure for the isolation of highly purified lysosomes from normal rat liver. *J. Biochem.* **95**, 1155–1160.

9. Nagle, D. L., Karim, M. A., Woolf, E. A., Holmgren, L., Bork, P., Misumi, D. J., et al. (1996) Identification and mutation analysis of the complete gene for Chediak-Higashi syndrome. *Nat. Genet.* **14**, 307–311.

10. Karim, M. A., Suzuki, K., Fukai, K., Oh, J., Nagle, D. L., Moore, K. J., et al. (2002) Apparent genotype-phenotype correlation in childhood, adolescent, and adult Chediak-Higashi syndrome. *Am. J. Med. Genet.* **108**, 16–22.

11. Cornillon, S., Dubois, A., Bruckert, F., Lefkir, Y., Marchetti, A., Benghezal, M., et al. (2002) Two members of the beige/CHS (BEACH) family are involved at different stages in the organization of the endocytic pathway in Dictyostelium. *J. Cell Sci.* **115**, 737–744.

12. Warnock, D. E., Fahy, E. and Taylor, S. W. (2004) Identification of protein associations in organelles, using mass spectrometry-based proteomics. *Mass Spectrom. Rev.* **23**, 259–280.

13. Brunet, S., Thibault, P., Gagnon, E., Kearney, P., Bergeron, J. J. M. and Desjardins, M. (2003) Organelle proteomics: looking at less to see more. *Trends Cell Biol.* **13**, 629–838.

14. Bell, A. W., Ward, M. A., Blackstock, W. P., Freeman, H. N. M., Choudary, J. S., Lewis, A. P., et al. (2001) Proteomics characterization of abundant Golgi membrane proteins. *J. Biol. Chem.* **276**, 5152–5165.

15. Leighton, F., Poole, B., Beaufay, H., Baudhuin, P., Coffey, J. W., Fowler, S., et al. (1968) The large-scale separation of peroxisomes, mitochondria, and lysosomes from the livers of rats injected with triton WR-1339. Improved isolation procedures, automated analysis, biochemical and morphological properties of fractions. *J. Cell Biol.* **37**, 482–513.

16. Bradford, M. (1976) A rapid and sensitive method for the quantitation of microgram quantities of protein utilizing the principle of protein-dye binding. *Anal. Biochem.* **72**, 248–254.

17. Mortz, E., Krogh, T. N., Vorum, H. and Gorg, A. (2001) Improved silver staining protocols for high sensitivity protein identification using matrix-assisted laser desorption/ionization-time of flight analysis. *Proteomics* **1**, 1359–1363.

17

Affinity Purification of Soluble Lysosomal Proteins for Mass Spectrometric Identification

Sylvie Kieffer-Jaquinod, Agnès Chapel, Jérôme Garin, and Agnès Journet

Summary

This chapter describes the process of production, purification, separation, and mass spectrometry identification of soluble lysosomal proteins. The rationale for purification of these proteins resides in their characteristic sugar, the mannose-6-phosphate (M6P), which allows an easy purification by affinity chromatography on immobilized M6P receptor (MPR). The secretion of M6P proteins (essentially soluble lysosomal proteins) from cells in culture is induced by adding a weak base in the culture medium. Secreted proteins are ammonium sulfate precipitated, dialyzed, and loaded onto the immobilized MPR column. After specific elution and collection of the M6P proteins, these are resolved by either bidimensional or monodimensional gel electrophoresis (designated as 2-DE or 1-DE, respectively). Mass spectrometry analysis is performed on spots excised from the 2-DE gel, or on discrete bands covering altogether the whole length of the 1-DE gel lane: these spots or bands are in-gel digested with trypsin and protein identification is obtained, thanks to peptide mass fingerprints [provided by analysis of the digests by matrix-assisted laser desorption ionization-mass spectrometry (MALDI-MS)] or peptide amino acid sequences (provided by analysis of the digests by the coupling between liquid chromatography and tandem mass spectrometry, LC-MS/MS).

Key Words: Lysosome; hydrolases; mannose-6-phosphate; bidimensional electrophoresis; SDS–PAGE; mass spectrometry; proteome.

1. Introduction

In the past 15 years, lysosomes and their content have been implicated in multiple specialized functions, besides their well-known role in basic cell

From: *Methods in Molecular Biology, vol. 432: Organelle Proteomics*
Edited by: D. Pflieger and J. Rossier © Humana Press, Totowa, NJ

catabolism. Although more than 50 hydrolases have been identified up to now *(1)*, new lysosomal hydrolases are still likely to be discovered. To search for these putative new lysosomal proteins, we and others performed proteomic studies of human soluble lysosomal proteins *(2–7)*. Such subcellular proteomic strategies, combined with improvements in mass spectrometry instruments, are a method of choice to discover proteins of very low abundance among total cellular proteins but enriched in particular subcellular fractions.

Mammalian soluble lysosomal proteins are specifically labeled by N-linked mannose-6-phosphate (M6P) *(8)*, which allows their lysosomal targeting through recognition by the M6P receptors (MPRs) *(9)*. Release of the M6P proteins in the late endolysosomal compartments occurs because of their luminal acidic pH. An increase in this pH after addition of a weak base to the extracellular medium leads to saturation of the MPRs and to consti- tutive secretion of the newly synthesized M6P proteins precursors *(10,11)*. In vitro, this M6P tag provides the rationale for affinity purification of M6P proteins. However, one has to keep in mind that a significant number of non-lysosomal proteins may be mannose-6-phosphorylated at low levels, as demonstrated very recently by Sleat et al. *(7)*. Therefore, ultimately, the subcel- lular localization of newly identified M6P proteins requires to be carefully assessed.

We describe here the whole process of M6P protein purification and identi- fication by mass spectrometry analysis. The secretion of M6P proteins from human monocytic U937 cells in culture is induced by adding NH_4Cl in the culture medium. Secreted proteins are ammonium sulfate precipitated, dialyzed, and loaded onto a column made of immobilized MPR. After specific M6P elution and collection of the M6P proteins, these are resolved by either bidimen- sional or monodimensional gel electrophoresis. Prior to mass spectrometry analysis, in-gel trypsin digestion is performed on the proteins present in spots excised from the 2-DE gel or in discrete bands covering altogether the entire length of a 1-DE gel lane. Proteins are finally identified through peptide mass fingerprints (MALDI-MS) or peptide sequences (LC-MS/MS). Following this protocol, the 2-DE separation of M6P proteins led to the identification of 26 proteins, including 19 well-known or supposed to be lysosomal proteins and 7 potentially new lysosomal proteins or contaminants. By 1-DE separation, we identified 93 proteins, among which about 40 are well-known or supposed to be lysosomal proteins. The relevance of the remaining proteins as potentially novel lysosomal proteins needs to be assessed according to criteria such as a demonstrated or predicted signal peptide and a known or potential function consistent with a role in the lysosome.

2. Materials

2.1. Preparation of the Immobilized MPR Affinity Chromatography Column

1. Soluble cation-independent MPR (sCI-MPR), stored in aliquots at −80°C (stable).
2. Coupling buffer: 0.1 M HEPES–NaOH, pH 7.5 (store at 4°C).
3. Amicon Ultra-15 (MWCO 100,000) (Millipore, Billerica, MA, USA).
4. Affigel-10 (Bio-Rad Hercules, CA, USA).
5. C16/20 column with A16 adaptors (GE Healthcare, Orsay, France).
6. Storage buffer: 50 mM imidazole-HCl, pH 6.5, 150 mM NaCl, 5 mM β-Na glycerophosphate, 5 mM EDTA, 0.05% (v/v) Triton X-100, and 0.02% (w/v) NaN$_3$. Prepare a 10× stock solution and store at 4°C.

2.2. Cell Culture and Induction of M6P Proteins Secretion

1. Human monocytic cell line U937 (American Type Culture Collection, Rockville, MD).
2. Culture flasks of 25 cm^2, 75 cm^2, and 175 cm^2 (BD Falcon, Franklin Lakes, NJ, USA).
3. RPMI 1640-GlutaMaxI, DMEM-GlutaMaxI, fetal calf serum (FCS), and PBS without Ca^{++} and without Mg^{++} (Invitrogen, Carlsbad, CA, USA).
4. Growth medium: RPMI 1640 supplemented with 10% (v/v) heat-inactivated FCS (56°C for 30 min) (stable for 1 month at 4°C).
5. Secretion medium: RPMI 1640-GlutaMAXI mixed with DMEM-GlutaMAXI in a 1:1 ratio, supplemented with 1% Nutridoma-SP (Roche, Indianapolis, IN, USA) (*see* **Note 1**) and 10 mM NH$_4$Cl. Prepare right before use.

2.3. Mannose-6-Phosphate Protein Purification

1. Buffer I: 50 mM imidazole-HCl, pH 6.5, 150 mM NaCl, 5 mM β-Na glycerophosphate, and 5 mM EDTA. Prepare a 10× stock and store at 4°C.
2. Buffer IT: buffer I supplemented with 0.05% (v/v) Triton X-100. Prepare a 10× stockand store at 4°C.
3. Dialysis tubing, MWCO 12 kDa (Sigma, St. Louis, MO, USA).
4. Protease inhibitors: Complete (with EDTA; Roche).
5. G6P elution buffer: 5 mM glucose-6-phosphate (G6P) in buffer IT. Store aliquots at −20°C.
6. M6P elution buffer: 5 mM mannose-6-phosphate in buffer IT. Store aliquots at −20°C.
7. Acetate buffer: 0.1 M sodium acetate-acetic acid, pH 4.5, and 0.5 M NaCl. Store at 4°C.
8. Tris buffer: 0.1 M Tris–HCl, pH 8.5, and 0.5 M NaCl. Store at 4°C.

2.4. Sodium Dodecyl Sulphate–Polyacrylamide Gel Electrophoresis Analysis

1. Molecular weight markers: SigmaMarker™ wide range (Sigma).
2. 5× reducing sample buffer: 125 mM Tris–HCl, pH 6.8, 20% (w/v) glycerol, 10% (w/v) sodium dodecyl sulphate (SDS), 25% (v/v) β-mercaptoethanol, and 0.1% (w/v) bromophenol blue. Store aliquots at –20°C.
3. XCell SureLock™ minicell, NuPAGE® Novex 4–12% Bis-Tris gels (10 wells, 1.5 mm thick), 20× NuPAGE® MES SDS running buffer, and SilverQuest™ silver staining kit (Invitrogen).
4. SE1200 Easy Breeze Air Gel Dryer and cellophane sheets (GE Healthcare).

2.5. Preparative Mannose-6-Phosphate Proteins Separation

1. 2-DE buffer: 8 M urea, 2 M thiourea, 4% (w/v) CHAPS, 40 mM dithioerythrol (DTE), 20 mM Tris–HCl, and 0.005% (w/v) bromophenol blue.
2. Bio-Lyte 3/10 Ampholyte (Bio-Rad).
3. Multiphor II electrophoresis system (GE Healthcare).
4. Immobiline DryStrip reswelling tray (GE Healthcare).
5. ReadyStrip IPG strips, pH 5–8, 17 cm long (Bio-Rad).
6. Isoelectric focusing (IEF) electrode strips (GE Healthcare).
7. Mineral oil (GE Healthcare).
8. Glycerol/urea buffer: 30% (w/v) glycerol, 6 M urea, 0.15 M Bis-Tris, 0.1 M HCl, and 2.5% (w/v) SDS. Heat at 37°C to dissolve the urea, and store as aliquots at –20°C.
9. Strip-reducing buffer: glycerol/urea buffer supplemented with 0.8% of dithiothreitol (DTT). Prepare freshly.
10. Strip-alkylating buffer: glycerol/urea buffer supplemented with 4% (w/v) iodoacetamide. Prepare freshly.
11. Protean™ II XL cell; Protean® II XL Ready Gel precast gels, Tris–HCl, and 10 or 12% acrylamide (Bio-Rad).
12. Running buffer: 10× Tris/glycine/SDS (Bio-Rad). Store at room temperature.
13. Sealing agarose: 1% (w/v) low-melting agarose in running buffer and 0.005% (w/v) bromophenol blue.
14. Bio-Safe™ Coomassie stain (Bio-Rad).

2.6. Preparation of the Samples for Mass Spectrometry Analysis

Solutions should be prepared freshly from commercial or stock solutions.

1. Bicarbonate-acetonitrile solution: 25 mM NH_4HCO_3/acetonitrile, 50/50 (v/v).
2. H_2O_2 [30% (w/v) in water; Fluka].
3. Trypsin, sequencing grade V5111 (Promega, Madison, WI, USA).

2.7. Mass Spectrometry Analysis

Solutions should be prepared freshly from commercial or stock solutions.

2.7.1. MALDI-MS

1. MALDI-TOF mass spectrometry is performed with an Autoflex (Bruker Daltonik, Bremen, Germany). Any other equivalent MALDI-TOF mass spectrometer may be used.
2. Matrix solution: α-cyano-4-hydroxycinnamic acid at 1 mg/mL in acetonitrile/water/TFA, 60/40/0.1 (v/v/v) (Aristar, VWR, Poole, England).

2.7.2. LC-MS/MS Analyses

1. Reverse-phase HPLC is performed with a CapLC system (Waters, Milford, MA, USA). Other HPLC systems capable of delivering a ~200 nL/min flow rate are suitable.
2. CapLC Autosampler (Waters).
3. A Q-TOF Ultima (Waters) is used in our laboratory. Other tandem mass spectrometers capable of automated acquisition of tandem mass spectra are suitable.
4. Columns: C18 PepMap, 75 μm ID, 15 cm length, containing silica beads of 3.5 μm diameter and 100 Å pore size (Dionex, Sunnyvale, CA, USA).
5. Guard columns: C18 PepMap, 300 μm ID, 5 cm length, containing silica beads of 5 μm diameter and 100 Å pore size (Dionex).
6. Buffer A: acetonitrile/water/formic acid, 2/98/0.1 (v/v/v).
7. Buffer B: acetonitrile/water/formic acid, 80/20/0.08 (v/v/v).
8. Buffer C: acetonitrile/water/formic acid, 5/95/0.2 (v/v/v).

3. Methods

Because of space limitations and the number of methods required to perform the whole process, commercial options will be suggested when available.

3.1. Preparation of the Immobilized MPR Affinity Chromatography Column

Whenever possible, proteins should be kept at 4°C, either on ice or in a cold room.

1. Transfer bovine sCI-MPR (*see* **Note 2**) in coupling buffer by ultrafiltration over an Amicon Ultra-15 device (MWCO 100,000) (*see* **Note 3**) and concentrate the protein to a final concentration of 2–5 mg/mL.
2. Couple the sCI-MPR to the Affigel-10 matrix (*see* **Note 4**) at a ratio of 2–5 mg of sCI-MPR/mL of matrix according to the manufacturer's protocol. Use ~20 mg of sCI-MPR and 10 mL of the matrix slurry (about 6 mL of drained matrix).

3. Degas the sCI-MPR matrix under a slight vacuum. Pour it gently into the C16/20 column. Avoid trapping air. With a peristaltic pump, equilibrate the column with 10 bed volumes of storage buffer at a rate of 40 mL/h (*see* **Notes 5** and **6**).

3.2. Culture of U937 Cells and Secretion of M6P Proteins

All manipulations implying cell culture and growth must be carried out under sterile conditions using sterile media, reagents and containers.

1. Grow human monocytic U937 cells (*see* **Note 7**) in suspension in culture flasks in 10–15 mL of growth medium per 25 cm^2 under 5% CO_2 at 37°C. Subculture them twice a week by a 1:6 dilution in fresh medium (*see* **Note 8**).
2. For secretion of M6P proteins, culture the cells to a density of more than 10^6 cells/mL. Collect them by centrifugation at 500 g for 5 min at 20°C, wash the pellet in prewarmed PBS (one-third of the initial volume), and repeat the centrifugation. Rinse the flasks that contained the cells with PBS to get rid of any remaining medium and FCS. Resuspend the cells at a density of 1×10^6 cells/mL in secretion medium, transfer the suspension back to the rinsed flasks (*see* **Note 9**), and let grow for a further 24 h. For proteomic analysis of the M6P proteins separated on a 2-DE gel, prepare 2×10^9 cells in 175 cm^2 flasks (100 mL per flask) (*see* **Note 10**).

3.3. Mannose-6-Phosphate Proteins Purification

From this point, sterility is not required any more. Apart from the first cell centrifugation, all steps of the purification and protein manipulation prior to separation should be performed at 4°C.

1. After the secretion step, centrifuge the cell suspension at 500 g for 5 min at 20°C. Transfer the supernatant in new tubes and centrifuge it again at 3200 g for 10 min at 4°C to get rid of any remaining cells or cell fragments. Add progressively ammonium sulfate under gentle agitation to a final concentration of 0.5 mg/mL. Allow protein precipitation to occur during the night with slow stirring. Pellet the precipitate by centrifugation at 10,000 g for 40 min (*see* **Note 11**).
2. Dissolve the ammonium sulfate pellet in cold buffer I (~2 mL/100 mL of the initial culture supernatant) and transfer the sample in a dialysis tubing (MWCO 12 kDa) (*see* **Note 12**). Dialyze extensively, with gentle stirring, the protein solution against cold buffer I in a large beaker containing at least 50 times the sample volume. Change the bath at least twice (after 2 h and after a night) (*see* **Note 13**). Transfer the retentate to a centrifugation tube. Add 0.05% Triton X-100 (*see* **Note 14**) and protease inhibitors. Centrifuge at 200,000 g for 30 min for clearing (*see* **Note 15**).
3. Equilibrate the affinity column by 10 bed volumes of buffer IT at 50 mL/h.

4. Load the protein sample onto the column at 25 mL/h. Repeat once with the flow through.
5. Wash the column extensively with buffer IT (50 bed volumes) at 25 mL/h.
6. Elute the non-specifically bound proteins with 2 bed volumes of G6P elution buffer at 6 mL/h (*see* **Note 16**).
7. At the same time, start collecting the eluate as 500 µL fractions.
8. Elute specifically the M6P proteins with 1.5 bed volume of M6P elution buffer at 6 mL/h.
9. After elution, wash the column with 10 bed volumes of buffer IT. Regenerate the matrix by three series of alternate washes of acetate or Tris buffers (3 bed volumes each) at 25 mL/h.
10. Equilibrate the column with 10 bed volumes of storage buffer and store at 4°C.

3.4. SDS–PAGE Analysis

The analysis of the eluted material is carried out by SDS–polyacrylamide gel electrophoresis (PAGE) on every second fraction. These instructions assume the use of an XCell SureLock™ minicell and NuPAGE® Novex 4–12% Bis-Tris gels (*see* **Note 17**). Follow the manufacturer's instructions.

1. Mix 20 µL (1/25) of each fraction to be analyzed with 10 µL of 5× reducing sample buffer (*see* **Note 18**). Heat for 5 min at 90°C in a water bath to denature the proteins.
2. Load the samples and molecular weight markers on the gel.
3. For electrophoresis, use 1× MES SDS running buffer. Run the gel at 200 V until the blue dye reaches the bottom of the gel (about 45 min).
4. Once the run is achieved, dissociate the electrophoresis unit and separate carefully the plates to remove the gel. Cut one angle of it to keep track of its orientation.
5. Silver stain the gel to visualize the proteins according to the manufacturer's protocol.
6. Incubate the gel in 2% glycerol for 30 min (*see* **Note 19**) and air dry it between stretched cellophane sheets for long storage.

3.5. Preparative Mannose-6-Phosphate Proteins Separation

1. Pool the fractions enriched in specifically eluted proteins (about 10 fractions, ~200 µg of proteins for 2×10^9 cells) in 50-mL Falcon tubes. Add 4 volumes of acetone and precipitate the proteins by overnight incubation at –20°C.
2. Centrifuge for 15 min at 4°C and 8000 g. A solid white pellet should be visible. Discard the supernatant. To get rid of the precipitated salts, rinse and dissociate the pellet by adding 3 mL of cold 10% TCA and pipetting up and down repeatedly. Transfer this suspension in two 2-mL microfuge tubes. Centrifuge for 15 min at 4°C and maximum speed (*see* **Note 20**). Wash the invisible pellet with 900 µL of cold (–20°C) methanol/900 µL of NH_3 vapors (*see* **Note 21**). Vortex. Centrifuge. Wash with 900 µL of cold (–20°C) ether/900 µL of NH_3 vapors to eliminate the

previous solvent and achieve neutralization of the sample pH. Vortex. Centrifuge. Wash with 900 μL of cold (–20°C) ether. Vortex. Centrifuge. Air dry the pellet for a few minutes, until all liquid has disappeared. Proteins can be kept at –20°C for several years, either in acetone or as dry precipitates.

3.5.1. Bidimensional Gel Electrophoresis for Mass Spectrometry Analysis

1. Solubilize the M6P proteins from each tube in 250 μL of 2-DE buffer + 2.5 μL (0.2% final) of Bio-Lyte 3/10 Ampholyte. Agitate for at least 1 h at room temperature (see **Note 22**). Pool both samples and centrifuge at 14,000 g for 15 min at room temperature.
2. The first dimension (IEF) will be performed on a Multiphor II electrophoretic system, with ReadyStrip IPG strips (17 cm long, pH 5–8).
3. For passive rehydration of a strip, load one chamber of the reswelling tray with the 500 μL sample. Take off the plastic film that protects the IPG strip and lay the strip, gel downwards, in contact with the sample. Remove the bubbles by slightly pressing on the strip back with a pipette tip. Check that the strip slides freely in the chamber by pushing it gently with the tip. Wait for 10 min before covering it with mineral oil (starting from its extremities) to avoid drying. Close the reswelling tray and allow the strips to rehydrate overnight at room temperature.
4. Connect a thermostatic circulator to the Multiphor device. Switch it on and set up the temperature at 20°C (see **Note 23**) 15 min in advance. Place the tray and electrode holder on the cooling plate. Align the drained IPG strip, gel upwards. Pay attention to its orientation (acidic extremity toward the anode). Place a piece of paper on each extremity (IEF electrode strip), 2 cm long, humidified with deionized water (see **Note 24**). This paper needs to cover a few millimeter of the gel only. Place both electrodes on the outward extremities of the paper pieces. Cover with mineral oil. Program the power supply as follows: 150 V, 30 min; 300 V, 1.5 h; 700 V, 1.5 h; 1000 V, 1.5 h; 1500 V, 1.5 h; 2000 V, 2 h; and 3000 V, 20 h (see **Note 25**). A total of 65,000–70,000 Vh is suitable. Take care to restrict the amperage at 50 μA/strip to avoid excessive heating and subsequent precipitation of the proteins.
5. After IEF, incubate the strips at room temperature for 15 min in strip-reducing buffer followed by 15 min in strip-alkylating buffer for reduction and alkylation of the proteins (see **Note 26**). Cysteines should become carbamidomethylated by this treatment.
6. Perform the second dimension (SDS–PAGE) in a Protean® II XL Cell on Protean II® Ready Gel precast gels (10 or 12%) (see **Note 27**). Pour melted sealing agarose (kept at 60°C) in the 2-DE well. Insert immediately the IPG strip (always in the same orientation) and push it down gently against the gel. Avoid trapping bubbles under the strip. Assemble the electrophoresis unit. Pour 1× Tris/glycine/SDS running buffer in the lower and upper tanks. Run the gel for 1 h at 25 V, then

overnight at 90 V. Once the bromophenol blue has reached the bottom of the gel, stop the run, separate the plates, remove the gel, and proceed to **step 3.5.3.**

3.5.2. Monodimensional Separation of the Proteins for Mass Spectrometry Analysis

Starting from washed and dried acetone precipitates, resolubilize M6P proteins corresponding to 1.5×10^8 cells in 30 μL of 2× reducing sample buffer by pipetting up and down. Heat at 90°C for 15 min for complete resolubilization and denaturation of the proteins. Load the sample onto a NuPAGE™ Novex 4–12% Bis-Tris gel and carry out the migration for a distance of 1.5 cm at 200 V (about 10 min). Separate the plates, remove the gel, and proceed to **step 3.5.3.**

3.5.3. Colloidal Blue Staining

Stain the gel with Bio-Safe Coomassie Stain, according to the manufacturer's protocol. **Figure 1** displays a 2-DE M6P proteins separation, achieved on a home-made IPG strip (4–8 pH gradient) (*see* **Note 28**).

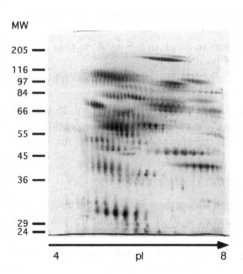

Fig. 1. Typical bidimensional electrophoresis (2-DE) of U937 soluble lysosomal proteins. Mannose-6-phosphate (M6P) proteins were purified from 2.5×10^9 U937 cells, separated on a linear pH gradient (4–8) followed by a 10% SDS–PAGE, and stained with colloidal blue (*see* **Note 28**). Soluble lysosomal proteins display a characteristic string profile. This figure was adapted from (*3*) with permission of Wiley-VCH.

3.6. Preparation of the Samples for Mass Spectrometry Analysis

3.6.1. Preparation of the Gel Spots or Bands Before Digestion

(see *Note 29*)

1. Excise the bands (*see* **Note 30**) or spots from the gel with respectively a scalpel or a pipette tip shortened with a scalpel to adjust its extremity to the spot size. Cut each band in two pieces only to avoid further losses during washings.
2. Wash the gel pieces with 200 μL of bicarbonate-acetonitrile solution for 15 min with shaking. Remove excess liquid.
3. Wash the gel pieces with 200 μL of 25 mM NH_4HCO_3 for 15 min with shaking. Remove excess liquid.
4. Wash the gel pieces with 200 μL of ultrapure HPLC grade water for 15 min with shaking. Remove excess liquid.
5. Add 200 μL of ultrapure HPLC grade acetonitrile to shrink the gel pieces for 15 min with shaking. Remove excess liquid. Dry the gel pieces in a vacuum centrifuge (*see* **Notes 31** and **32**).
6. Add 200 μL of H_2O_2 7% (v/v) for 15 min (*see* **Note 33**). Discard excess liquid.
7. Repeat steps 4 and 5 (*see* **Note 34**).

3.6.2. "In-Gel" Trypsin Digestion of the Proteins Contained in the Gel Spots/Bands

1. Prepare a 0.4 μg/μL stock solution of trypsin in 50 mM acetic acid (*see* **Note 35**). Store aliquots at −20°C.
2. Rehydrate gel pieces in 25 μL of a 6 ng/μL trypsin solution in 25 mM NH_4HCO_3 at 4°C for 15 min (*see* **Note 36**).
3. Add 30 μL of 25 mM NH_4HCO_3 for complete rehydration.
4. Shake the samples at 37°C for 30 min.
5. If necessary, cover gel pieces with additional 25 mM NH_4HCO_3 to keep them wet during enzymatic digestion and to favor peptide exit to the solution.
6. Leave samples in a thermostatic oven at 37°C for 3–5 h.

3.6.3. Extraction of the Peptides from the Gel

1. Withdraw the digest supernatant and keep it in a separate 0.5-mL Eppendorf tube.
2. Incubate the gel pieces in 30 μL of acetonitrile/water (50/50, v/v) for 15 min with shaking.
3. Spin gel pieces down. Collect the supernatant and pool it with the supernatant from **step 1**.
4. Incubate the gel pieces in 30 μL of an aqueous solution containing 5% formic acid for 15 min with shaking.
5. Spin gel pieces down. Collect the supernatant and pool it with the two previous ones.
6. Incubate the gel pieces in 30 μL of acetonitrile for 15 min with shaking.

7. Spin gel pieces down and collect the supernatant to pool it with the three previous ones.

8. Evaporate the pooled supernatants in a vacuum centrifuge.

3.7. Mass Spectrometry Analysis

3.7.1. MALDI-MS (see Note 37)

1. For MALDI-TOF analyses, take 0.5 μL aliquot out of ~30 μL of the peptide mixture (right after digestion, without peptide extraction) and mix with 0.5 μL of the matrix solution on the target plate. Air dry the sample and rinse with 2 μL of 0.1% TFA in water to remove the salts.

2. Generate a peptide mass fingerprint with a MALDI-TOF mass spectrometer. With the Autoflex, work in reflector/delayed extraction mode over a mass range of 0–4200 Da. For each sample, make an automatic acquisition (FlexControl software); use a spiral pattern with an average of 200 laser shots per position after external calibration.

3. Make a database search in the Swissprot-Trembl database with each peptide mass fingerprint using MH+ monoisotopic peaks. Allow cysteine carbamidomethylation, protein N-terminal acetylation and methionine oxidation as variable modifications, as well as one tryptic missed cleavage. The tolerance limit for the peptide masses will depend on the instrument specifications. With the Autoflex instrument, we used a precision of 50 ppm. If working with Mascot (http://www.matrixscience.com/cgi/search_form.pl?FORMVER=2& SEARCH=PMF), accept automatically proteins with a score higher than 70 and a sequence coverage of more than 20%. If one of these parameters is not respected, manual inspection of the data can be helpful to improve the coverage. If no or unclear identification is obtained after a MALDI-TOF analysis, perform an LC-MS/MS analysis of the extracted sample.

3.7.2. LC-MS/MS Analyses

1. Resuspend the dried sample obtained in **step 8** of **Subheading 3.6.3** in 5 μL of 5% TFA in water. Sonicate and add 10 μL of buffer C (*see* **Note 38**). Centrifuge the sample and take 10 μL from the upper part to introduce into a vial adapted to the autosampler.

2. Set up the injection system for an injection of 6.4 μL in the microliter pickup mode.

3. Set up the LC method in the MassLynx 4.0 CapLC diagnostic window (*see* **Note 39**).

4. Set up the MS/MS method for the Q-TOF Ultima instrument (*see* **Note 40**).

5. Launch the LC-MS/MS analysis.

6. Process the raw data files from the ProteinLynx set up window of MassLynx to generate a peak list file (*see* **Note 41**).

7. Make a database search using the obtained pkl. File with the Mascot search engine. Use SpTrembl as the database, human as the taxonomy, and consider N-terminal acetylation of the proteins, oxidation or sulphone of methionine, and cysteic acid as variable modifications for 1D bands sample, because oxygen peroxide was used during band treatments and cysteine was not alkylated with iodoacetamide. Enzyme is specified as Trypsin/P; to accept K–P cleavages, two tryptic missed cleavages are allowed; 2+ and 3+ are the possible charge states, monoisotopic MH+ masses are given; and a tolerance of ±0.3 Da is specified on the mass measurement of precursors (MS) and fragments (MS/MS).

8. Proteins are validated using these criteria: proteins that are identified with at least two peptides both showing a score higher than the identity significant threshold given by Mascot are validated without any manual validation. For proteins identified by only one peptide having a score higher than this significant threshold, the peptide sequence is checked according to the corresponding spectrum. Peptides with scores higher than 20 and lower than the significant threshold are systematically checked and/or interpreted manually to confirm or cancel Mascot suggestion (*see* **Note 42**). In manual validation, we consider an MS/MS spectrum to be reliable when at least three consecutive y or b ions can be read with an S/N ratio superior to 3 (thus providing a sequence tag of two amino acids).

4. Notes

1. 1. Nutridoma-SP is a serum substitute, which contains very low amounts of protein and thus makes purification easier.
2. sCI-MPR was purified from FCS according to Causin et al. *(12)* and Journet et al. *(2)*.
3. Any kind of similar concentration device is suitable.
4. Affigel 15 (Bio-Rad) has also been used *(4,5)*.
5. This column can be stored for years at 4°C in the presence of 0.02% NaN_3, which prevents bacterial growth.
6. The flow rates depend on the type of matrix and column.
7. The starting sample may be any other kind of cell or brain tissue that is specifically enriched in M6P proteins *(13)*. When using cells deficient in both MPRs *(14,15)*, M6P proteins are spontaneously secreted in the medium, rendering NH_4Cl treatment useless.
8. In these conditions, U937 cells commonly reach a density of $1.2–1.5 \times 10^6$ cells/mL after 3 days.
9. The secretion step is performed in the initial flasks to avoid any loss of proteins by adsorption on "clean" flask walls.
10. For proteomic analysis after 1-DE separation, proteins from 1.5×10^8 cells are a sufficient amount. The various steps of the protocol should be consequently adapted.

11. This step may be repeated once, allowing the additional recovery of about 5–10% of the precipitate.
12. Be aware that the volume of the sample will increase to about 1.5 times the initial volume during dialysis.
13. The disappearance of the orange-red color of the sample, due to the pH indicator of the culture medium, provides some indication on the dialysis progress.
14. As undiluted Triton X-100 precipitates when added to a cold solution, use a stock solution of 5% Triton X-100 in buffer I.
15. A small brown pellet is always observed after the centrifugation step, but it does not contain proteins, as checked by SDS–PAGE analysis.
16. This elution might be avoided, because almost no proteins are eluted as checked by SDS–PAGE analysis.
17. Any kind of minigel, home-made or commercial, is suitable.
18. The final concentration in Laemmli buffer is 1.6×, providing an excess of SDS over the 0.05% Triton X-100 present in the sample. Otherwise, Triton X-100 might interfere with protein migration.
19. The glycerol bath avoids cracking of the gel during drying.
20. As TCA washes the salts out of the pellet, only traces along the tube wall may remain visible. Centrifuge the tubes with their hinge upwards to keep track of the position of the pellet.
21. Take the 900 μL of NH_3 vapors by pipetting gas above a solution at 30% (w/v) of NH_3. NH_3 is used to neutralize the remaining TCA.
22. The solubilization step is essential for the quality of the isoelectrofocalization, and a total time of 1 h is a minimum. The subsequent centrifugation eliminates any remaining precipitate.
23. If the temperature drops below this limit, urea will precipitate.
24. During IEF, salts migrate toward the electrodes and accumulate on the paper pieces. If those are too short, the accumulated salts may interfere with migration.
25. This is an example. The principle is to increase progressively the voltage.
26. IPG strips can be stored at –20°C either before this step, wrapped in Saran plastic film without previous draining, or after treatment in the alkylation buffer, in hybridization glass tubes containing 25 mL of glycerol/urea buffer.
27. Any kind of gel, home-made or commercial, is suitable.
28. This home-made 4–8 pH gradient was prepared according to reference *(16)*. Because such a gradient is not commercially available, we recommend using of a 5–8 pH gradient, which will result in further horizontal spreading of the spot strings.
29. The procedure described for destaining applies to bands or spots excised from both Coomassie blue- or silver nitrate-stained gels. Caution must be taken not to use the cross-linking reagent formaldehyde for silver staining of gels *(17)*.
30. To obtain protein identification for the whole sample, cut out bands of ∼1–2 mm from the top to the bottom of the gel. Otherwise, just cut out discrete bands of interest.
31. Repeat steps 2–4 if staining is intense.

32. Stop the washes at this step if working with 2-DE spots stained with Coomassie blue.

33. Treatment with H_2O_2 allows both gel destaining and oxidation of free (not alkylated) cysteine residues from 1D gels. During this treatment, methionine can also be oxidized. Consequently, when running database searches, cysteic acid, oxidized or sulphone methionine must be specified as variable modifications.

34. If trypsin digestion is not performed right away, gel pieces must be stored at $-20°C$.

35. If the trypsin is not from Promega, refer to the manufacturer's instructions.

36. The ratio enzyme/proteins must be between 1/50 and 1/20. When adding 25 µL of a 6 ng/µL trypsin solution, we consider that 3–5 µg of proteins are present in the gel band. The concentration of trypsin must be adapted to the protein amount and gel volume.

37. MALDI-MS analysis is carried out on samples from 2-DE gels only.

38. It is sometimes necessary to dilute the sample, in order not to saturate the LC-MS system. Refer to your instrument specifications to check saturation issues.

39. Before sample loading, the HPLC system must be adjusted to a nanoflow rate adapted to the used column. The master flow rate of the CapLC system is set at 4–5 µL/min and a split is set upstream of the column to get a final 200–300 nL/min flow rate through the nano-LC column. Optionally, a guard column is set close to the injection system to concentrate, filter, and desalt the sample before the column. A 20 µL/min flow rate of buffer C is run on the guard column to load the sample. A typical chromatographic separation uses the following gradients: (1) from 10 to 40% buffer B in 40 min, (2) from 40% to 90% buffer B in 5 min, (3) from 90% to 10% buffer B in 10 min, and (4) 10% buffer B for 5 min for further column equilibration. The remaining percentage of the elution solvent is made of buffer A. For maintenance purposes and checking of the LC-system (flow rate, pressure, and leakages), refer to the manufacturer's instructions.

40. Prior to the mass spectrometry analysis, the calibration and the sensitivity of the instrument used for MS/MS analysis must be checked according to the manufacturer's instructions. For automatic LC-MS/MS analysis, the Q-TOF Ultima instrument is run in data-dependent mode (DDA) with the following parameters: 1 s scan time and 0.1 s interscan delay for MS survey scans; 400–1600 and 50–2000 m/z mass ranges for the survey and the MS/MS scans, respectively; selection for MS/MS of the five ions detected with highest intensities in the latest MS scan; MS/MS to MS switch after 5 s for each ion with a switchback threshold of 10 counts/scan; and include charge states 2, 3, and 4 with the corresponding optimized collision energy profiles. Optionally, a list of the m/z corresponding to the most intense peptides of trypsin autolysis can be set as an exclusion list.

41. To generate a peak list file with ProteinLynx, we use the following parameters: Electrospray instrument type; QA threshold: 10; Mass Measure option with no background subtract; Savitzky Golay Smooth window: 3; number of smooths:

2; 4 channels; and 80% centroid top. Alternatively, other processing tools can be used such as Mascot Distiller with equivalent parameters.

42. Depending on the data quality, the MS instrument used, the processing parameters and tools, and the database searching tool, the criteria for validating protein identifications can be different and relies on the expertise of the scientists controlling the results.

Acknowledgments

This work was supported by the Commissariat à l'Energie Atomique, the Institut National de la Santé et de la Recherche Médicale, and the Réseau des Génopoles.

References

1. Holtzman, E. (1989) Lysosomes. In *Cellular Organelles*. Plenum Press, New York.
2. Journet, A., Chapel, A., Kieffer, S., Louwagie, M., Luche, S., and Garin, J. (2000) Towards a human repertoire of monocytic lysosomal proteins. *Electrophoresis* **21**, 3411–3419.
3. Journet, A., Chapel, A., Kieffer, S., Roux, F., and Garin, J. (2002) Proteomic analysis of human lysosomes: application to monocytic and breast cancer cells. *Proteomics* **2**, 1026–1040.
4. Sleat, D. E., Lackland, H., Wang, Y., Sohar, I., Xiao, G., Li, H., et al. (2005) The human brain mannose 6-phosphate glycoproteome: a complex mixture composed of multiple isoforms of many soluble lysosomal proteins. *Proteomics* **5**, 1520–1532.
5. Kollmann, K., Mutenda, K. E., Balleininger, M., Eckermann, E., von Figura, K., Schmidt, B., et al. (2005) Identification of novel lysosomal matrix proteins by proteome analysis. *Proteomics* **5**, 3966–3978.
6. Czupalla, C., Mansukoski, H., Riedl, T., Thiel, D., Krause, E., and Hoflack, B. (2006) Proteomic analysis of lysosomal acid hydrolases secreted by osteoclasts: implications for lytic enzyme transport and bone metabolism. *Mol. Cell. Proteomics* **5**, 134–143.
7. Sleat, D. E., Wang, Y., Sohar, I., Lackland, H., Li, Y., Li, H., et al. (2006) Identification and validation of mannose 6-phosphate glycoproteins in human plasma reveals a wide range of lysosomal and non-lysosomal proteins. *Mol. Cell. Proteomics* **5**(10), 1942–1956.
8. von Figura, K. and Hasilik, A. (1986) Lysosomal enzymes and their receptors. *Annu. Rev. Biochem.* **55**, 167–193.
9. Hille-Rehfeld, A. (1995) Mannose 6-phosphate receptors in sorting and transport of lysosomal enzymes. *Biochim. Biophys. Acta* **1241**, 177–194.
10. Hasilik, A. and Neufeld, E. F. (1980) Biosynthesis of lysosomal enzymes in fibroblasts. Synthesis as precursors of higher molecular weight. *J. Biol. Chem.* **255**, 4937–4945.

11. Gonzalez-Noriega, A., Grubb, J. H., Talkad, V., and Sly, W. S. (1980) Chloroquine inhibits lysosomal enzyme pinocytosis and enhances lysosomal enzyme secretion by impairing receptor recycling. *J. Cell Biol.* **85**, 839–852.

12. Causin, C., Waheed, A., Braulke, T., Junghans, U., Maly, P., Humbel, R. E., et al. (1988) Mannose 6-phosphate/insulin-like growth factor II-binding proteins in human serum and urine. Their relation to the mannose 6- phosphate/insulin-like growth factor II receptor. *Biochem. J.* **252**, 795–799.

13. Sleat, D. E., Sohar, I., Lackland, H., Majercak, J., and Lobel, P. (1996) Rat brain contains high levels of mannose-6-phosphorylated glycoproteins including lysosomal enzymes and palmitoyl-protein thioesterase, an enzyme implicated in infantile neuronal lipofuscinosis. *J. Biol. Chem.* **271**, 19191–19198.

14. Kasper, D., Dittmer, F., von Figura, K., and Pohlmann, R. (1996) Neither type of mannose 6-phosphate receptor is sufficient for targeting of lysosomal enzymes along intracellular routes. *J. Cell Biol.* **134**, 615–623.

15. Ludwig, T., Munier-Lehmann, H., Bauer, U., Hollinshead, M., Ovitt, C., Lobel, P., et al. (1994) Differential sorting of lysosomal enzymes in mannose 6-phosphate receptor-deficient fibroblasts. *EMBO J.* **13**, 3430–3437.

16. Rabilloud, T., Valette, C., and Lawrence, J. J. (1994) Sample application by in-gel rehydration improves the resolution of two-dimensional electrophoresis with immobilized pH gradients in the first dimension. *Electrophoresis* **15**, 1552–1558.

17. Richert, S., Luche, S., Chevallet, M., Van Dorsselaer, A., Leize-Wagner, E., and Rabilloud, T. (2004) About the mechanism of interference of silver staining with peptide mass spectrometry. *Proteomics* **4** (4), 909–916.

18

Purification and Proteomic Analysis of Synaptic Vesicles

Holly D. Cox and Charles M. Thompson

Summary

Synaptic vesicles store and subsequently release neurotransmitters into the synaptic cleft thereby regulating chemical neurotransmission in the brain. Proteins present in synaptic vesicles vary greatly in structure and function and have been identified primarily by genetic knock-out analysis in *C. elegans, Drosophila,* and mice *(1,2)*. However, knock-out methods are not useful for the identification of proteins when a detectable phenotype is not created. Further, certain knocked-out proteins have function(s) that could be compensated for by another protein or cause a lethal phenotype when deleted. Additionally, some transporters and enzymes that appear to copurify with synaptic vesicles have not been characterized and confirmed *(3–5)*. We have determined the proteins associated with purified synaptic vesicles using 2-D polyacrylamide gel electrophoresis (PAGE) protein separation followed by identification by mass spectrometry *(6)*. Some of the new proteins identified were evaluated by western blot and confocal immunofluorescence analysis.

Key Words: Synaptic vesicles; membrane proteins; 2-D PAGE; 16-BAC/SDS–PAGE; mass spectrometry.

1. Introduction

To identify proteins strictly associated with synaptic vesicles, a new method of vesicle isolation and purification was developed, which produces large amounts of vesicles that are of high purity. Proteins derived from the purified vesicles can be solubilized in detergents and run on 2-D polyacrylamide gel electrophoresis (PAGE) gels. After in-gel digestion with trypsin, proteins in each gel spot can be identified by mass spectrometry. The presence and identity

From: *Methods in Molecular Biology, vol. 432: Organelle Proteomics*
Edited by: D. Pflieger and J. Rossier © Humana Press, Totowa, NJ

of new proteins on synaptic vesicles can be confirmed by western blot and confocal microscopy.

2. Materials

2.1. Synaptic Vesicle Purification

1. Solution A: 0.32 M sucrose, 1 mM sodium bicarbonate buffer, pH 7.2 (adjusted with a concentrated HCl solution), 1 mM magnesium acetate, and 0.5 mM calcium acetate.
2. Solution B: 6 mM Tris-maleate, pH 8.1.
3. Solution C: 0.32 M sucrose and 10 mM HEPES–KOH, pH 7.4.
4. 100 mM PMSF: 0.174 g phenylmethylsulfonyl fluoride (PMSF) in 10 mL of isopropanol. Store in single-use aliquots at –80°C (toxic, wear gloves when handling solutions containing PMSF).
5. 55-mL Wheaton mortar with Teflon pestle, drill that can hold Teflon pestle, and 5-mL Wheaton mortar with Teflon pestle.
6. Dounce homogenizer with tight-fitting pestle.
7. Sucrose solutions: 0.2 M, 0.4 M, 0.6 M, and 0.8 M sucrose in 10 mM HEPES–KOH, pH 7.4.
8. Protease Inhibitor Cocktail Mini-Pellets, ethylenediaminetetraacetic acid (EDTA)-free (# 1 836 170; Roche Applied Science).
9. Capless, 30-mL polyallomer ultracentrifuge tubes (326823; Beckman Instruments).
10. 10× HEPES-buffered saline: 100 mM HEPES–KOH, pH 7.4, and 1.5 M NaCl.
11. 5 and 25% buffered glycerol: 5 or 25% (w/v) glycerol in 10 mM HEPES–KOH, pH 7.4, and 150 mM NaCl.
12. Hoefer 50-mL gradient maker (Amersham Life Sciences, Piscataway, NJ).
13. Mini-peristaltic pump with variable flow (Fisher).

2.2. Western Blot of Glycerol Gradient Fractions

1. 100% TCA: 100 g of trichloroacetic acid brought to a final volume of 100 mL with nanopure water.
2. Ethanol/ether, 1/1 (v/v).
3. 10% acrylamide Tris–HCl sodium dodecyl sulphate (SDS)–PAGE minigels (Bio-Rad).
4. 10× Tris/glycine/SDS buffer (Bio-Rad).
5. Laemmli sample buffer (Bio-Rad).
6. Transfer buffer: 100 mL of 10× Tris/glycine buffer, pH 8.3 (Bio-Rad), 700 mL of nanopure water, and 200 mL of methanol.
7. PBS-T: 11.9 mM phosphate buffer, pH 7.4, 137 mM NaCl, 2.7 mM KCl (sold as 10× solution, Fisher) + 0.1% (w/v) Tween-20.
8. Blocking Solution: PBS-T + 5% (w/v) bovine serum albumin (Calbiochem).

9. Primary antibody: anti-synaptophysin monoclonal antibody (Synaptic Systems, Göttingen, Germany).
10. Secondary antibody: HRP-conjugated anti-mouse IgG antibody (Cell Signaling Systems, Beverly, MA).
11. Enhanced chemiluminescence reagents (Amersham Biosciences).
12. Imager for chemiluminescence: Versadoc Imager (Bio-Rad).
13. Imaging software: Quantity One (Bio-Rad).

2.3. Concentration and Collection of Vesicles

1. 0.5-mL glass Wheaton mortar and Teflon pestle.
2. 25 mM Tris–HCl, pH 8.0.
3. Ultracentrifuge equipped with a fixed angle Ti-70 rotor.

2.4. Assay for Endoplasmic Reticulum Contamination

1. 96-Well microtiter plate.
2. Assay buffer: 50 mM KH_2PO_4, pH 7.7, and 0.1 mM EDTA (refrigerated at 4°C).
3. NADPH solution: 2 mg/mL NADPH (N-1630 Sigma) in assay buffer. Wrap in foil. Aliquot in single use amounts and freeze at –20°C.
4. Cytochrome c solution: 25 mg/mL cytochrome c (C-3131, bovine heart; Sigma) in assay buffer. Wrap in foil. Aliquot in single use amounts and freeze at –20°C.
5. Positive control: dilute 500 µL of the synaptic vesicle containing 0.2 M/0.4 M sucrose interface layer with 4.5 mL of 1× HEPES-buffered saline or with a larger volume to fill an ultracentrifuge tube. Pellet the membranes at 200,000 g for 1 h. Resuspend the pellet in 50 µL of 25 mM Tris–HCl, pH 8.0.
6. Blank: 25 mM Tris–HCl, pH 8.0.

2.5. Two-Dimensional IEF/SDS–PAGE

1. 10% (w/v) ammonium persulfate (APS) in water.
2. 1.5 M Tris–HCl, pH 8.8.
3. Acrylamide solution: 30% acrylamide and 0.8% Bis-acrylamide. Toxic, wear gloves when handling acrylamide.
4. 17-cm isoelectric focusing (IEF) strip, pH 3–10 NL (Bio-Rad).
5. IEF rehydration tray (Bio-Rad).
6. Protean IEF apparatus (Bio-Rad).
7. IEF sample buffer: 2 M thiourea, 7 M urea, 2% (w/v) CHAPS, 2% (w/v) ASB-14, 5 mM tributylphosphine (TBP), 0.2% (w/v) ampholytes (pH 3–10) (Bio-Rad), and 25 mM Tris–HCl, pH 8.8. Prepare solution fresh immediately before use. Do not heat solution (*see* **Note 1**).
8. IEF strip reducing solution: 6 M urea, 2% (w/v) SDS, 20 mg/mL dithiothreitol (DTT), 375 mM Tris–HCl, pH 8.8, 20% (v/v) glycerol, and 0.05 mg/mL bromophenol blue.

9. IEF strip alkylating solution: 6 M urea, 2% (w/v) SDS, 25 mg/mL iodoacetamide, 375 mM Tris–HCl, pH 8.8, 20% (v/v) glycerol, and 0.05 mg/mL bromophenol blue (*see* **Note 2**).

10. IEF overlay solution: 0.5% (w/v) agarose in 1× Tris/glycine/SDS buffer. Heat agarose solution in boiling waterbath until agarose is completely dissolved and solution is clear. Keep heated until use, always make fresh.

11. Precision Plus Protein Standards (Bio-Rad).

12. BioSafe Coomassie blue (Bio-Rad).

2.6. Two-Dimensional 16-BAC/SDS–PAGE

1. 80 mM ascorbic acid.
2. 5 mM ferrous sulfate.
3. Urea.
4. 300 mM KH_2PO_4 buffer, adjusted to pH 2.1 using 1 M HCl.
5. 30% (v/v) H_2O_2.
6. 500 mM KH_2PO_4 buffer, pH 4.1.
7. 1.7% Bis-acrylamide.
8. 10× 16-BAC running buffer: 1.5 M glycine, 500 mM phosphoric acid, and 25 mM 16-BAC detergent (Sigma).
9. 16-BAC sample buffer: 6 M urea, 55 mM DTT, 7.5% (w/v) 16-BAC, 7.5% (v/v) glycerol, and 0.05% (w/v) Pyronin Y (*see* **Note 3**).
10. Destain solution: isopropanol/glacial acetic acid/nanopure water, 3.5/1/5.5 (v/v/v).
11. Coomassie blue staining solution: 0.15% (w/v) Coomassie blue R-250 in destain solution.

2.7. In-Gel Digestion

1. 50 mM ammonium bicarbonate, pH 8.0 (*see* **Note 4**).
2. Gel Spot Destain Solution: 50 mM ammonium bicarbonate/acetonitrile, 50/50 (v/v).
3. Trypsin, sequencing grade (Promega).
4. Peptide extraction solution: trifluoroacetic acid/acetonitrile/water, 0.1/60/40 (v/v/v).

3. Methods

Determination of the protein components of synaptic vesicles by 2-D PAGE separation followed by protein identification by mass spectrometry requires a rigorous purification method that can produce a large amount of vesicle protein typically exceeding 0.5 mg/2-D gel. The purification strategy used is an adaptation of a method in which crude washed synaptosomes are prepared from adult rat brain followed by hypotonic lysis, removal of large membranes, and collection of small organelles and small membrane fragments by ultra-centrifugation *(7)*. The small membrane components are then separated on a sucrose density gradient. Membrane components of equal density, containing the synaptic vesicles, are then separated on the basis of size using a glycerol velocity gradient. Alternatively, controlled-pore glass chromatography has been used in the final purification step to separate vesicles on the basis of size *(7)*. Glycerol gradient fractions containing synaptic vesicles are identified by detecting the presence of the vesicle-specific protein, synaptophysin, by western blot. Contamination from membrane fragments of the endoplasmic reticulum (ER) is present in the lower gradient fractions and should be monitored by measuring the activity of the ER-specific enzyme, NADPH-cytochrome c reductase.

Purified synaptic vesicles contain a significant number of integral membrane proteins, which can be solubilized using a mixture of strong zwitterionic detergents prior to the isoelectric focusing step for 2-D PAGE analysis. Regardless of the detergents used, most integral membrane proteins in synaptic vesicles cannot be resolved on IEF gels and require the use of the cationic detergent system 16-BAC/SDS–PAGE gels (16-BAC is 16-benzylhexadecylmethonium chloride) *(6,8)*. 16-BAC gels solubilize and separate proteins in the first dimension, and standard SDS–PAGE is used to separate proteins in the second dimension. While protein separation in both the first and second dimensions is based upon molecular weight, the two detergents bind to proteins differently, causing different migration patterns. Proteins in each spot are in-gel digested using trypsin, and the resulting peptide samples are analyzed by mass spectrometry to identify the original proteins. Poor digestion of membrane proteins and low recovery of their proteolytic peptides from gels require identification of the proteins using MS/MS peptide sequence analysis. Newly identified proteins not previously described as being associated with synaptic vesicles can be validated by measuring the migration of the protein in the glycerol gradient fractions by western blot. The staining intensity of a putative vesicle protein should peak in the same fractions as the known synaptic vesicle protein, synaptophysin. Further confirmation can be obtained by demonstrating colocalization of the new protein with synaptophysin in cultured granule neurons by confocal microscopy.

3.1. Purification of Synaptic Vesicles

1. All steps are performed on ice using solutions at 4°C unless otherwise stated. Ten adult rat brains are rinsed in 100 mL of solution A and minced with scissors in another 100 mL of fresh solution A. Minced brains are homogenized, 33 mL at a time, in a 55-mL glass Wheaton mortar with a Teflon pestle fitted into a drill.

2. The 100-mL brain homogenate is diluted to 600 mL with solution A in a large flask and mixed briefly using a stir bar. The solution is divided into four large centrifuge tubes and spun at 1500 g for 15 min to remove large neuronal cell bodies, glial cells, red blood cells, and myelin.

3. The supernatant containing crude synaptosomes is carefully removed using a large serological pipet and transferred to a fresh centrifuge tube. Crude synaptosomes are pelleted by centrifugation at 20,000 g for 30 min.

4. The pellet is resuspended in 200 mL of solution A, and steps 2 and 3 are repeated without dilution to 600 mL.

5. Each of the four crude synaptosome pellets is vigorously resuspended in 30 mL of solution B for a total volume of 120 mL. Lysis of the synaptosomes in hypotonic buffer is improved by homogenization of the solution in a glass dounce homogenizer with tight-fitting pestle, 10 strokes. The centrifuge tubes are rinsed with an additional 80 mL of solution B, which is added to the homogenized pellet solution for a final volume of 200 mL. The last wash with 80 mL of buffer does not add much more cell material but serves to properly dilute the sample. Two milliliters of 100 mM PMSF are added to the solution and lysis is allowed to proceed for 45 min on ice with stirring.

6. The lysed synaptosome solution is divided into eight centrifuge tubes and spun at 47,000 g for 15 min to remove the large membrane components.

7. The supernatant is carefully removed using a serological pipet and transferred to eight fresh ultracentrifuge tubes. The small membrane components, including synaptic vesicles, are pelleted by centrifugation at 200,000 g for 1 h.

8. Solution C (2 mL) is added to a single pellet that is dislodged from the side of the tube with a Teflon pestle. The pellet solution is then transferred to a second pellet to dislodge, and this solution is transferred to the remaining six pellets resulting in a final volume of ~3 mL. The vesicles should be completely homogenized with a 5-mL glass Wheaton mortar and Teflon pestle so that no membrane pieces are visible. Vesicles can be stored at this stage at –80°C.

9. Two sucrose step gradients are prepared in two topless, polypropylene tubes for swinging bucket rotors. Each step gradient consists of 6 mL of 0.8 M sucrose followed by 8 mL of 0.6 M sucrose, 8 mL of 0.4 M sucrose, and 8 mL of 0.2 M sucrose.

10. One tablet of Roche Protease Inhibitors is added to the 3 mL of crude vesicle solution, and the solution is homogenized by hand in a small glass Wheaton mortar with a Teflon pestle. The solution is then diluted to 5 mL with nanopure water and mixed gently.

11. The vesicle solution (2.5 mL) is carefully layered on top of the 0.2 M sucrose layer of the step gradient. The vesicles are centrifuged in a swinging bucket rotor at 100,000 g for 2 h. The vesicles will migrate to the 0.2 M/0.4 M sucrose interface producing a cloudy layer. Approximately 2.25 mL of the cloudy layer at the interface of each tube is removed and transferred to a fresh tube to provide 4.5 mL total volume. 1.5 mL of this solution is saved to use as a positive control in **Subheading 3.4** (*see* **Subheading 2.4.**, **step 5**).

12. While the sucrose density gradient is running, the 5–25% glycerol gradients can be prepared for the next step. Four 30-mL gradients of 5–25% buffered glycerol are prepared using a 50-mL Hoefer gradient maker fitted with 3/32" internal diameter tubing that is threaded through a mini-peristaltic pump and ends inside a glass 9" Pasteur pipet. The gradients are layered from the bottom by placing the long tip of the glass pipet at the bottom of a topless, polypropylene ultracentrifuge tube on ice and secured by hand against the side of the tube. After pouring the gradient, the Pasteur pipet tip is lifted up along the side of the tube so that the gradient is not disturbed.

13. The vesicle-containing sucrose layer (4 mL) is diluted with 6.8 mL of nanopure water plus 1.2 mL of 10× HEPES-buffered saline, and the solution is gently mixed. Three milliliters of this solution is carefully layered on the top of each of the four glycerol gradients. The gradients are centrifuged at 100,000 g in a swinging bucket, SW28 rotor for 2 h.

14. A small hole is punched in the bottom of one of the gradient tubes with a 26 gauge syringe needle to produce a slow drop-wise release of the gradient, and 30 1-mL fractions are collected from each tube. The same fraction number from each gradient tube can be combined or collected in the same tube resulting in a 4-mL solution for each fraction if four gradient tubes are run.

3.2. Determination of Vesicle-Containing Gradient Fractions by Western Blot

1. Prepare a 10% acrylamide pre-made mini-gel by removing the comb, rinsing the wells with 1× Tris/glycine/SDS buffer and removing the plastic strip from the bottom of the gel. Assemble into the mini-gel box apparatus and add 1× Tris/glycine/SDS buffer according to the manufacturer's instructions.

2. Transfer a 100-µL aliquot from each gradient fraction to a fresh microfuge tube and add 11 µL of 100% TCA. Incubate the protein solution for 30 min at −20°C and collect the protein pellet by centrifugation at 15,000 g for 5 min. Wash the protein pellet with an ethanol/ether solution, 1/1 (v/v), and collect the pellet by centrifugation as described. Decant and completely remove the supernatant. Resuspend protein pellets in 20 µL of Laemmli sample buffer, mix and boil for 5 min. Spin down samples in a microcentrifuge.

3. Load each sample into the wells of the mini-gel and run at 100 V for 1.5 h.

4. Remove the gel from the apparatus, rinse it in transfer buffer and assemble it in the blotting apparatus with pre-wetted PVDF membrane according to the manufacturer's instructions. Allow proteins to transfer for 1.5 h at 100 V.

5. Rinse the PVDF membrane with PBS-T and incubate in 10 mL of blocking solution for 1 h.

6. Rinse the PVDF membrane with an excess of PBS-T and incubate in 10 mL of PBS-T containing the primary antibody at a 1:500,000 dilution for 1 h.

7. Rinse the PVDF membrane with an excess of PBS-T, three times for 5 min. Incubate with 10 mL of PBS-T containing the secondary antibody at a 1:4000 dilution for 1 h.

8. Rinse the membrane with an excess of PBS-T, three times for 5 min. Develop the membrane using enhanced chemiluminescence reagents and visualize bands using an imager for chemiluminescence.

9. Quantitate the intensity of synaptophysin staining in each fraction using an imaging software. The staining intensity should peak near fraction 12 (*see* **Fig. 1**).

3.3. Concentration and Collection of Vesicles

1. To obtain a sufficient signal for the ER assay described in **Subheading 3.4** and to collect the vesicle membranes for 2-D PAGE, vesicle fractions are combined in threes and pelleted by ultracentrifugation. Vesicle fractions 1–3, 4–6, 7–9, and so on are combined resulting in a 12-mL solution for each combined fraction.

2. Six milliliters of each combined fraction is diluted with 12 mL of nanopure water plus 2.0 mL of 10× HEPES-buffered saline, and the solution is vortexed. The solution is transferred to a polyallomer, capped ultracentrifuge tube for a fixed angle, Ti-70 rotor. Purified vesicle membranes are collected by centrifugation at 200,000 g for 2 h.

3. The supernatant is removed and **step 1** is repeated with the remaining 6 mL of each fraction added to the corresponding centrifuge tube still containing the first pellet. Each final vesicle pellet is resuspended in 100–200 μL of 25 mM Tris–HCl buffer, pH 8.0, using a small 0.5-mL glass Wheaton mortar and Teflon pestle. Protein amount in each fraction is determined by BCA assay (Bio-Rad).

3.4. Measurement of Endoplasmic Reticulum Contamination in Gradient Fractions

1. Combined fractions are tested in duplicate, including one blank and one positive control resulting in 24 samples per microtiter plate.

2. Add 5 μL of cytochrome c to each of the 24 wells in a microtiter plate.

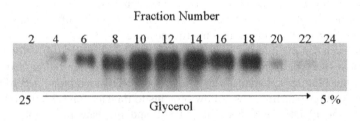

Fig. 1. Synaptophysin staining detected by western blot of glycerol gradient fractions. Synaptic vesicles were purified from rat brain as described in **Subheading 3.1**, and 30 1-mL fractions were collected from the bottom of the 5–25% glycerol velocity gradient. Proteins in 100 μL of each fraction were TCA precipitated and probed for the vesicle-specific protein, synaptophysin, by western blot. Synaptic vesicle concentration peaks near fraction 12.

3. Add 100 µL of assay buffer to each well.
4. Add 5 µL of each gradient fraction, blank or positive control, to a well, mix gently. Incubate at room temperature for 3 min.
5. Add 10 µL of NADPH to each well, mix gently.
6. Read plate at 550 nm at 30-s intervals for 10 min. Using enzyme kinetics software often provided with the microplate reader, Microsoft Excel, or GraphPad Prizm, the absorbance values are plotted versus time for each sample.
7. The slope of this line is divided by the extinction coefficient for cytochrome c (indicated in the Sigma product sheet) to determine the rate of reduced cytochrome c production, typically expressed in nmol/min.
8. The enzyme concentration is directly proportional to the rate of reduced cytochrome c production. Given the protein concentration in each fraction, the concentration of enzyme in nanomoles of product/minute/milligram protein is calculated. The peak of the NADPH–cytochrome c reductase activity should appear between fractions 7 and 9, below the peak of synaptophysin staining determined in section 3.2 (*see* **Fig. 2**).
9. The glycerol gradient fractions collected for 2-D PAGE analysis, typically fractions 12–18, should contain the highest concentration of synaptic vesicles and a very low concentration of ER membranes.

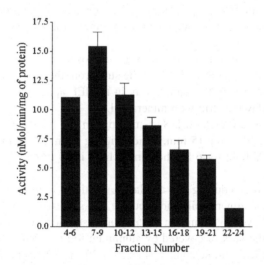

Fig. 2. NADPH-cytochrome c reductase (NCR) activity in pooled glycerol gradient fractions. The presence of contamination by endoplasmic reticulum (ER) membranes in the glycerol gradient fractions was tested by measuring NCR activity. Synaptic vesicles were purified as described in **Subheading 3.1**, and the resulting glycerol velocity gradient fractions were pooled in sets of three. Membranes in the pooled fractions were collected by centrifugation and assayed for NCR activity. The graph demonstrates that the concentration of ER membranes peaks near fractions 7–9 and decreases in the vesicle-containing fractions 12–18.

3.5. Vesicle Protein Separation by IEF 2-D PAGE

1. Wear gloves during subsequent steps to avoid contact with acrylamide solutions and to prevent sample contamination with keratin. Prepare a fresh solution of 10% (w/v) APS in water. Assemble two large glass plates, one long and one short, with a 1 mm spacer on each side according to the manufacturer's directions (*see* **Note 5**).

2. For one large 20 × 20 cm, 1-mm thick, 10% acrylamide SDS–PAGE gel: combine 16 mL of nanopure water, 10 mL of 1.5 M Tris–HCl, pH 8.8, and 13.3 mL of acrylamide solution in a small flask and swirl to mix. Add 200 µL of 10% APS and 20 µL of TEMED, swirl to mix, and slowly pour the solution between the two glass plates to the top. Gently tap plates to remove air bubbles and insert the IEF strip sample comb (wear eye protection to prevent acrylamide exposure). Cover plates with plastic wrap (*see* **Note 6**).

3. Remove a volume of purified vesicle solution to equal 500 µg of protein and pellet membranes by centrifugation at 200,000 *g* for 1 h. Resuspend the pellet in 300 µL of IEF sample buffer by vigorous vortexing. Further resuspension using a 0.5-mL glass mortar and pestle may be required. Centrifuge the sample at 15,000 *g* for 5 min to remove debris and layer supernatant into the IEF rehydration tray.

4. Place a 17-cm IEF strip, pH 3–10 NL, on top of the sample solution and rehydrate for 1 h before overlaying with mineral oil. Place a cover on the strips and allow rehydration overnight.

5. Wet paper wicks with water and place one over each electrode in the sample tray of the IEF apparatus. Remove the strip from the rehydration tray and place it, gel side down, in the sample tray in the IEF apparatus on top of the paper wicks. Overlay the strip with mineral oil.

6. For the Bio-Rad Protean IEF System, program IEF focusing using the following conditions: 250 V for 15 min, rapid ramping from 250 V to 10,000 V in 5 h, and 10,000 V hold for 6 h. The current limit is 50 µA/gel. The gels can be run overnight.

7. Prepare the IEF reducing and alkylating solutions. Remove the strip, blot away the excess of mineral oil, and incubate the strip for 30 min in the reducing solution followed by 30 min in the alkylating solution. Incubate in the alkylation solution in the dark or covered by aluminum foil (*see* **Note 2**).

8. Remove the IEF sample comb from a 20 × 20 cm, 10% acrylamide SDS–PAGE gel and rinse the gel with 1× Tris/glycine/SDS buffer. Secure the gel into the gel box apparatus according to the manufacturer's instructions.

9. Gently slide the IEF strip between the two glass plates of the gel so that it is lying flat along the top of the gel. Heat the IEF overlay solution until clear and pour it over the IEF strip until it is completely covered. Allow the overlay solution to gel for a few minutes and then fill the inner and outer chambers of the gel box with 1× Tris/glycine/SDS buffer. The inner chamber must be completely full, and the outer chamber must cover the bottom of the glass plates

(about 1.5 L total). Load 20 µL of Precision Plus Protein Standards into the empty well.

10. Run the gel at 28 mA/gel for 4–5 h or until the blue tracking dye has migrated to the bottom of the gel.

11. Rinse the gel three times with nanopure water and stain it with BioSafe Coomassie blue overnight (*see* **Fig. 3**).

3.6. Vesicle Protein Separation by 16-BAC/SDS–PAGE

1. Assemble two large glass plates, one long and one short, with a 0.75-mm spacer on each side according to the manufacturer's directions. The plates should be tightly blocked at the bottom.

2. Prepare fresh aqueous solutions of 80 mM ascorbic acid and 5 mM ferrous sulfate.

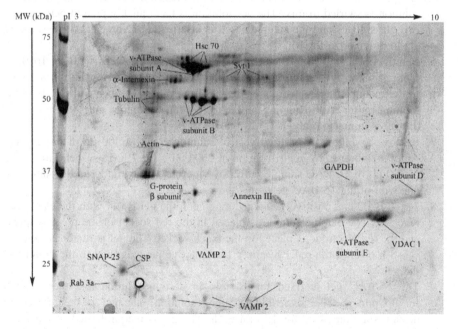

Fig. 3. Vesicle proteins resolved by IEF/SDS–PAGE. Synaptic vesicles collected from glycerol gradient fractions 12–18 were solubilized in 7 M urea, 2 M thiourea, 2% (w/v) CHAPS, 2% (w/v) ASB-14, 5 mM TBP, 25 mM Tris–HCl, pH 8.8, and 0.2% (w/v) ampholytes. The proteins were resolved on a 17-cm IEF strip, pH 3–10 NL, followed by SDS–PAGE separation in the second dimension. The gel was stained with colloidal Coomassie blue, and proteins were identified using in-gel digestion (*see* **Subheading 3.7.**) followed by peptide sequence analysis by ESI-MS/MS. Reproduced with permission from reference (*6*).

3. Wear gloves when handling acrylamide solutions. For one large 16 × 20 cm, 0.75-mm thick, 7.5% acrylamide 16-BAC separating gel: 5.4 g of urea is dissolved in 7.5 mL of acrylamide solution (*see* **Subheading 2.5.**), 7.5 mL of 300 mM KH_2PO_4 buffer, pH 2.1, 7.5 mL of nanopure water, and 1.1 mL of 1.7% Bis-acrylamide. To start polymerization, 1.5 mL of 80 mM ascorbic acid, 48 μL of ferrous sulfate, and 1.2 mL of 30% (v/v) hydrogen peroxide are added and the solution is swirled. The gel solution is then quickly poured between the two glass plates while leaving room for a stacking gel.

4. Isopropanol is poured over the gel, and the gel is wrapped in plastic wrap. The gel should polymerize overnight.

5. Pour off the isopropanol, rinse the top of the gel with nanopure water, and blot dry. To prepare the 16-BAC stacking gel: 1.0 g of urea is dissolved in 1.33 mL of acrylamide solution (*see* **Subheading 2.5.**), 2.5 mL of 500 mM KH_2PO_4 buffer, pH 4.1, 3.0 mL of nanopure water, and 1.38 mL of 1.7% Bis-acrylamide. To start polymerization, 500 μL of 80 mM ascorbic acid, 8.5 μL of 5 mM ferrous sulfate, and 500 μL of 30% (v/v) hydrogen peroxide are added and the solution is swirled to mix. The gel solution is then quickly poured between the two glass plates and the sample comb is inserted (*see* **Note 7**). Allow stacking gel to polymerize for 1 h.

6. Remove the comb from the stacking gel and rinse the wells with 1× 16-BAC running buffer. Assemble the gel in the large format gel apparatus according to the manufacturer's instructions and fill the inner chamber with 1× 16-BAC running buffer. Add Running buffer to the outer chamber until the bottom of the gel is immersed (about 1.5 L total).

7. Remove a volume of purified vesicle solution equal to 300 μg of protein and pellet membranes by centrifugation at 200,000 *g* for 1 h. Resuspend the pellet in 100 μL of 16-BAC sample buffer by vigorous vortexing. Load the sample in one well of the 16-BAC gel.

8. As 16-BAC is a positively charged detergent, the gel is run with the electrodes reversed. The black electrode is placed in the red socket of the power supply, and the red electrode is placed in the black socket. The gel is run at 28 mA for ~7–8 h or until the red Pyronin Y tracking dye has migrated to the bottom of the gel.

9. The 16-BAC gel is fixed for 15 min in destain solution and then stained for 30 min in Coomassie blue staining solution. The gel is then rinsed in destain solution twice for 15 min each. The gel is reequilibrated in nanopure water.

10. Prepare the second dimension, a 10% acrylamide SDS–PAGE gel, the day before use as described in steps 1 and 2, **Subheading 3.5** and assemble as described in **step 8, Subheading 3.5.**

11. Each lane in the 16-BAC gel is cut vertically into a single strip using a razor blade (one lane = one strip). The gel strips are incubated for 30 min in reducing solution followed by 30-min incubation in alkylating solution in the dark (*see* **Note 8**).

12. Each gel strip is placed on a piece of parafilm, and the gel strip is gently transferred along the top of the SDS–PAGE gel. Folded parafilm can be used to push the gel strip into place. 1× Tris/glycine/SDS buffer is added to fill the inner chamber, and the buffer is added to outer chamber until the bottom of the gel is immersed. Twenty microliters of Precision Plus Protein Standards are loaded into the well.

13. The gel is run at 28 mA for 4–5 h. The gel is rinsed three times for 5 min each in nanopure water and stained overnight in BioSafe Coomassie blue (*see* **Fig. 4**). The gel is destained three times in an excess of nanopure water.

3.7. In-gel Digestion of Vesicle Proteins

1. Wear gloves during all steps to prevent keratin contamination. Place the destained gel on a clean glass plate. Cut off the tip of a 1000-µL pipet and use

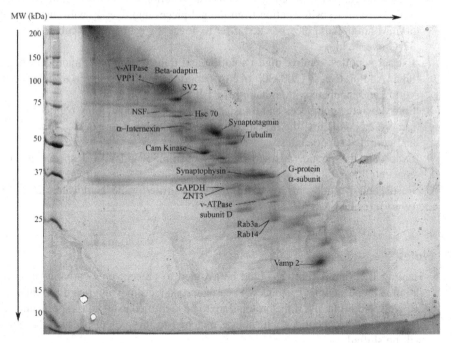

Fig. 4. Vesicle proteins resolved by 16-BAC/SDS–PAGE. Synaptic vesicles collected from glycerol gradient fractions 12–18 were solubilized in 6 M urea, 55 mM DTT, 7.5% (w/v) 16-BAC, 7.5% (v/v) glycerol, and 0.05% (w/v) Pyronin Y. Proteins were separated in the cationic, 16-BAC detergent system in the first dimension followed by SDS–PAGE in the second dimension. The gel was stained with colloidal Coomassie blue, and proteins were identified using in-gel digestion (*see* **Subheading 3.7.**) followed by peptide sequence analysis by ESI-MS/MS. Reproduced with permission from reference *(6)*.

it to cut out stained protein spots from the gel. Large spots should be cut into smaller pieces of about 1 mm in diameter. Transfer each gel spot to a siliconized, 0.6-mL microcentrifuge tube.

2. Add enough Gel Spot Destain Solution to cover the spot, typically 200–300 µL. Vortex the tubes and incubate the gel pieces in the solution with gentle shaking for 1 h at room temperature.

3. Spin down the gel pieces and discard the supernatant. Repeat **step 2**, two to three more times or until all color has been removed from the gel spots.

4. Dry the gel pieces thoroughly in a speed vacuum concentrator. They should be shrunk to tiny squares when done.

5. Digest protein: reconstitute trypsin (20 µg/vial) in 40 µL of supplied resuspension buffer (Promega). Transfer the trypsin solution to a larger tube and add 1.6 mL of 50 mM ammonium bicarbonate buffer (12.5 ng trypsin/µL). Add 50 µL of this solution to each gel spot (*see* **Note 9**).

6. Allow the gel pieces to rehydrate in trypsin solution at 4°C for 15 min. Remove any excess of trypsin solution from the gel pieces and overlay with just enough 50 mM ammonium bicarbonate buffer to cover the gel pieces. Incubate tubes at 37°C overnight.

7. Extract peptides: remove digest solution and transfer to a fresh microcentrifuge tube. Add an excess (200–300 µL) of peptide extraction solution to gel pieces, vortex and incubate with agitation for 1 h at room temperature.

8. Spin down the gel pieces and transfer the supernatant containing peptides to the same tube containing the digest solution. Repeat **step 7** one or two more times and combine supernatants.

9. Evaporate the peptide extraction solution from peptides in a speed vacuum concentrator.

10. Resuspend peptides in 20–40 µL of acetonitrile/formic acid/water, 2/0.1/98 (v/v/v), for ESI-MS/MS analysis.

4. Notes

1. Heating IEF sample buffer should be avoided to prevent urea carbamylation of proteins in the sample. Proteins with carbamyl adducts may migrate differently on IEF gels. Additionally, if the proteins are to be identified using mass spectrometry, the carbamylated peptides will not be matched to the protein because the mass will be shifted.

2. Iodoacetamide is an alkylating agent and toxic. Wear gloves when handling solutions containing iodoacetamide. Iodoacetamide is light sensitive, store solutions in the dark.

3. Pyronin Y is a DNA intercalator and a carcinogen. Wear gloves when handling Pyronin Y.

4. The pH should be around 8.0. Do not adjust the pH of this solution as this will increase the amount of non-volatile salt in the buffer, which may interfere with peptide analysis by mass spectrometry.

5. The plates should be perfectly aligned at the bottom and contain no chips in the glass. Secure the plates tightly with a foam pad at the bottom so that fluid will not leak.

6. The gels should be prepared the day before use to allow complete polymerization to occur overnight.

7. Wear lab goggles, acrylamide solution can splash out while inserting comb. Acrylamide is toxic. Have paper towels on hand to clean up any excess acrylamide solution that is displaced by the comb.

8. Wear gloves when handling gel strips. This will avoid keratin contamination of the gel and prevent contact with alkylating solutions that contain iodoacetamide.

9. Promega sequencing grade trypsin has been methylated to avoid autolysis. However, trypsin autolytic peptides are commonly observed in gel spots digested with this method upon peptide analysis by mass spectrometry. The autolysis does not interfere with the activity of the enzyme, and the autolytic peptides can be used for MS calibration.

Acknowledgments

This work was supported in part by NIH/NINDS NS38248 (CMT) and NIH/NCRR RR15583. Additional support was provided by the National Science Foundation (MCB9808372 and EPS). The mass spectrometers used in this study were purchased with support from the Murdock Charitable Trust (Vancouver, WA), the National Science Foundation (99-77757), and through NIH/NCRR RR15583.

References

1. Vijayakrishnan, N. and Broadie, K. (2006) Temperature-sensitive paralytic mutants: insights into the synaptic vesicle cycle. *Biochem. Soc. Trans.* **34**, 81–87.

2. Fernandez-Chacon, R. and Sudhof, T. C. (1999) Genetics of synaptic vesicle function: toward the complete functional anatomy of an organelle. *Annu. Rev. Physiol.* **61**, 753–776.

3. Gualix, J., Pintor, J., and Miras-Portugal, M. T. (1999) Characterization of nucleotide transport into rat brain synaptic vesicles. *J. Neurochem.* **73**, 1098–1104.

4. Goncalves, P. P., Meireles, S. M., Neves, P., and Vale, M. G. (2000) Distinction between Ca(2+) pump and Ca(2+)/H(+) antiport activities in synaptic vesicles of sheep brain cortex. *Neurochem. Int.* **37**, 387–396.

5. Ikemoto, A., Bole, D. G., and Ueda, T. (2003) Glycolysis and glutamate accumulation into synaptic vesicles. Role of glyceraldehyde phosphate dehydrogenase and 3-phosphoglycerate kinase. *J. Biol. Chem.* **278**, 5929–5940.

6. Coughenour, H. D., Spaulding, R. S., and Thompson, C. M. (2004) The synaptic vesicle proteome: a comparative study in membrane protein identification. *Proteomics* **4**, 3141–3155.

7. Huttner, W. B., Schiebler, W., Greengard, P., and De Camilli, P. (1983) Synapsin I (protein I), a nerve terminal-specific phosphoprotein. III. Its association with synaptic vesicles studied in a highly purified synaptic vesicle preparation. *J. Cell Biol.* **96**, 1374–1388.

8. Hartinger, J., Stenius, K., Hogemann, D., and Jahn, R. (1996) 16-BAC/SDS-PAGE: a two-dimensional gel electrophoresis system suitable for the separation of integral membrane proteins. *Anal. Biochem.* **240**, 126–133.

19

Purification and Proteomics Analysis of Pancreatic Zymogen Granule Membranes

Xuequn Chen and Philip C. Andrews

Summary

Pancreatic zymogen granules (ZGs) are specialized for digestive enzyme storage and regulated secretion in exocrine pancreas and are a classical model for studying secretory granule function. To understand the function of this organelle, we have conducted a proteomic study to identify the ZG membrane (ZGM) proteins from ZGs purified by Percoll gradient centrifugation. By combining multiple separation strategies including two-dimensional gel electrophoresis and two-dimensional HPLC with tandem mass spectrometry, we identified 101 proteins from purified ZGMs including a large number of proteins previously unknown on ZGMs. To distinguish intrinsic membrane proteins from soluble and peripheral membrane proteins, a quantitative proteomics strategy was used to measure the enrichment of intrinsic membrane proteins through the purification steps by labeling crude, KBr-, and Na_2CO_3-washed ZGMs with multiplexed isobaric tags (iTRAQ™), 114, 116, and 117, respectively. The proteins with 117:114 ratios greater than one correlated well with intrinsic membrane proteins that contain either known or predicted transmembrane domains.

Key Words: Pancreatic zymogen granule; Percoll gradient; 2D gel electrophoresis; 2D HPLC; tandem mass spectrometry; iTRAQ; membrane proteins; quantitative proteomics.

1. Introduction

The primary function of pancreatic acinar cells is to synthesize, package, and secrete digestive enzymes. This process is regulated by gastrointestinal hormones and neurotransmitters (1,2). In acinar cells, digestive enzymes are stored in zymogen granules (ZGs). Stimulation of acinar cells by secretagogues

From: *Methods in Molecular Biology, vol. 432: Organelle Proteomics*
Edited by: D. Pflieger and J. Rossier © Humana Press Inc., Totowa, NJ

triggers fusion of the ZG membrane (ZGM) with the apical plasma membrane leading to exocytosis. In addition to fulfilling this important physiological role, ZGs have been used as a model system for studying secretory granule function in general. Identification of the component proteins of the ZGM is an essential first step in understanding the molecular architecture of the ZG and its function. It is believed that ZGs share common mechanisms with other secretory organelles such as synaptic vesicles and chromaffin granules, in that two families of proteins, SNARE and Rabs, govern vesicular trafficking from the Golgi network to the plasma membrane *(3,4)*. However, many key components have not yet been identified on the ZGM. Early sodium dodecyl sulphate–polyacrylamide gel electrophoresis studies indicated a relatively simple protein composition for the ZGM because of limited resolution and sensitivity. Subsequently, a number of low-abundance proteins including Rab3D *(5,6)* and several SNARE proteins have been localized on ZGM by immunoblotting and immunocytochemistry *(7–10)*. Not surprisingly, our thorough proteomics analysis indicates a more complex scenario for ZG function.

Organellar proteomics combines biochemical fractionation and comprehensive protein identification, which results in reduced sample complexity and provides a functional context for proteins. Taking advantage of an established protocol to obtain high-purity ZGs *(11,12)*, we conducted an organellar proteomic analysis of ZGM proteins. By combining the complementary techniques of 2D gel electrophoresis and 2D LC with tandem mass spectrometry, over 100 proteins have been identified on ZGMs, among which 73 have been localized on ZGs for the first time including multiple small GTP-binding proteins, SNARE proteins, and molecular motor proteins. These observations bring new insights into the molecular mechanisms of ZG functions.

Recently, quantitative proteomics technologies have been applied to organellar proteomics to identify the dynamic compositions of the human nucleolus *(13)* and distinguish bona fide centrosome proteins from copurified proteins *(14)*. In this study, we also developed a relative quantitative proteomics strategy to monitor intrinsic ZGM protein enrichment during purification; the intrinsic membrane proteins characterized based on iTRAQ™ ratios correlate well with known or TMHMM-predicted membrane proteins *(15)*.

2. Materials

2.1. Isolation of ZGs and Purification of ZGMs

2.1.1. Isolation of ZGs

1. Typically, pancreases are obtained from 10–12 Sprague–Dawley rats with body weight between 250 and 300 g.

2. Homogenization buffer: 0.25 M sucrose, 25 mM 2-morpholinoethanesulfonic acid (MES, monohydrate), pH 6.0 (adjusted with 1 N HCl), 2 mM EGTA, and 0.1 mM phenylmethylsulfonyl fluoride (*see* **Note 1**). It is made from stock solutions (2 M sucrose and 1 M MES, pH 6.0) stored at 4°C.
3. Teflon glass homogenizer with tight pestle (Thomas Scientific, Swedesboro, NJ).
4. Percoll (Amersham Biosciences, Uppsala, Sweden).
5. Beckman ultracentrifuge with a Ti 70.1 rotor.

2.1.2. Purification of ZGMs

1. Nigericin (Sigma, St. Louis, MO) is dissolved in ethanol at a concentration of 10 mg/mL and stored at 4°C.
2. Protease inhibitors cocktail (Roche, Indianapolis, IN) is used according to the company's instruction.
3. ZG lysis buffer: 150 mM sodium acetate, 10 mM MOPS, pH 7.0, 27 µg/mL nigericin, 0.1 mM $MgSO_4$, and 0.1 mM phenylmethylsulfonyl fluoride supplemented with protease inhibitors cocktail. The stock solutions of 1 M sodium acetate and 1 M MOPS, pH 7.0 (adjusted with 1 N HCl), are stored at 4°C.
4. Beckman ultracentrifuge with a Ti 70.1 rotor.
5. 250 mM KBr.
6. 0.1 M Na_2CO_3, pH 11.0 (adjusted with 1 N HCl).

2.2. Two-Dimensional Gel Electrophoresis and In-Gel Protein Digestion

2.2.1. Two-Dimensional Separation by IEF/SDS–PAGE

1. 18-cm IPG strips of pH 4–7 and IPG buffer are from Amersham Pharmacia Biotech (Pittsburgh, PA).
2. Amidosulfobetaine-14 (ASB-14) and ASB-C8φ are from Calbiochem (San Diego, CA).
3. Isoelectric focusing (IEF) rehydration buffer: 7 M urea, 2 M thiourea, 1% (w/v) ASB-14, 1% (w/v) ASB-C8φ, 1% (v/v) Triton X-100, and 1% (w/v) CHAPS. It is aliquoted in 1.5-mL Eppendorf tubes and stored at –80°C.
4. Reducing solution: 2.5 mM TriButylPhosphine (TBP) in 6 M urea, 2% (w/v) SDS, 20% (v/v) glycerol, and 50 mM Tris–HCl, pH 8.8.
5. Alkylating solution: 55 mM iodoacetamide in 6 M urea, 2% (w/v) SDS, 20% (v/v) glycerol, and 50 mM Tris–HCl, pH 8.8.
6. Acrylamide gels: 12.5% acrylamide, 1.5 mm thick, 20 × 20 cm.
7. A Multiphor II electrophoresis unit (Amersham Biosciences) is used for IEF separation.
8. A PROTEAN Plus Dodeca cell is used for SDS–PAGE.
9. SYPRO Ruby is from Molecular Probes (Eugene, OR), stored at room temperature and can be reused for up to three times.
10. 2D-gel imaging: Molecular Imager PharosFX Plus System (Bio-Rad, Herculus, CA).

2.2.2. In-Gel Protein Digestion

1. All solutions used for in-gel digestion and for ZipTip cleaning are made in Optima H₂O (Fisher Scientific, Pittsburgh, PA).
2. For cutting 2D-gel spots, use a 2-mm dermal punch (Sklar, West Chester, PA).
3. Sequencing grade-modified trypsin is from Promega (Madison, WI) and stored at –20°C.
4. MassPrep sample digestion robot (Waters, Billerica, MA).
5. ZipTip$_{C18}$ tips are from Millipore, Foster City, CA.

2.3. iTRAQ™ Labeling and 2D LC Separation of Tryptic Peptides

1. iTRAQ™ reagents are from Applied Biosystems and stored at –80°C.
2. Strong cation exchange (SCX) buffer: 10 mM KH_2PO_4, and 25% (v/v) acetonitrile with pH adjusted to 3.0 (adjusted with 1 N HCl). It is stored at 4°C.
3. MicroSpin™ columns (PolyLC) are stored at room temperature.
4. Zorbax 300 SB C18 column, 75 μm x 150 mm, 3.5-μm particles (Agilent Technologies, Santa Clara, CA) are used for reversed-phase HPLC.
5. For reversed-phase HPLC separation, solvent A is 0.1% (v/v) TFA in water and solvent B is acetonitrile/water/TFA, 90/10/0.1% (v/v).
6. Agilent 1100 HPLC system for reversed-phase separation.

2.4. Tandem Mass Spectrometry

1. Matrix-assisted laser desorption ionization (MALDI) matrix solution: 10 mg/mL of α-cyano-4-hydroxycinnamic acid (Sigma Chemical) in acetonitrile/water/trifluoroacetic acid, 50/50/0.1 (v/v/v), is stored at –20°C.
2. The diluted MALDI matrix solution used for automated LC spotting on a target plate is at 2 mg/mL α-cyano-4-hydroxycinnamic acid in the same solution as above.
3. MALDI-TOF/TOF instrument: Applied Biosystems 4700 Proteomics Analyzer (Applied Biosystems/MDX Sciex, Foster City, CA).
4. GPS explorer software (Applied Biosystems) for MS/MS data interpretation.

3. Methods

3.1. Preparation of Purified ZGMs

3.1.1. Isolation of ZGs

1. In a typical experiment, 10–12 rats are killed by decapitation following CO_2 anesthesia. The blood is drained (see **Note 2**), and the pancreases are removed, minced, and homogenized (two pancreases at a time) in 10 mL of ice-cold homogenization buffer. The homogenization is conducted for nine strokes on an electronic motor using a Teflon glass homogenizer.

2. Homogenates are combined and distributed in six 15-mL tubes and then centrifuged first at 300 g for 10 min at 4°C to remove unbroken cells and nuclei, and the supernatant is transferred to six new tubes and spun down at 2000 g to generate a white particulate enriched in ZGs covered by a tan layer containing mainly mitochondria (*see* **Note 3**).

3. The particulate is gently resuspended in 40 mL of homogenization buffer and mixed with 60 mL of Percoll solution containing 50 mL of Percoll, 9.4 mL of 2 M sucrose and 600 μL of 1 M MES, pH 6.0.

4. The mixture is distributed in 10 10-mL ultracentrifugation tubes and centrifuged in a Beckman ultracentrifuge at 30,000 rpm (60,000 g) for 20 min using a Ti 70.1 rotor.

5. The dense white ZG bands near the bottom of the centrifuge tube are collected and diluted in 60 mL of homogenization buffer. To remove excess Percoll, the suspension is centrifuged to pellet ZGs at 1000 g for 10 min.

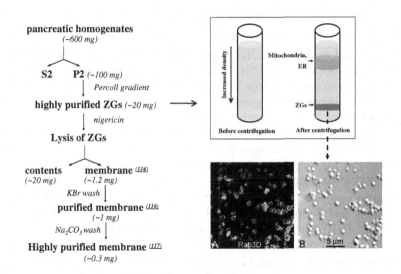

Fig. 1. **Outline of ZG membrane purification.** *Left*, rat pancreases were homogenized and then centrifuged in two low speed steps to generate a crude particulate fraction (P2). The ZGs were purified by Percoll gradient. To purify ZG membrane, the isolated ZGs were lysed and centrifuged. The membrane pellet was then washed sequentially with 250 mM KBr and 0.1 M Na_2CO_3 (pH 11.0). The proteins were extracted from crude, KBr- and Na_2CO_3-washed ZG membrane and the digested peptides were labeled with 114, 116 and 117 iTRAQ™ reagents respectively. The amount of protein recovered at each step is also indicated. *Right*, Top shows a cartoon illustrating the Percoll gradient ultracentrifugation; at the bottom are Nomarski and fluorescent images of purified ZGs to demonstrate the purity of ZGs and the positive staining of a previously established ZG marker, Rab3D. Note essentially all granules stain for Rab3D. This figure is modified from Figure 1 in our recent publication *(15)*.

3.1.2. Purification of ZGMs

1. To purify ZGM, the above ZG pellets are resuspended in 20 mL of ZG lysis buffer and incubated at 37°C for 15 min. The lysate, which becomes clear at the end of the incubation, is centrifuged at 38,000 rpm (100,000 g) for 1 h in a Beckman ultracentrifuge using a Ti 70.1 rotor to pellet the ZGM.
2. To remove absorbed content proteins, the ZGM pellet is washed with 10 mL of 250 mM KBr. To further purify the membrane and remove some of the peripheral membrane proteins, the ZGM pellet is resuspended in 0.1 M Na_2CO_3 (pH 11.0) and incubated on ice for 30 min, then centrifuged for 1 h at 100,000 g (*see* **Notes 4** and **5**).
3. The procedure for ZGM purification and the protein recovery at each step have been illustrated in a flow chart diagram (*see* **Fig. 1**).

3.2. Two-Dimensional Electrophoresis and In-Gel Protein Digestion

3.2.1. Two-Dimensional Separation by IEF/SDS–PAGE

1. The ZGM pellets (300 μg) obtained from the previous purification steps are resuspended in 360 μL of IEF rehydration buffer with 1% (v/v) corresponding IPG buffer and 2.5 mM tributyl phosphine (TBP) added to the buffer immediately before use and solubilized for 30 min at 30°C (*see* **Note 6**). The insoluble material, which is minimal in this protocol, is removed by centrifugation at 14,000 rpm (20,000 g) for 10 min in a tabletop centrifuge.
2. Sample loading of solubilized ZGM is performed by in-gel rehydration on 18-cm IPG strips (pH 4–7) overnight at room temperature.
3. Isoelectric focusing is conducted at 20°C, 110,000 Vh at 6000 V maximum voltage on the electrophoresis unit.
4. After the strips are equilibrated in buffer [6 M urea, 2% (w/v) SDS, 20% (v/v) glycerol, and 50 mM Tris–HCl, pH 8.8] at room temperature, first containing 2.5 mM TBP (reduction) for 15 min and then 55 mM iodoacetamide (alkylation) for another 15 min, the strips are run for the second dimension separation on 1.5-mm thick, 20 × 20 cm, 12.5% polyacrylamide gels at 70 V overnight at 4°C.
5. The gels are fixed and stained with SYPRO Ruby according to the methods provided by the manufacturers.

3.2.2. In-Gel Protein Digestion by Trypsin

1. For in-gel digestion, SYPRO Ruby-stained 2D gels are scanned and digital images analyzed with Quantity One software to detect protein spots.
2. All detected protein spots are excised using a 2-mm dermal punch and placed individually in a 96-well plate.
3. A sample digestion robot is used to perform the in-gel digestion. The dry gel pieces are rehydrated in a volume equal to the volume of the wet gel pieces with

50 mM ammonium bicarbonate buffer containing 12.5 ng/μL sequencing grade trypsin, and digestion is carried out for 5 h at 37°C.

4. The tryptic peptides are extracted with 30 μL of an aqueous solution containing 1% (v/v) formic acid and 2% (v/v) acetonitrile for at least 2 h at room temperature.

5. The peptides are desalted on ZipTip$_{C18}$ tips. Bound peptides are directly eluted with MALDI matrix solution and spotted on MALDI target plates.

3.3. iTRAQ™ Labeling and 2D LC Separation of Tryptic Peptides

3.3.1. iTRAQ™ Labeling

1. Multiplexed isobaric tags (iTRAQ™ reagents) 114, 116, and 117 are used to label tryptic peptides from crude, KBr- and Na$_2$CO$_3$-washed ZGMs, respectively. The labeling procedure is according to the manufacturer as described previously *(16)*, and more details are given below.

2. Proteins are extracted from crude, KBr- and Na$_2$CO$_3$-washed ZGM by incubating with 0.5 M triethylammonium bicarbonate and 0.1% (w/v) SDS on ice for an hour following three 10-s sonications. The protein concentrations are determined using a Bio-Rad protein assay kit based on the Bradford method.

3. Twenty micrograms of protein in 20 μL from each fraction is reduced by adding 5 mM Tris-(2-carboxyethyl)phosphine and incubated at 60°C for 1 h, and cysteines are blocked by adding 10 mM methyl methanethiosulfonate (MMTS) and incubated in dark at room temperature for 10 min.

4. The proteins are digested with trypsin overnight (1:10, w/w, 37°C).

5. Dissolve the iTRAQ™ reagents by adding 70 μL of ethanol. The tryptic peptides are then mixed with 114, 116, and 117 iTRAQ™ reagents at room temperature for an hour.

6. The three iTRAQ™-labeled peptide samples are mixed and diluted over 10-fold in SCX buffer containing 10 mM KH$_2$PO$_4$, pH 3.0, and 25% (v/v) acetonitrile (*see* **Note 7**).

3.3.2. 2D LC Separation of Tryptic Peptides

1. The iTRAQ™-labeled peptide mixture is fractionated on a SCX MicroSpin™ column with sequential elution of bound peptides in eight salt steps: 50 μL of a 10 mM potassium phosphate buffer, pH 3.0, containing 25% (v/v) acetonitrile and 20, 50, 75, 100, 125, 150, 200, and 500 mM NaCl are used successively.

2. The volume of the eluted material from each salt step is reduced in a SpeedVac and reconstituted with 40 μL of 0.1% (v/v) TFA in water.

3. Each reconstituted sample is separated by reversed-phase chromatography using a Zorbax 300 SB C18 column on an Agilent 1100 HPLC system. The flow rate is set at 300 nL/min, and the following binary gradient is run: 0 min, 6.5% B; 9 min, 6.5% B; 12 min, 15% B; 92 min, 45% B; 97 min, 60% B; 102 min, 100% B; 104 min, 100% B; 105 min, 6.5% B; and 115 min, 6.5% B.

5. The column effluent is mixed with diluted MALDI matrix solution (at 2 mg/mL α-cyano-4-hydroxycinnamic acid) (*see* **Note 8**) through a 25-nL mixing tee and spotted on 192-well MALDI target plates using an Agilent 1100 series microcollection/spotting system.

6. The matrix is delivered to the mixing tee by an external infusion pump at a rate of 800 nL/min.

3.4. Tandem Mass Spectrometry

1. All MS and MS/MS spectra for both 2D gel and 2D LC samples are acquired on a MALDI-TOF/TOF instrument in positive ion reflection mode with a 200 Hz Nd : YAG laser operating at 355 nm. Accelerating voltage is 20 kV with 400 ns extraction delay. For MS/MS spectra, the collision energy is 1 keV, and the collision gas is air.

2. Typical spectra are obtained by averaging 3000 laser shots with the minimum possible laser energy to maintain the best resolution. Single-stage MS peptide mass fingerprints for the entire samples are collected first, and in each sample well, MS/MS spectra are acquired from the five most intense peaks above the signal to noise (S/N) ratio threshold of 30. Both MS and MS/MS data are acquired using the instrument default calibration, without applying internal or external calibration.

3. In the 2D LC-MALDI experiments, a MS survey scan is first performed for each salt fraction across the entire plate. After applying the exclusion list (*see* **Note 9**), the 10 most intense peaks above S/N ratio of 60 are selected from each well for MS/MS analysis.

3.5. Database Search and Data Analysis

3.5.1. Database Search

1. Both MS and MS/MS spectra are processed in 4700 Explorer™ software (v2.0; Applied Biosystems) with Gaussian smoothing at filter width of 7 points.

2. For MS spectra, a S/N threshold of 30 whereas for MS/MS spectra, a threshold of 20 is used to detect peaks.

3. Monoisotopic peak lists are generated in Applied Biosystem's GPS Explorer™ v2.0 and submitted to the GPS Explorer™ v2.0 search tool (based on MASCOT) for protein identification.

4. The Non-Redundant Protein Database, NCBInr, with 2,419,798 mammalian or 109,660 rodent sequences (National Center for Biotechnology Information, USA) is searched using the following parameters for 2D gels: 0 or 1 missed cleavage by trypsin, carboxyamidomethylation of cysteines as fixed modification, and methionine oxidations, N-terminal protein acetylation, Pyro-glu (N-term E), and Pyro-glu (N-term Q) as variable modifications.

5. For iTRAQ™-labeled samples, iTRAQ modification of N-terminal and lysine is selected as fixed, together with MMTS for cysteines; methionine oxidation is selected as a variable modification and 0 or 1 missed tryptic cleavage is allowed.

Table 1
Summary of iTRAQ 117:114 Ratios from Identified ZG Proteins

Protein name	Accession number	Number of spectra	117:114 ratios (mean ± SD)	TM domains[a]
ATP synthase beta chain	54792127	6	0.24 ± 0.09	0
GP3 (PLRP-2)	17105374	21	0.43 ± 0.15	0
Colipase	203503	11	0.44 ± 0.16	0
Pancreatic lipase	1865644	21	0.44 ± 0.05	0
Sterol esterase	1083805	17	0.47 ± 0.21	0
Anionic trypsin precursor	67548	4	0.49 ± 0.16	0
Syncollin	20806121	9	0.52 ± 0.10	1
Alpha-amylase	62644218	26	0.60 ± 0.17	0
ATP synthase alpha chain	114523	8	0.69 ± 0.20	0
Carboxypeptidase A1 precursor	8393183	8	0.70 ± 0.30	0
Carboxypeptidase A2 precursor	61556903	3	0.70 ± 0.25	0
ZG16	19705541	6	0.70 ± 0.15	0
Elastase 3B precursor	62649890	6	0.73 ± 0.22	0
Caldecrin precursor (chymotrypsin c)	1705913	2	0.79 ± 0.31	0
Protein disulfide isomerase precursor	129731	2	0.85 ± 0.16	0
Pancreatic lipase related protein 1 (PLRP-1)	14091772	29	0.87 ± 0.22	0
Clusterin	46048420	3	0.93 ± 0.10	0
RAB27B	16758202	7	1.35 ± 0.13	1 (prenyl)
Rac1	54607147	3	1.40 ± 0.15	1 (prenyl)
21-kDa transmembrane trafficking protein	3915137	2	1.40 ± 0.19	1
Similar to osmotic stress protein	34856875	6	1.48 ± 0.17	5
Ubiquitin	1050930	5	1.54 ± 0.32	0

(Continued)

Table 1
(Continued)

Protein name	Accession number	Number of spectra	117:114 ratios (mean ± SD)	TM domains[a]
Vacuolar-type H⁺-ATPase 115 kDa subunit, a1 isoform	13928826	7	1.74 ± 0.14	6
Lysosomal-associated membrane protein 2	40254785	2	1.85 ± 0.13	1 (GPI)
Gamma-glutamyl transpeptidase	16758696	19	2.02 ± 0.33	1
Rab8A	49522647	6	2.12 ± 0.76	1 (prenyl)
Rab6	62654200	6	2.35 ± 0.73	1 (prenyl)
Rab1	56605816	12	2.37 ± 0.75	1 (prenyl)
Signal sequence receptor, alpha	57114346	3	2.68 ± 1.20	1
RAB3D	18034781	8	2.94 ± 0.10	1 (prenyl)
GP2	121538	28	3.06 ± 1.11	1 (GPI)
Protein transport protein SEC61 beta subunit	27714473	4	3.13 ± 0.60	1
Integral membrane-associated protein 1(Itmap 1)	5916203	11	3.24 ± 0.35	2
Dipeptidase 1	16758372	7	3.30 ± 1.10	1
Voltage-dependent anion channel 2 (VDAC 2)	13786202	2	3.48 ± 0.48	1 (β-sheet)
Voltage dependent anion channel 1 (VDAC 1)	48734887	5	3.58 ± 0.62	1 (β-sheet)
Myosin Vc	62653910	7	3.67 ± 0.71	0

[a]The number of transmembrane (TM) domains are listed in the column. The TM domains are either predicted by TMHMM or known from database or literature. The known TM domains or posttranslational modifications are included in parentheses.

This table a modified version of Table 2 in our recent publication (15).

6. For all searches, precursor ion mass tolerance is 100 ppm and fragment ion mass tolerance is 0.6 Da. For both 2D gel and 2D LC experiments, only MS/MS data are used for identification. Protein identifications are considered significant when they are based on at least two unique peptides, each of which has a Mascot score corresponding to $p < 0.05$.

3.5.2. Data Analysis

1. In a representative 2D gel *(15)*, about 130 spots were visualized, among which 61 were identified by tandem mass spectrometry. In 2D LC experiments, a total of 2498 peptides were selected for MS/MS analysis, and 630 peptides were assigned to 84 non-redundant proteins.
2. The peak areas of low-mass reporter ions from the isobaric tags, 114, 116 and 117, are extracted from the spectra using 4700 Explorer™ and matched in an Excel datasheet to the identified peptides retrieved from the MS/MS summary table in GPS Explorer™ (*see* **Note 10**).
3. The complete list of identified peptides with reporter ion peak areas is then grouped into proteins for calculation of average ratios and standard deviations. Although the same peptides identified from multiple wells in the LC-MALDI experiments are counted as only one unique peptide, the iTRAQ™ values obtained from the fragmentation of these peptides are considered as independent measurements for the calculation of average ratios and standard deviations.
4. Abundance ratio calculation includes corrections for overlapping isotopic contributions (both natural and enriched ^{13}C components).
5. The iTRAQ 117:114 ratios of identified ZG proteins are summarized in **Table 1**. Proteins with 117:114 ratios greater than one (the minimal ratio measured on a known membrane protein, Rab27B, is 1.35) correlate well with intrinsic membrane proteins that contain either known or predicted transmembrane domains, whereas proteins with ratios less than one do not have known or predicted membrane domains or posttranslational modifications for membrane insertion.

4. Notes

1. Because phenylmethylsulfonyl fluoride is not stable, it is added to the buffer from a 0.2 M stock solution stored at –20°C immediately before use and every hour thereafter.
2. The blood needs to be drained as completely as possible. Large amount of residual blood in pancreas tissue may contaminate ZG preparation and is apparent as a reddish pellet at the very bottom of the white pellet after the second low-speed centrifugation. This blood contaminant is hard to separate efficiently by Percoll gradient centrifugation.
3. The loose layer of tan pellet containing mainly mitochondria can be largely removed by rinsing the pellet with homogenization buffer several times. The residual mitochondria appear in the top band of the centrifuge tubes after Percoll gradient ultracentrifugation.

4. The high salt wash alone is not efficient to remove absorbed ZG content proteins. In Coomassie-stained 2D gels of 300 μg of KBr-washed ZGM proteins, only predominant membrane proteins such as GP2 and dipeptidase and contaminant content proteins could be detected. In contrast, small GTPase such as Rab27B and Rab11 became apparent in Coomassie-stained 2D gels of 300 μg of Na_2CO_3-washed ZGM proteins.

5. ZGM pellets can be used immediately after preparation or alternatively stored in the centrifuge tubes at –80°C with the caps sealed with parafilm.

6. In our experience, ASB-C8ϕ is a key detergent in the buffer to improve the solubility of ZGM proteins. As an alternative to dissolve ZGM pellets directly in IEF buffer, we have tried to dissolve ZGM pellets in Tris–HCl buffer with 0.5% (v/v) Triton X-100 and precipitate ZGM proteins by acetone/TCA precipitation. The precipitant pellets were washed and then dissolved in IEF buffer. Direct solubilization in IEF buffer seemed to give better result.

7. If the SCX step is not planned immediately after tagging, the labeled samples can be kept separately or mixed and diluted immediately in SCX buffer. The iTRAQ tagging reaction proceeds rapidly and to completion under the conditions described here, so a specific quenching step is generally not required; however, effective quenching may be attained, if desired, by addition of 50 μL of 100 mM ammonium bicarbonate.

8. A reduced concentration of α-cyano-4-hydroxycinnamic acid is used to prevent matrix crystals from clogging the capillary in the nanoflow pump.

9. Prior to the 2D LC-MALDI MSMS experiments, a 1D LC-MALDI MSMS run is performed using 5 μg of the same sample mixture to identify the most abundant proteins in the sample. From this experiment, an exclusion list is generated containing 70 peptides from the abundant proteins in the sample including GP2, GP3, syncollin, colipase, pancreatic lipase, sterol esterase, and amylase.

10. GPS Explorer™ v3.0 allows the iTRAQ ratios to be calculated automatically without the need to export the peak areas of low-mass reporter ions.

Acknowledgments

We gratefully acknowledge Dr. John A. Williams for helping us to prepare the manuscript. We thank the Michigan Proteome Consortium for performing some of the analyses. This research was supported by the National Resource for Proteomics and Pathways (P41 RR018627) to Philip C. Andrews and also by NIH grants DK 41122 to John A. Williams.

References

1. Williams, J. A. (2001) Intracellular signaling mechanisms activated by cholecystokinin-regulating synthesis and secretion of digestive enzymes in pancreatic acinar cells. *Annu. Rev. Physiol.* **63**, 77–97.

2. Jamieson, J. D. and Palade, G. E. (1971) Synthesis, intracellular transport, and discharge of secretory proteins in stimulated pancreatic exocrine cells. *J. Cell Biol.* **50**, 135–158.

3. Pfeffer, S. R. (2001) Rab GTPases: specifying and deciphering organelle identity and function. *Trends Cell Biol.* **11**, 487–491.

4. Ungar, D. and Hughson, F. M. (2003) SNARE protein structure and function. *Annu. Rev. Cell Dev. Biol.* **19**, 493–517.

5. Ohnishi, H., Ernst, S. A., Wys, N., McNiven, M., and Williams, J. A. (1996) Rab3D localizes to zymogen granules in rat pancreatic acini and other exocrine glands. *Am. J. Physiol.* **271**, G531–G538.

6. Valentijn, J. A., Sengupta, D., Gumkowski, F. D., Tang, L. H., Konieczko, E. M., and Jamieson, J. D. (1996) Rab3D localizes to secretory granules in rat pancreatic acinar cells. *Eur. J. Cell Biol.* **70**, 33–41.

7. Hansen, N. J., Antonin, W., and Edwardson, J. M. (1999) Identification of SNAREs involved in regulated exocytosis in the pancreatic acinar cell. *J. Biol. Chem.* **274**, 22871–22876.

8. Gaisano, H. Y., Ghai, M., Malkus, P. N., Sheu, L., Bouquillon, A., Bennett, M. K., et al. (1996) Distinct cellular locations of the syntaxin family of proteins in rat pancreatic acinar cells. *Mol. Biol. Cell* **7**, 2019–2027.

9. Gaisano, H. Y., Sheu, L., Grondin, G., Ghai, M., Bouquillon, A., Lowe, A., et al. (1996) The vesicle-associated membrane protein family of proteins in rat pancreatic and parotid acinar cells. *Gastroenterology* **111**, 1661–1669.

10. Wang, C. C., Ng, C. P., Lu, L., Atlashkin, V., Zhang, W., Seet, L. F., et al. (2004) A role of VAMP8/endobrevin in regulated exocytosis of pancreatic acinar cells. *Dev. Cell* **7**, 359–371.

11. De Lisle, R. C., Schulz, I., Tyrakowski, T., Haase, W., and Hopfer, U. (1984) Isolation of stable pancreatic zymogen granules. *Am. J. Physiol.* **246**, G411–G418.

12. Burnham, D. B., Munowitz, P., Thorn, N., and Williams, J. A. (1985) Protein kinase activity associated with pancreatic zymogen granules. *Biochem. J.* **227**, 743–751.

13. Andersen, J. S., Lam, Y. W., Leung, A. K., Ong, S. E., Lyon, C. E., Lamond, A. I., et al. (2005) Nucleolar proteome dynamics. *Nature* **433**, 77–83.

14. Andersen, J. S., Wilkinson, C. J., Mayor, T., Mortensen, P., Nigg, E. A., and Mann, M. (2003) Proteomic characterization of the human centrosome by protein correlation profiling. *Nature* **426**, 570–574.

15. Chen, X., Walker, A. K., Strahler, J. R., Simon, E. S., Tomanicek-Volk, S. L., Nelson, B. B., et al. (2006) Organellar proteomics: analysis of pancreatic zymogen granule membranes. *Mol. Cell. Proteomics* **5**, 306–312.

16. Ross, P. L., Huang, Y. N., Marchese, J. N., Williamson, B., Parker, K., Hattan, S., et al. (2004) Multiplexed protein quantitation in saccharomyces cerevisiae using amine-reactive isobaric tagging reagents. *Mol. Cell. Proteomics* **3**, 1154–1169.

20

Isolation and Proteomic Analysis of *Chlamydomonas* Centrioles

Lani C. Keller and Wallace F. Marshall

Summary

Centrioles are barrel-shaped cytoskeletal organelles composed of nine triplet micro-tubules blades arranged in a pinwheel-shaped array. Centrioles are required for recruitment of pericentriolar material (PCM) during centrosome formation, and they act as basal bodies, which are necessary for the outgrowth of cilia and flagella. Despite being described over a hundred years ago, centrioles are still among the most enigmatic organelles in all of cell biology. To gain molecular insights into the function and assembly of centrioles, we sought to determine the composition of the centriole proteome. Here, we describe a method that allows for the isolation of virtually "naked" centrioles, with little to no obscuring PCM, from the green alga, *Chlamydomonas*. Proteomic analysis of this material provided evidence that multiple human disease gene products encode protein components of the centriole, including genes involved in Meckel syndrome and Oral-Facial-Digital syndrome. Isolated centrioles can be used in combination with a wide variety of biochemical assays in addition to being utilized as a source for proteomic analysis.

Key Words: Centrioles; basal bodies; *Chlamydomonas*; proteomics; MudPIT; nephronophthisis; Meckel Syndrome; Oral-Facial-Digital Syndrome; primary cilia; PACRG.

1. Introduction

The centriole *(1)*, which is at the heart of the centrosome, is composed of nine triplet microtubule blades arranged in a pinwheel-like array. The biological significance of centrioles is demonstrated by the fact that they are required for

From: *Methods in Molecular Biology, vol. 432: Organelle Proteomics*
Edited by: D. Pflieger and J. Rossier © Humana Press, Totowa, NJ

recruitment of pericentriolar material (PCM) during centrosome formation and that they act as basal bodies, which are necessary for the outgrowth of cilia and flagella. Other roles for centrioles in cytokinesis and cell-cycle regulation have been proposed but remain controversial. The remarkable centriole duplication cycle in which new daughter centrioles form at an intriguing right angle to the pre-existing mother centriole, along with the precise function of centrioles have perplexed researchers for decades.

To fully understand large macromolecular structures, such as the centriole, it is essential to identify the protein composition in its entirety. We therefore determined the first published centriole proteome, using material isolated from the green alga *Chlamydomonas reinhardtii (2)*. *Chlamydomonas* has become a prominent model organism to study centrioles for several reasons: *Chlamydomonas* has genetics very similar to yeast, but unlike yeast, it has centrioles similar to animal cells. Many molecular and genomic techniques are now routine in *Chlamydomonas*, including RNA interference (RNAi), Green Fluorescent Protein (GFP) tagging, and microarrary analysis. Furthermore, *Chlamydomonas*, unlike other genetic model organisms including *Drosophila* and *Chlamydomonas elegans*, has centrioles with triplet microtubules that are virtually identical to mammalian centrioles in both structure and duplication cycle. Unlike mammalian centrioles, which are surrounded by a complex mesh work of PCM containing over 300 proteins, *Chlamydomonas* has virtually "naked" centrioles allowing for convenient large-scale biochemical isolation and direct analysis in the absence of the obscuring PCM *(3)*. These features make *Chlamydomonas* an ideal model organism to study the composition of centrioles through proteomics.

Previously, comparative-genomic analyses were used to unveil genes conserved in species with cilia and flagella *(4)*. In addition, proteomic analyses were reported on enriched preparations of centrosomes, but that analysis was unable to distinguish the bona fide centriolar proteins from the surrounding PCM *(5)*. The exploitation of *Chlamydomonas* allows direct proteomic analysis of isolated centrioles that lack PCM, presenting a huge advantage over mammalian systems *(2)*. Our method, adapted from an earlier procedure developed in the Rosenbaum laboratory *(6)*, utilizes large-scale isolation of centrioles from *Chlamydomonas* cells. Briefly, *Chlamydomonas* cells are deflagellated and lysed in detergent before enriching for centrioles through multiple rounds of velocity sedimentation and a final round of equilibrium centrifugation. The isolated centrioles are then examined proteomically using *m*ultidimensional *p*rotein *i*dentification *t*echnology (MudPIT), a mass spectrometry-based method in which complex mixtures of proteins can be analyzed without prior electrophoretic separation *(7)*.

2. Materials

2.1. Cell Culture

1. *Chlamydomonas* (cell-wall-less strain), *Chlamydomonas Genetics Center* Duke University. Web site for the center: http://www.chlamy.org.
2. 100× Tris: dissolve 24.2 g of Tris in 100 mL of distilled H_2O (dH_2O).
3. 100× Tris-acetate-phosphate (TAP) salts: dissolve 18.75 g of NH_4Cl, 5 g of $MgSO_4 \cdot 7H_2O$, 2.5 g of $CaCl_2 \cdot 2H_2O$ in 500 mL of dH_2O.
4. Phosphate solution: dissolve 21.6 g of K_2HPO_4 and 10.8 g of KH_2PO_4 in 200 mL of dH_2O.
5. Hutner trace elements (*see* **Subheading 3.1.**).
6. Cell culture carboy, 2 gallon (for large-scale isolation), and Nalgene #2551.
7. Two-liter glass bottle (for small-scale isolation), Fisher, Pittsburgh, PA, USA.
8. Foam corks.
9. Boring tool.
10. Stir plate large enough for carboy, with large-size magnetic stir bar.
11. Fish tank pump with regulation valve (*see* **Note 1**).
12. Rubber latex tubing.
13. Glass filter unit (Bellco) plugged with cotton.

2.2. Deflagellation and Cell Lysis

1. Deflagellation buffer: 0.5 M acetic acid.
2. Recovery buffer: 0.5 M KOH.
3. TE: 10 mM Tris adjusted to pH 8.0 with HCl solution and 1 mM ethylenediaminetetraacetic acid (EDTA).
4. 25% (w/v) sucrose in TE.
5. Nonidet P-40 (NP-40) (*see* **Note 2**).
6. Protease inhibitors (final concentrations): 2 μg/mL aprotinin, 1 μg/mL pepstatin, 1 μg/mL leupeptin, 1 mM phenylmethylsulfonyl fluoride (PMSF), and 10 μg/mL soybean trypsin inhibitor (STBI).

2.3. Centriole Enrichment

1. TE: 10 mM Tris adjusted to pH 8.0 with HCl and 1 mM EDTA.
2. 25% (w/v) sucrose in TE.
3. 40% (w/v) sucrose in TE.
4. 50% (w/v) sucrose in TE.
5. 60% (w/v) sucrose in TE.
6. 70% (w/v) sucrose in TE.
7. 80% (w/v) sucrose in TE.
8. 40% (w/v) Nycodenz in TE.
9. 2 M NaCl.

10. Dounce homogenizer (40 mL).
11. Light microscope.
12. Coverslips (circle and square).
13. 12-well dish.
14. Swinging bucket rotor.
15. Plate spinner.
16. High-speed centrifuge that can reach 14,000 *g*.

2.4. Monitoring and Evaluating Centriole Enrichment

2.4.1. Immunofluorescence

1. Anti-acetylated tubulin antibody (#T6693, clone 6-11B-1; Sigma, St. Louis, MO, USA).
2. Anti-mouse secondary antibody (#115-095-003; Jackson ImmunoResearch Laboratories, West Grove, PA, USA).
3. Polylysine solution (Sigma).
4. PBST: phosphate-buffered saline (PBS) + 0.1% (w/v) Tween-20.
5. Normal goat serum.
6. Vectashield (#H-1000; Vector, Burlingame, CA, USA).

2.4.2. Western Blot

1. 0.45-μm nitrocellulose (#162-0115; Bio-Rad, Hercules, CA, USA).
2. Peroxidase-conjugated donkey anti-mouse IgG (#115-035-003; Jackson Immuno-Research Laboratories).
3. Chemiluminescence detection reagents (#RPN2132; Amersham Biosciences, Piscataway, NJ, USA).

3. Methods

The use of a flagellated cell type (*Chlamydomonas*) as a source of centrioles creates a potential pitfall in that flagella share many proteins in common with centrioles and are of a similar radius and density. This makes it difficult to separate flagella from centrioles if both are present in the starting lysate. For this reason, it is critical to carefully remove the flagella prior to cell lysis. Fortunately, the natural biology of *Chlamydomonas* provides a simple solution. *Chlamydomonas* cells exposed to a transient decrease in pH will spontaneously shed their flagella, which can then be easily separated from the cell bodies as the cell bodies are much larger than the flagella. For this strategy to work, it is critical to verify, by microscopy, that virtually all cells have undergone flagellar loss prior to beginning cell lysis. It is also important to keep the cells in the cold after deflagellation to prevent flagellar regrowth.

The major contaminant in all biochemical preparations from *Chlamydomonas* cell bodies is the chloroplast. Because the chloroplast contains naturally pigmented chlorophyll, it is possible to visually assess the degree of chloroplast contamination at each stage simply by observing the green color.

3.1. Preparation of Media

1. To make 1 L of TAP media, add 10 mL of 100× Tris, 10 mL of 100× TAP salts, 1 mL of phosphate solution, 1 mL Hutner Trace Elements, and 1 mL of glacial acetic acid to a bottle and add water to 1 L.

2. To make 1 L of Hutner Trace Elements, dissolve each compound in the volume of dH_2O indicated in **Table 1**. Mix all solutions together in a large flask, except for the EDTA stock solution. Bring this mixture to a boil before adding the EDTA. This mixture should be green in color. After everything has dissolved, cool to 70°C and maintain this temperature while adjusting the pH to 6.7 with ~85 mL of hot KOH (20%, w/v). The pH-meter should first be standardized with buffer of the same temperature. Bring this mixture to a final volume of 1 L. Put a cotton plug in the flask and let stand for 1–2 weeks at room temperature, swirling it once per day. The initially clear-green solution should turn purple eventually and have a rust-brown precipitate settled at the bottom of the flask. The precipitate is filtered through two layers of Whatman #1 filter paper two times. This clear-purple solution can be kept at 4°C for years *(8)*.

3.2. Cell Culture

1. A small pea-sized amount of a cell-wall-less strain of *Chlamydomonas* is inoculated into a 200-mL flask containing 100 mL of TAP media with a foam cork that has a hole bored through it. Allow this culture to grow until reaching a dark green color by bubbling air through a disposable pipet, sticking through the foam cork, hooked up to a standard fishtank air pump. A dark green color (demonstrated in **Fig. 1**) indicates an approximate cell density of 10^7 cells/mL. For best results, bubble at a constant but slow rate so as not to lyse the cells (a sterile stir bar can also be used to keep the cells from settling).

Table 1
Hutner's Trace Elements

	Grams salt	Milliliters of dH_2O
EDTA, disodium salt	50	250
$ZnSO_4 \cdot 7H_2O$	22	100
H_3BO_3	11.4	200
$MnCl_2 \cdot 6H_2O$	5.06	50
$CuSO_4 \cdot 5H_2O$	1.57	50
$((NH4)_6Mo_7)_{24} \cdot 4H_2O$	1.10	50
$FeSO_4 \cdot 7H_2O$	4.99	50

Fig. 1. Photograph of 100 mL of culture of *Chlamydomonas* illustrating the correct density of culture for procedure.

2. Prepare the 8-L culture carboy of TAP media. Autoclave TAP media in container with foam cork that has a glass pipet through it attached to a glass filter unit filled with cotton and latex rubber tubing (*see* **Fig. 2** for photograph of apparatus). Sterilization of entire apparatus plus tubing significantly decreases the chance of contamination. Once the 100-mL culture is dark green, pour this into the 8-L carboy and allow growth with aeration and stirring until obtaining a dark green color. For proteomic analysis, it is recommended to start with at least 32 L of *Chlamydomonas* culture (*see* **Note 3**).

3.3. Deflagellation and Cell Lysis

1. Concentrate *Chlamydomonas* from 32 L of culture by low-speed centrifugation (400 *g* for 3 min without brake). Aspirate media off cell pellet. Add TAP media to each of the cell pellets and combine them until a final volume of 400 mL is

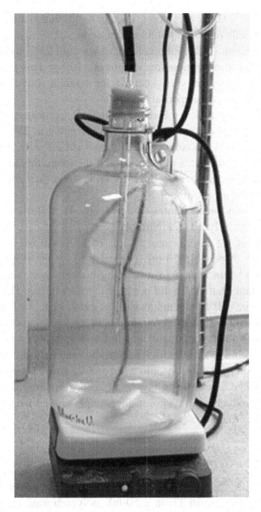

Fig. 2. Photograph of 8-L carboy apparatus showing configuration of foam plug, rubber hose, and glass filter holding unit. This should be assembled, filled with TAP media, a stir-bar added, and then entire apparatus autoclaved.

reached. Let the cells bubble and stir for 90 min. Check that cells are alive and motile by looking with a compound microscope.

2. After 90 min of aeration, which ensures that cells have recovered from the first spin and will be able to robustly deflagellate in response to pH shock, concentrate the cells again by centrifugation at 400 g for 3 min without brake. Aspirate the supernatant and resuspend in 100 mL of TAP media.

3. Put the 100 mL on ice and reduce the pH to 4.5–4.7 with 0.5 M acetic acid for 1 min while stirring and monitoring with a pH-meter. Then add 0.5 M KOH to

return the pH to 7.0. At this point, check that deflagellation was successful by looking at the cells on a compound microscope.

4. After deflagellation, overlay the cells onto ice-cold 25% sucrose in TAP media. Spin cells on sucrose cushion for 30 min at 1200 *g*. This spin should pellet the cells through the sucrose while keeping the flagella above the sucrose cushion. Aspirate and discard the supernatant.

5. Resuspend cell pellet in 150 mL of ice-cold TAP. Centrifuge for 20 min at 1200 *g* at 4°C. Aspirate supernatant and resuspend the pellet in 25 mL of ice-cold TE.

6. Add NP-40 to 10% of the volume (in this case add 2.5 mL). Also add the protease inhibitors. Let this stir for 5 min on ice and then take this sample and put it in a 40-mL dounce homogenizer. Dounce 15 strokes and then put the sample back on ice for stirring for an additional 10 min.

7. Check cell lysis by phase contrast microscopy. It is absolutely essential that all cells are completely lysed (*see* **Note 4**).

8. Layer the 25 mL of lysate onto 25 mL of 25% sucrose in TE cushion. This can be done in disposable 50-mL conical tubes. Centrifuge the lysate/cushion at 1500 *g* for 15 min at 4°C in a hanging bucket rotor. Then collect the entire green supernatant (that which did not go into the cushion). This first spin removes most of the large cell fragments as well as much of the cellular starch granules; however, this sample will still contain lots of vesicles and starch granules visible by microscopy, so we recommend an additional spin over a fresh sucrose cushion to completely remove large particles from the preparation. Layer the supernatant from the first sucrose cushion onto a second 25-mL cushion of 25% sucrose in TE and centrifuge at 1500 *g* in hanging bucket rotor for 15 min at 4°C. Collect the entire supernatant. At this point, the sample can be frozen in liquid nitrogen until future use. Keep at –80°C.

3.4. Enrichment of Centrioles by Sedimentation

1. Layer the 25 mL of lysate onto 10 mL of 50% sucrose in TE overlayed with 10 mL of 40% sucrose in TE in a 50-mL polycarbonate tube. Spin at 14,000 *g* in hanging bucket rotor for 1 h. Collect the entire 40% sucrose fraction. The 50% fraction and the pellet forming below it are green in color, suggesting that a portion of the chloroplasts is removed during this step. Layer this 40% sucrose fraction onto a step gradient in a polycarbonate tube consisting of (from the bottom) 5 mL of 80%, 5 mL of 70%, 5 mL of 60%, and 5 mL of 50% sucrose in TE. Centrifuge this sample at 14,000 *g* in hanging bucket rotor for 3 h. After centrifugation, collect 2-mL fractions starting from the top and freeze in liquid nitrogen. Save each of these fractions at –80°C for western blot analysis and immunofluorescence.

2. During analysis of the fractions from the 80–50% sucrose in TE spin (*see* **Subheadings 3.5.** and **3.6.**), prepare a Nycodenz gradient. Make 40% Nycodenz in TE and freeze-thaw the solution in a polycarbonate tube three times. This freeze-thaw method creates a continuous density gradient of Nycodenz.

3. Combine the two fractions having the highest enrichment of centrioles (as determined by western blot and/or immunofluorescence) from the 80–50% sucrose, spin them down at 35,000 g, and wash them by resuspending in 4 mL of TE containing 2% (v/v) NP-40 and 500 mM NaCl for 10 min to remove any peripherally associated proteins. Soluble material from this wash will not enter the next gradient and therefore does not need to be removed at this stage. Add the mixture of the two fractions suspended in 4 mL TE plus NP-40 and NaCl to the top of the pre-made Nycodenz gradient and centrifuge for 18 h at 14,000 g in hanging bucket rotor to reach equilibrium. At this point, there should be a visible band where the fractions migrated during the spin. Take 1-mL fractions from the top of the tube and proceed with determination of centriole enrichment by both Western blot and immunofluorescence.

3.5. Tracking Centriole Enrichment by Immunofluorescence

1. It is important to be able to track where the centrioles are during the course of the isolation to select the fractions from each step that contain the majority of centrioles. Using antibodies directed against centriolar proteins, one can determine which of the fractions collected throughout the preparation are enriched for centrioles. We have used commercially available acetylated tubulin antibodies for this procedure.
2. Add 200 μL of polylysine to coverslips and let sit for 5 min. Rinse the coverslips in dH$_2$O and let air-dry. To speed procedure up, excess of dH$_2$O can be aspired. Pipet ~20 μL of sample onto polylysine-coated coverslip and let sit for 10 min to adhere (see below for method to increase the number of centrioles per coverslip). After the 10 min, rinse coverslips in PBST once quickly followed by a 1-min and a 5-min incubation in PBST. Block coverslips with normal goat serum diluted 1:10 in PBST for 15 min. After blocking, wash coverslips with PBST for 5 min.
3. To stain coverslips with acetylated tubulin, dilute antibody 1:500 in PBST and put 100-μL drops on a piece of Parafilm. Then take coverslips and invert them onto the 100-μL drops of diluted antibody. Let coverslips sit for 20 min. After antibody staining, wash the coverslips twice with PBST for 10 min each. During these washes, dilute anti-mouse secondary antibody 1:1000 in PBST and make 100-μL drops on Parafilm again. After washing, invert the coverslips onto the secondary antibody. Let coverslips incubate in secondary antibody for 15 min. Wash coverslips with PBST twice for 15 min each and mount coverslips with a small drop of Vectashield. For the best visualization, aspirate excess Vectashield around coverslip and polish edges.
4. To increase the number of centrioles per coverslip, a spin-down method can be used. This method involves a 12-well dish and 18-mm circle coverslips. Circle coverslips are first washed with water and soap and then rinsed well with dH$_2$O. Coverslips are then polylysine-coated by adding 200 μL of polylysine for 10 min. Rinse coverslips with H$_2$O and let air-dry. Put the coverslips into the bottom of the 12-well dish (polylysine-coated side up). Dilute sample 1:4 mL in TE and add 4.8 mL of this mixture into each well of the 12-well dish. Spin this sample

(make sure there is another 12-well plate for balancing) in a centrifuge with plate holders at 750 *g* for 10 min with NO brake. Stain coverslips as stated above.

3.6. Evaluating Enrichment by Western Blotting

1. The immunofluorescence method described above provides a rapid way to assess the progress of the isolation. Once the procedure is completed, progressive enrichment can be evaluated throughout the procedure by analyzing frozen samples on western blots using centriole-specific antibodies. Samples with equal amounts of protein (determined by Bradford assay) are run on a 10% acrylamide sodium dodecyl sulphate–polyacrylamide gel electrophoresis (SDS–PAGE) gel under normal conditions. The gels are then transferred onto 0.45-μm nitrocellulose.

2. Western blot analysis is then performed by using a mouse monoclonal anti-acetylated tubulin IgG at a dilution of 1:1000 and a peroxidase-conjugated donkey anti-mouse IgG at a dilution of 1:10,000. Detection by chemiluminescence can be performed using commercially available reagents.

3. We also analyzed our final preparation using two-dimensional gel electrophoresis to estimate the complexity of the mixture. Starting with 32 L of *Chlamydomonas* cells, we ended up with a 200 μL sample that contained ~100 μg of total protein. Two-dimensional gel electrophoresis was then performed on this final sample by Kendrick Labs (Madison, WI) with ampholines spanning pH values of 3.5–10 (Amersham Pharmacia Biotech, Piscataway, NJ). The second dimension was run in a 10% acrylamide gel (0.75-mm thick) for 4 h at 12.5 mA. This analysis indicated that at least 50–100 proteins were present by silver stain.

3.7. Proteomic Analysis by MudPIT Mass Spectrometry

1. LC-LC-MS/MS (two-dimensional liquid chromatography followed by tandem mass spectrometry) was carried out by MudPIT analysis in the Yates laboratory, Scripps Research Institute (see **Note 5**). Each LC-LC-MS/MS analysis was done with a starting material of ~100 μg of protein. Mass spectrometry was performed using a biphasic column (strong cation exchange, SCX, and reversed phase, RP) coupled to an LCQ ion trap mass spectrometer (ThermoFinnigan, San Jose, CA, USA). Fourteen salt steps were performed to displace the initial sample loaded on the SCX column onto the RP column. Peptides were then eluted into the mass spectrometer using a linear gradient of 5–60% reversed-phase solvent B (acetonitrile/acetic acid/water, 80/0.5/19.5, v/v/v) over 60 min. The collection of resulting MS^2 spectra was searched against the *C. reinhardtii* predicted protein database (JGI Assembly October 2003, release 2.0) using the SEQUEST algorithm *(9)*.

2. To compensate for high levels of chloroplast contamination throughout the final Nycodenz gradient (as judged by uniform green color in all fractions), we analyzed not only the peak centriole fraction but also two additional fractions above and below the peak in the gradient. Proteins identified in these fractions were subtracted from the protein list identified in the peak fraction. This resulted in a

final list of 194 proteins. This subtracted list still contained a substantial number of known chloroplast proteins, suggesting that more extensive analysis of the control fractions could be beneficial.

3. The final list of proteins contained 8 out of 11 previously known *Chlamydomonas* centriole proteins. In addition, 45 new proteins were identified, whose localization in the centriole was supported by bioinformatic cross-validation, including presence in the human centrosome proteome *(5)*, conservation of the genes in species that contain centrioles but not those missing centrioles *(4)*, or upregulation of the genes during flagellar regeneration *(10)*, a property shared by many genes encoding basal-body-localized proteins. Among the new genes were the *Chlamydomonas* homologs of four mammalian ciliary disease genes: oral-facial-digital syndrome type I, nephronophthisis NPHP-4, Meckel syndrome MKS1, and the Parkin coregulated gene PACRG. These results suggest that mutations in these genes cause ciliary disease because of defects in centrioles/basal bodies that nucleate cilia and indicate that further proteomic analysis of centrioles may identify additional ciliary disease genes.

4. Notes

1. Any fish tank pump with adjustable air-flow rate will be adequate. We used Air-Tech #VAT-5.5.
2. It is essential that name-brand Nonidet P-40 be used. Other supposedly equivalent detergents fail to completely lyse *Chlamydomonas* cell walls.
3. Smaller-scale centriole isolations may be carried out to examine centriole ultrastructure. We recommend a no less than 2-L starting volume be used for this purpose. Proceed as suggested above but scale down appropriately. Western blots and immunofluorescence may be done on fractions to determine approximate enrichment. These fractions can then be processed for electron microscopy to investigate ultrastructure. Centrioles from mutant strains of *Chlamydomonas* can be processed in this fashion to examine ultrastructural defects in centrioles. Remember that this procedure is designed for cell-wall-less strains, so mutant lines may have to be crossed into a cell-wall-less background. It is often cumbersome to deal with large volumes so we recommend doing a maximum of 8-L of centriole preparation at one time if larger samples are necessary.
4. If cells visualized by phase contrast microscopy are not completely lysed, allow sample to stir on ice for an additional 5 min and recheck.
5. We strongly encourage proteomics to be done in collaboration with established experts in the field.

Acknowledgments

The authors thank Edwin Romijn and John R. Yates III for a highly productive collaboration and for many helpful discussions. This work was supported by NSF grant MCB0416310, NIH grant R01 GM077004-01A1, the Searle Scholars Program, a Hellman Family Award for Early Career Faculty, and a UCSF REAC award.

References

1. Marshall, W. F. (2001) Centrioles take center stage. *Curr. Biol.* **11**, R487–R496.
2. Keller, L. C., Romijn, E. P., Zamora, I. Yates, J. R., and Marshall, W. F. (2005) Proteomic analysis of isolated Chlamydomonas centrioles reveals orthologs of ciliary disease genes. *Curr. Biol.* **15**, 1090–1098.
3. Doxsey, S., Zimmerman, W., and Mikule, K. (2005). Centrosome control of the cell cycle. *Trends Cell Biol.* **15**, 303–311.
4. Li, J. B., Gerdes, J. M, Haycraft, C. J., Fan, Y., Teslovich, T. M., May-Simera, H., et al. (2004). Comparative genomics identifies a flagellar and basal body proteome that includes the BBS5 human disease gene. *Cell* **117**, 541–552.
5. Andersen, J. S., Wilkinson, C. J., Mayor, T., Mortensen, P., Nigg, E., and Mann, M. (2003). Proteomic characterization of the human centrosome by protein correlation profiling. *Nature* **426**, 570–574.
6. Snell, W. J, Dentler, W. L., Haimo, L. T., Binder, L. I., Rosenbaum, J. L. (1974). Assembly of chick brain tubulin onto isolated basal bodies of *Chlamydomonas reinhardtii*. *Science* **185**, 357–360.
7. Washburn, M. P., Wolters, D., and Yates, J. R. III (2001). Large-scale analysis of the yeast proteome by multidimensional protein identification technology. *Nat. Biotechnol.* **19**, 242–247.
8. Harris, E. H. (1989). Procedures and Resources *The Chlamydomonas Sourcebook: A Comprehensive Guide to Biology and Laboratory Use.* Academic Press Inc., San, Diego, CA, USA pp. 578–579.
9. Eng, J., McCormack, A., and Yates, J. R. (1994). An approach to correlate tandem mass-spectral data of peptides with amino-acid sequences in a protein database. *J. Am. Soc. Mass Spectrom.* **5**, 976–989.
10. Stolc, V., Samanta, M. P., Tongprasit, W., and Marshall, W. F. (2005). Genome-wide trancriptional analysis of flagellar regeneration in *Chlamydomonas reinhardtii* identifies orthologs of ciliary disease genes. *Proc. Natl. Acad. Sci. U.S.A.* **102**, 3703–3707.

21

Purification and Proteomic Analysis of 20S Proteasomes from Human Cells

Marie-Pierre Bousquet-Dubouch, Sandrine Uttenweiler-Joseph, Manuelle Ducoux-Petit, Mariette Matondo, Bernard Monsarrat, and Odile Burlet-Schiltz

Summary

The 20S proteasome is a multicatalytic protein complex present in all eukaryotic cells. Associated to regulatory complexes, it plays a major role in cellular protein degradation and in the generation of Major Histocompatibility Complex (MHC) class I antigenic peptides. In mammalian cells, this symmetrical cylindrical complex is composed of two copies of 14 distinct subunits, three of which possess a proteolytic activity. The catalytic standard subunits can be replaced by immunosubunits to form the immunoproteasome, which possesses different proteolytic efficiencies. Both types of 20S proteasomes can be present in cells in varying distributions. The heterogeneity of 20S proteasome complexes in cells leads to different protein degradation patterns. The characterization of the subunit composition of 20S proteasomes in cells thus represents an important step in the understanding of the effect of the heterogeneity of proteasome complexes on their activity. This chapter describes the use of proteomic approaches to study the subunit composition of 20S proteasome complexes purified from human cells. An immunoaffinity purification method is presented. The separation of all 20S proteasome subunits by 2D gel electrophoresis and the subunit identification by matrix-assisted laser desorption ionization time-of-flight mass spectrometry (MALDI-TOF MS) analysis and database search are then described. These methods are discussed with the study of 20S proteasomes purified from two human cancer cell lines.

Key Words: Immunoaffinity chromatography; 2D gel electrophoresis; MALDI-TOF mass spectrometry; peptide mass fingerprint; human colorectal Caco2 cells; monoblastic U937 cells; human 20S proteasome; catalytic protein complex.

From: *Methods in Molecular Biology, vol. 432: Organelle Proteomics*
Edited by: D. Pflieger and J. Rossier © Humana Press, Totowa, NJ

1. Introduction

The 20S proteasome constitutes the catalytic core complex of the 26S proteasome, which is the main machinery responsible for intracellular protein degradation. In mammalian cells, the 20S proteasome is composed of 14 distinct subunits arranged in four stacked rings of seven subunits each, $\alpha_7\beta_7\beta_7\alpha_7$ (1). The catalytic activity is attributed to three β-type subunits, β1, β2, and β5, whose proteolytic specificity is characterized by a peptidylglutamyl-peptide hydrolyzing (PGPH) activity, a trypsin-like activity, and a chymotrypsin-like activity, respectively (2). The 20S proteasome subunit composition may vary as a function of cellular environment, and the three catalytic subunits can be exchanged by three interferon γ-inducible catalytic subunits, β1i, β2i, and β5i, leading to the immunoproteasome. Standard 20S proteasome and immunoproteasome differ in their protein degradation efficiency, resulting in the production of different sets of peptides. This may, for example, dramatically affect the generation of MHC class I antigenic peptides (3–6). In the literature, evidence has accumulated that the proteasome subunit composition is heterogeneous and that the distribution of the different proteasome forms varies in different cells and tissues (7–9). The precise characterization of 20S proteasome forms in cells thus represents a major challenge to investigate the structure/activity relationships of this multicatalytic complex.

Proteomic approaches represent an efficient strategy to study 20S proteasome subunit composition heterogeneity. The separation of 20S proteasome subunits by 2D gel electrophoresis followed by their identification by mass spectrometry (MS) has been used successfully to identify subunits of 20S proteasomes purified from different species (10–15). These analyses reveal, in addition to various proportions of standard catalytic subunits and immunosubunits, the presence of isoforms for most subunits, which suggests an increased heterogeneity of 20S proteasome complexes. Further MS/MS analyses may allow the characterization of these isoforms, which may be due to post-translational modifications and splicing variants. The effect of subunit modifications and of the subunit composition heterogeneity in general on the 20S proteasome activity remains to be investigated.

This chapter focuses on the proteomic analysis of 20S proteasomes purified from human cells. The first part describes an immunoaffinity purification method used to purify human 20S proteasomes from various cell lines. This method can also be applied to human cells from tissue. The second part presents proteomic analyses based on 2D gel electrophoresis separation of purified 20S proteasome subunits and subunit identification by matrix-assisted laser desorption ionization time-of-flight mass spectrometry (MALDI-TOF MS). The whole procedure is illustrated in a third part by the analysis of purified 20S proteasomes from two human cancer cell lines.

2. Materials

2.1. Production and purification of MCP21 antibodies

1. MCP21 cell line (ECACC Hybridoma collection, No. 96030418).
2. Culture medium: Dulbecco's modified Eagle's medium (DMEM), 10 mM Hepes, 2 mM L-glutamine, and 5% fetal calf serum (FCS). The pH of this medium is not adjusted.
3. 0.2 µm filter.
4. Sepharose HiTrap Protein G column (5 mL; GE Healthcare, Uppsala, Sweden).
5. 2 M Tris base.
6. Phosphate-buffered saline (PBS).
7. Elution buffer: 0.1 M citric acid (pH must be around 3.0).
8. Dialysis membrane (10-kDa cut-off membrane).
9. Immobilization buffer: 0.1 M NaHCO$_3$, and 0.5 M NaCl. The pH of this solution is adjusted to 8.3 with 2 N NaOH.

2.2. Covalent immobilization of MCP21 antibodies on Sepharose

1. CNBr-activated Sepharose™ 4 Fast Flow beads (GE Healthcare).
2. Glass filter (pore size <40 µm to retain sepharose beads).
3. Hydration solution: 1 mM HCl.
4. Glycine solution: 0.2 M glycine in immobilization buffer.
5. Immobilization buffer: 0.1 M NaHCO$_3$ and 0.5 M NaCl. The pH of the solution is adjusted to 8.3 with 2 N NaOH.
6. Wash buffer ❶: 0.1 M sodium acetate, pH 4.0 (mix 0.1 M sodium acetate and 0.1 M acetic acid to reach pH 4.0), and 0.5 M NaCl.
7. Wash buffer ❷: 0.1 M Tris–HCl, pH 8.0, and 0.5 M NaCl.

2.3. Purification of 20S proteasome from human cell lines

1. Lysis buffer: 20 mM Tris–HCl, pH 7.6, 50 mM NaCl, 10 mM ethylenedi-aminetetraaceticacid (EDTA), protease inhibitors (2 tablets for 100 mL buffer; Roche, Mannheim, Germany, reference: 1873580), and phosphatase inhibitors (1 mM sodium orthovanadate, 10 mM sodium pyrophosphate, and 20 mM sodium fluoride) (*see* **Note 1**).
2. Liquid nitrogen.
3. Water bath at 30°C.
4. FPLC system equipped with a UV detector and a pump.
5. FPLC column (10 mm ID, 10 cm length) fitting to the FPLC system.
6. Equilibration buffer: 20 mM Tris–HCl, pH 7.6, 150 mM NaCl, and 1 mM EDTA.
7. Elution buffer: 20 mM Tris–HCl, pH 7.6, 2 M NaCl, and 1 mM EDTA.

2.4. Salt elimination and TCA/acetone protein precipitation

1. Dialysis membrane (10-kDa cut-off membrane).
2. 1.5 mL microcentrifuge tubes.

3. Floating stand for microcentrifuge tubes.
4. 40% trichloroacetic acid solution, stored at 4°C.
5. Acetone, stored at –20°C.

2.5. 2D gel electrophoresis separation

1. Resuspending buffer: 9 M urea, 2.2% (w/v) CHAPS, and 0.003% (w/v) bromophenol blue. Aliquots can be stored at –20°C.
2. 2 M dithiothreitol (DTT).
3. Bio-lyte ampholytes 3–10 buffer (Bio-Rad, Hercules, CA, USA).
4. One- and two-dimensional gel electrophoresis equipment (Bio-Rad or GE Healthcare).
5. First dimension gel strips: ReadyStrip™ IPG Strips, 17 cm, pH 3–10 NL (Bio-Rad, reference: 163-2009) (*see* **Note 2**).
6. Mineral oil: biotechnology grade (use the oil recommended by the one-dimensional gel electrophoresis device's supplier).
7. Electrode wicks (Bio-Rad).
8. Equilibrating buffer: 50 mM Tris–HCl, pH 8.8, 6 M urea, 30% (v/v) glycerol, 2% (w/v) sodium dodecyl sulphate (SDS), and 0.25% (w/v) bromophenol blue. Prepare aliquots of 2.5 mL and store at –20°C.
9. 12–12.5% acrylamide gel for SDS–polyacrylamide gel electrophoresis (PAGE) second dimension of separation (*see* **Note 3**).
10. Molecular weight markers (low range: ~10 kDa–120 kDa).

2.6. 2D gel staining

1. Fixation solution: EtOH/water/acetic acid, 40/50/10 (v/v/v).

2.6.1. Coomassie Blue staining

1. Staining solution: 1 tablet of PhastGelBlue (GE Healthcare, reference: 17-0518-01) in 500 mL of EtOH/water/acetic acid, 40/50/10 (v/v/v).
2. Destaining solution: EtOH/water/acetic acid, 25/67/8 (v/v/v).

2.6.2. Colloidal Coomassie Blue staining

1. Prestaining solution: 17% (w/v) ammonium sulphate, 34% (v/v) methanol, and 3% (w/v) phosphoric acid in water.
2. Staining solution: prestaining solution supplemented with 0.1% (w/v) colloidal blue G 250 (*see* **Note 4**).

2.6.3. Silver nitrate staining

1. Sensitizing solution: 3 g/L potassium tetrathionate, 0.5 M potassium acetate, and 30% (v/v) ethanol.
2. Staining solution: 2 g/L silver nitrate.

3. Development solution: 30 g/L potassium carbonate, 0.012% (v/v) formaldehyde, and 0.00125% (w/v) sodium thiosulfate.
4. Destaining solution: 15 mM potassium ferricyanide and 50 mM sodium thiosulfate.

2.7. In-gel digestion and peptide extraction

1. Trypsin, sequencing grade (Sequencing Grade Modified Trypsin; Promega, Madison, WI, USA reference: V511A).
2. Trypsin solution: 12.5 ng/μL of trypsin in 12.5 mM ammonium bicarbonate prepared from a stock solution of 0.1 μg/μL of trypsin in the Promega resolubilization buffer provided by the supplier.
3. Ammonium bicarbonate.
4. Acetonitrile.
5. Thermomixer.
6. Speed-vacuum centrifuge.
7. Ultrasonic bath.

2.8. Sample preparation for MALDI-TOF analysis

1. ZipTip$_{\mu C18}$ (Millipore, Bedford, MA, USA reference: ZTC18M960).
2. Equilibration solution: 0.1% (v/v) trifluoroacetic acid (TFA) in water.
3. Elution solution: H$_2$O/acetonitrile/TFA, 20/80/0.1 (v/v/v).
4. Matrix solution: 10 mg/mL of α-cyano-4-hydroxy-cinnamic acid (Sigma, Steinheim, Germany reference: C-2020) in H$_2$O/ACN/TFA (50/50/0.1, v/v/v). Vortex vigorously and let stand on the bench at room temperature for 30 min–1 h for maximum solubilization in the solution, whose concentration in matrix is close to saturation.
5. MALDI plate.

2.9. MALDI-TOF analysis and database search

1. MALDI-TOF mass spectrometer.
2. Database search software for "peptide mass fingerprint" data.

3. Methods

The methods developed in the following subheadings describe *(1)* the purification of 20S proteasome from human cell lines and *(2)* the proteomic analysis of purified 20S proteasome complexes based on 2D gel electrophoresis and "peptide mass fingerprint" analysis. The whole procedure is then illustrated with the proteomic analysis of purified 20S proteasomes from two human cancer cell lines.

3.1. Purification of 20S proteasomes from human cell lines

We perform immunoaffinity chromatography to purify human 20S proteasome complexes. This approach uses antibodies specifically interacting with one 20S proteasome subunit. The method is fast and provides highly purified active proteasomes even though the degree of purity may vary from one cell line to another. The 20S proteasome purity can be controlled by 2D gel electrophoresis analysis.

3.1.1. Production and purification of MCP21 antibodies

The hybridoma cell line MCP21 secretes antibodies (IgG1) reacting with the α2 subunit of human and rabbit proteasome complexes, which can either be native or denatured.

1. Maintain the culture between 5×10^4 and 1×10^6 cells/mL, under 5% CO_2 and at 37°C.
2. Culture cells until you reach a volume between 1.5 and 2 L of cell suspension and let the cells grow for 1 week without medium replacement to increase the antibody concentration in the medium.
3. Recover the supernatant by centrifugation (10 min, 1200 g) and clarify it by filtration on 0.2 μm filters.
4. Purify the antibodies by FPLC on a Sepharose HiTrap Protein G column:
    a. Equilibrate the column with PBS buffer (flow rate 1 mL/min).
    b. Load the supernatant containing antibodies onto the column (flow rate 0.5–1 mL/min, 4°C, overnight) (*see* **Note 5**).
    c. Wash the column with PBS buffer (flow rate 1 mL/min, 2h).
    d. Elute with the acidic elution buffer at a flow rate of 0.5 mL/min.
    e. Collect 4-mL fractions in tubes containing 200 μL of 2 M Tris-base to neutralize the pH.
5. Pool the most concentrated fractions (OD > 0.1) and dialyze them against immobilization buffer.
6. Calculate the MCP21 antibody concentration by measuring the solution absorbance at 280 nm ($\varepsilon_{280} = 1.4$ cm^{-1}/gL^{-1}) (*see* **Note 6**).

3.1.2. Covalent immobilization of MCP21 antibodies on Sepharose

1. Add 10 mL of hydration solution to 1 g of dehydrated CNBr-activated Sepharose™ 4 Fast Flow beads.
2. Wash the beads with 200 mL of hydration solution using a glass filter. Gently scrape the beads from the glass filter using a spatula (*see* **Note 7**).
3. Add 8 mg of 0.5–1 mg/mL purified MCP21 antibody solution prepared in immobilization buffer to the beads and incubate overnight at 4°C under gentle agitation (*see* **Note 8**).

4. Wash the beads with 200 mL of immobilization buffer using the glass filter.
5. Incubate the beads with 10 mL of glycine solution for 6 h at 4°C under gentle agitation. Glycine saturates the CNBr sites that did not react with antibodies.
6. Wash the beads successively with 100 mL of wash buffers ❶ and ❷. Repeat three more times.
7. Recover the beads in ~15 mL of 10 times diluted wash buffer ❷ and store at 4°C (*see* **Note 9**).

3.1.3. Purification of 20S proteasome from human cell lines

1. Grow cells until you obtain $0.5–1.0 \times 10^9$ cells. Harvest cells by centrifugation at 1500 g and wash them three times with PBS (*see* **Note 10**).
2. Suspend the pellet in 30–50 mL of lysis buffer and break the cell wall by three consecutive freeze/thaw cycles (liquid nitrogen/30°C water bath) to obtain cytoplasmic proteasome.
3. Centrifuge to eliminate cell debris (48,000 g, 2 h, 4°C).
4. Sediment or centrifuge the sepharose beads (1–5 min, low speed, 2000 g maximum to avoid bead damage) and discard the supernatant. Incubate immediately the lysis extract supernatant with sepharose beads coupled to MCP21 antibodies (overnight, 4°C, gentle agitation to enable permanent homogenization of the mixture).
5. Discard the major part of the supernatant by sedimentation or centrifugation (1–5 min, low speed, 2000 g maximum).
6. Fill the FPLC column with the sepharose beads (*see* **Note 11**) and connect the column to the FPLC system. Rinse it with the equilibration buffer for 30 min (0.5 mL/min, 4°C) or longer if necessary to obtain a stable UV baseline signal. Switch the mobile phase to the elution buffer (0.5 mL/min, 4°C) to elute the proteasome with a saline step. Collect 2-mL fractions (*see* **Note 12**).
7. Calculate the proteasome concentration in the fractions by measuring their absorbance at 280 nm ($\varepsilon_{280} = 0.865$ cm^{-1}/gL^{-1}) (*see* **Note 13**).

3.2. Proteomic analysis of purified 20S proteasomes

Proteomic analyses can be performed on a small quantity of proteasome sample (1–40 μg) to control the purity and thus the purification steps efficiency. It can also be performed on larger quantities (40–100 μg) to study the proteasome subunit composition in more detail. This involves the identification of all the proteins detected after 2D gel separation, including subunit isoforms, and possibly the characterization of post-translational modifications. The main steps to identify all subunits of a purified proteasome sample are described in this subheading and are as follows: (1) salt elimination and protein precipitation, (2) 2D gel electrophoresis separation and staining, (3) protein in-gel digestion and peptide extraction, and (4) sample cleanup and MALDI-TOF MS analysis.

3.2.1. Salt elimination and TCA/acetone protein precipitation

Purified proteasome samples are stored in a highly saline solution (2 M NaCl), which is prohibited to perform isoelectrofocusing, the first dimension of 2D gel electrophoresis separation. Before performing 2D gel electrophoresis separation, purified proteasome samples are thus desalted by dialysis against deionized water. We usually perform a "small-scale dialysis" adapted from the protocol of Orr et al. *(16)* for sample volumes ranging from 50 to 500 µL. A precipitation step using TCA/acetone is then performed to ensure complete salt elimination and to concentrate and resuspend the proteasome in the appropriate buffer for isoelectrofocusing separation.

1. To dialyze the proteasome fractions in 2 M NaCl solution against deionized water, prepare centrifuge tube caps by cutting off the middle part of the cap using a cutter blade (*see* **Note 14**). Load the sample into a centrifuge tube. Cut 2–3 cm² of single-thickness dialysis membrane, rinse it in deionized water, and remove excess liquid. Place the membrane on top of the centrifuge tube containing the sample (*see* **Fig. 1A**) and push down the perforated cap (*see* **Fig. 1B**). Place the tube upside down on a floating stand so as to position the membrane in the dialysis buffer. Gently tap the tube to ensure that the whole sample is in contact with the membrane interface and check that no air bubble is trapped under the membrane surface (*see* **Fig. 1C**). Gently stir the dialysis buffer overnight at 4°C.

2. Perform all precipitation steps in 1.5- or 2-mL centrifuge tubes at 4°C. Add 1 volume of cold 40% (w/v) trichloroacetic acid to the proteasome sample. After vigorous homogenization, keep at 4°C for 30 min. Centrifuge for 15 min at 14,000 *g*, 4°C (*see* **Note 15**). Discard the supernatant by gentle pipetting at the side opposite to the pellet. Wash with one volume of cold acetone, vortex, and centrifuge for 10 min at 14,000 *g*, 4°C. Discard the supernatant. Wash with acetone one more time. Leave the pellet for 1 h at room temperature to ensure complete acetone evaporation.

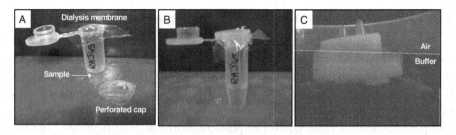

Fig. 1. Microdialysis of 20S proteasome samples. (**A**) A small piece of dialysis membrane is placed on top of the tube containing the proteasome sample and a perforated tube cap has been prepared. (**B**) The perforated cap is pushed down on top of the tube. (**C**) The tube is placed on a floating stand and put upside down with the dialysis membrane in the dialysis buffer.

3.2.2. 2D gel electrophoresis separation

The 17 constitutive 20S proteasome subunits (14 standard subunits and 3 immunosubunits) possess pI ranging from 4 to 9 and molecular weights ranging from 21 to 31 kDa. To separate all subunits in a single gel, IPG strips of 3–10 in pI scale and of at least 11 cm in length are necessary. However, to achieve isoform separation, IPG strips of 17 cm in length are necessary, and the use of a non-linear ampholyte gradient improves the separation resolution in the middle range of pI. To obtain confident mass spectrometric identifications, we usually perform the 2D gel electrophoresis separation with 40 μg of purified proteasome.

1. Resuspend the precipitated proteasome sample in 344 μL of resuspending buffer and add 5.3 μL of 2 M DTT to obtain a 30 mM final concentration. Vortex vigorously. Let stand for 1 h at room temperature and then overnight at 4°C.
2. Add 0.7 μL of Bio-Lyte ampholytes 3–10 buffer to the sample just before isoelectrofocalization. Rehydrate the IPG gel strip with the sample solution first in a passive way for 9 h at 20°C and second in an active way for 6 h at 50 V. Increase the voltage linearly from 50 to 8000 V in 2 h and maintain the high voltage until you reach a total of 65,000 Vh (*see* **Note 16**).
3. Incubate the IPG strip in the equilibrating buffer containing 30 mM DTT for 15 min at room temperature to reduce the proteins. This second reduction step is necessary because disulfure bridges might have spontaneously reformed during the isoelectrofocalization step. Repeat this step by replacing DTT with 135 mM iodoacetamide to alkylate the protein cysteines.
4. Perform the SDS–PAGE (second dimension of separation) according to the 2D electrophoresis system supplier's recommendations.

3.2.3. 2D gel staining

The staining method is first chosen according to the quantity of protein loaded onto the 2D gel. Roughly, for proteasome samples, quantities below 10 μg will require the use of the most sensitive staining methods, like Coomassie colloidal blue, silver nitrate, and Sypro staining. Otherwise, Coomassie blue staining is used. To perform mass spectrometric analysis of the 2D gel spots, Coomassie blue staining and colloidal Coomassie blue staining are fully compatible and are thus recommended. If silver nitrate staining has to be used for low quantities, a modified protocol compatible with MS needs to be followed. The use of the fluorescent Sypro dye requires special equipment for gel imaging and spot cutting and will therefore not be presented in this chapter.

3.2.3.1. COOMASSIE BLUE STAINING

1. Place the gel successively in fixation solution for 20 min, in staining solution for 1 h, and in destaining solution as long as needed to remove the maximum of background signal. The destaining solution can be replaced several times to speed up the process.

3.2.3.2. Coomassie Colloidal Blue Staining

1. Place the gel in the fixation solution for 20 min and incubate for 1 h in the prestaining solution.
2. Stain the gel for 2–3 days in the staining solution. Change the staining solution after several hours of incubation for better results.
3. Wash with deionized water to remove background signal.

3.2.3.3. Silver Nitrate Staining

The silver nitrate staining protocol given hereafter is compatible with mass spectrometric analyses: it avoids the use of glutardialdehyde in the sensitizing step and of formaldehyde in the staining step. Glutardialdehyde and formaldehyde probably induce some cross-linking between amino acid-reactive side chains (17).

1. Place the gel successively in fixation solution for 20 min and in sensitizing solution for 45 min. Wash three times for 10 min with deionized water.
2. Place the gel in staining solution for 50 min and rinse with deionized water for 1 min.
3. Place the gel in development solution for around 10 min (to be adapted to the quantity of proteins) and stop staining by adding 5% (v/v) acetic acid for 15 min (*see* **Note 17**).
4. Immediately excise the gel spots and cover them with the destaining solution with gentle shaking until no bands are visible, approximately 10 min. The gel pieces should turn yellow. Wash gel pieces four to five times for 15 min with deionized water until they are transparent and have no background color anymore. Now proceed to **step 2** of **Subheading 3.2.4**.

3.2.4. Spot excision, in-gel protein digestion, and peptide extraction

1. Excise gel spots using a clean scalpel blade (*see* **Note 18**) and wash the gel pieces in centrifuge tubes containing deionized water (100 μL).
2. Remove the water, add 30–50 μL of acetonitrile to dehydrate the gel pieces, and discard the supernatant.
3. Incubate in 100 mM ammonium bicarbonate (30–50 μL, 15 min, 37°C, under shaking). Add 30–50 μL of acetonitrile (15 min, 37°C, under shaking) to destain the gel pieces. Discard the supernatant.
4. Repeat the incubation step (**step 3**) until the gel pieces become colorless.
5. Lyophilize to dryness (around 10 min in a speed-vacuum centrifuge; gel pieces become white and hard).
6. Add 5–10 μL of trypsin solution (enough to cover the gel pieces) and incubate overnight at 37°C under shaking.
7. Sonicate for 5–10 min and transfer the peptide-containing supernatant to another tube.
8. Incubate the tube containing the gel pieces with 15 μL of 25 mM ammonium bicarbonate for 15 min at 37°C under gentle shaking and sonicate for 5 min. Add 15 μL

of acetonitrile and reincubate for 15 min at 37°C under gentle shaking. Resonicate and pool the supernatant with the previous one from **step 7**. Add 15 µL of 5% (v/v) formic acid solution to the tube containing the gel pieces and incubate for 15 min at 37°C under shaking. Sonicate for 5 min. Add 15 µL of acetonitrile and reincubate for 15 min at 37°C under shaking. Resonicate and combine all supernatants. Lyophilize to dryness using a speed-vacuum (*see* **Note 19**).

3.2.5. Sample preparation for MALDI-TOF analysis: desalting and plate spotting

After peptide extraction from 2D gel spots, a desalting step is necessary to achieve high-quality MALDI-TOF spectra. However, if the peptide gel extraction step is omitted (*see* **Subheading 3.2.4.**, **step 8**), then the tryptic digest supernatant can be directly analyzed, without performing the desalting step. In this supernatant, the salt concentration is indeed low enough (12.5 mM ammonium bicarbonate) to obtain good-quality MALDI-TOF spectra.

1. Wash the ZipTip$_{\mu C18}$ stationary phase successively with acetonitrile/water, 80/20 (v/v) (10 µL, three times), and acetonitrile/water, 50/50 (v/v) (10 µL, three times). Equilibrate with 0.1% (v/v) TFA (10 µL, three times). Load the peptide mixture onto the ZipTip$_{\mu C18}$ stationary phase by 10 admission/delivery cycles. Wash three times with 10 µL of 0.1% (v/v) TFA to eliminate residual salts. Elute with 1.5–3 µL of elution solution by applying several admission/delivery cycles to maximize the elution efficiency (*see* **Note 20**).
2. Dilute the supernatant of a freshly prepared matrix solution twice with a solution of H$_2$O/ACN/TFA, 50/50/0.1 (v/v/v).
3. Deliver 0.5 µL of desalted peptide sample onto the MALDI plate and let it dry. Add 0.3 µL of the diluted matrix solution. Let the mixture crystallize at room temperature.

3.2.6. MALDI-TOF MS analysis and database search

MALDI-TOF MS analysis allows a fast analysis of protein digests. It is well suited for the analysis of 2D gel spots, which often contain only one protein. This is the case of 20S proteasome samples, for which all subunits can be separated on a 2D gel. Protein digests are thus not too complex, and therefore suppression effects during ionization should be minimized. The mass spectra obtained are referred to as "peptide mass fingerprints" of the protein and are used to search databases. The confidence in protein identifications is directly related to the quality of the acquired data and to the parameters used to perform the database search. Protein identification is based on the comparison of the peptide masses given by the mass spectrometric analysis to the peptide masses obtained by theoretical tryptic digestion of each protein in the database. The mass accuracy and the number of peptides detected per protein are thus critical parameters. One of the limitations of the "peptide mass fingerprint" approach is the presence of unexpected modifications of peptide sequences, which change the peptide masses and prevent any match to the theoretical masses. The protein, however, may still be identified, but a precise characterization of the protein modifications requires MS/MS analyses.

1. Set the MALDI-TOF mass spectrometer in positive ion mode using the reflectron (*see* **Note 21**).
2. Adjust all necessary parameters and acquire mass spectra from *m/z* 700 to *m/z* 3000 (*see* **Note 22**).
3. Calibrate the spectra preferably by internal calibration using autodigestion peaks of trypsin ([M + H]$^+$ at *m/z* 842.510 and *m/z* 2211.105) or by external calibration using a mixture of known peptides if autodigestion peaks of trypsin are not present or not sufficiently resolved in the spectrum (*see* **Note 23**).
4. Extract a peak list from each acquired spectrum (*see* **Note 24**).
5. Perform the database search using dedicated software, such as Protein Prospector and Mascot (*see* **Note 25**). Select the human Swiss-Prot database (*see* **Note 26**), trypsin that is able to cleave before proline residues as the digestion enzyme, carbamidomethylation as a systematic cysteine modification, methionine oxidation and protein N-acetylation as possible modifications (*see* **Note 27**), a number of missed cleavages of 1 or 2, and a minimum number of peptides belonging to the same protein of 4. Set the mass accuracy according to the calibration used for data acquisition: the lower the value in ppm, the more stringent the database searches (*see* **Note 28**).

3.3. Proteomic analysis of 20S proteasomes purified from human monoblastic U937 and colorectal Caco2 cell lines

The methods described in **Subheadings 3.1** and **3.2** have been carried out to purify and analyze 20S proteasomes from two human cancer cell lines. The 2D gel images obtained after 2D gel electrophoresis separation and Coomassie blue staining of 20S proteasomes purified from colorectal Caco2 and monoblastic U937 cultured cells are presented in **Fig. 2**. MALDI-TOF analysis of each detected spot has been performed, and the identified 20S proteasome subunits are labeled on the 2D gels. These analyses show that all 20S proteasome subunits can be identified. The 2D gel image of the Caco2 20S proteasome sample reveals additional spots of higher molecular weights corresponding to copurified proteins. For U937 purified 20S proteasome, however, no other spot than those corresponding to proteasome subunits has been detected, and only the part of the 2D gel containing the 20S proteasome subunits is shown in **Fig. 2**. This indicates a high purity of the U937 20S proteasome sample and thus proves the efficiency of the purification method. The 2D gel images of the two purified 20S proteasomes shown in **Fig. 2** also reveal the presence

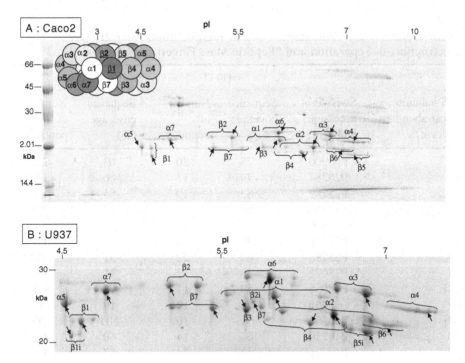

Fig. 2. 2D gel electrophoresis images of 20S proteasomes purified from Caco2 (**A**) and U937 (**B**) cells. The 2D gel images have been obtained with estimated amounts of 80 and 60 μg of 20S proteasomes purified from colorectal Caco2 (**A**) and monoblastic U937 (**B**) cells, respectively, after Coomassie blue staining. All labeled spots have been identified by the "peptide mass fingerprint" approach. The arrows indicate the major isoform of each identified 20S proteasome subunit, the detailed analyses of which are reported in **Table 1**. A schematic representation of the 20S proteasome complex showing the 28 subunits assembly is inserted in the top left corner of the figure. The catalytic subunits β1, β2, and β5 that are replaced by β1i, β2i, and β5i to form the immunoproteasome are underlined.

of isoforms of different pI associated to almost all subunits. This has previously been described for human 20S proteasome purified from erythrocytes *(10)* and placenta *(18)*. The full characterization of each isoform cannot be achieved using MALDI-TOF MS. It requires complementary MS/MS analyses to obtain peptide sequence information and more material to maximize the protein sequence coverage. The detailed results of the "peptide mass fingerprint" analyses are presented in **Table 1** for the most abundant isoform of each 20S proteasome subunit indicated by an arrow in **Fig. 2**. For all analyzed spots but one, at least nine distinct peptides were detected, leading to high sequence coverages ranging from 41 to 83% and to unambiguous protein identifications

Table 1
Purified Human 20S Proteasome Subunit Identification After 2D Gel
Electrophoresis Separation and "Peptide Mass Fingerprint" Analysis

20S subunit (most abundant isoform)[a]	Swiss-Prot accession number	U937		Caco2	
		Sequence coverage (%)	Number of peptides[b]	Sequence coverage (%)	Number of peptides[b]
α1	P60900	78	20	63	22
α2	P25787	79	13	66	12
α3	P25789	83	19	55	16
α4	O14818	80	24	72	20
α5	P28066	71	15	72	13
α6	P25786	76	24	52	17
α7	P25788	63	22	51	15
β1	P28072	56	12	41	10
β1i	P28065	59	11	ND	ND
β2	Q99436	53	13	19	4
β2i	P40306	51	10	ND	ND
β3	P49720	64	14	60	12
β4	P49721	73	24	57	12
β5	P28074	ND	ND	60	12
β5i	P28062	63	13	ND	ND
β6	P20618	76	14	54	12
β7	P28070	51	9	NA	NA

The database search has been performed using the following parameters: Swiss-Prot database, trypsin that cleaves before proline residues as the enzyme, two missed cleavages, carbamidomethyl-cysteine, possible oxidation of methionine and protein N-terminus acetylation, and 25 ppm tolerance on mass measurements.

[a] Indicated by an arrow on the 2D gels presented in **Fig. 2**.

[b] Oxidized peptide analogs are excluded.

ND, Not detected on the 2D gel and NA, not analyzed by MALDI-TOF MS but the subunit identification has been confirmed by ESI-MS/MS.

(see **Note 29**). The MALDI-TOF mass spectra of the tryptic digests of Caco2 20S proteasome β2 and α7 subunits are displayed in **Fig. 3**. An internal calibration using the autodigestion peaks of trypsin was applied, which allowed a mass accuracy below 10 ppm to be obtained over the acquisition mass range.

——————————————————————————————————————▶

Fig. 3. in parentheses below each matched mass. Amino acid modifications are also indicated when present. Mox indicates an oxidized methionine residue.

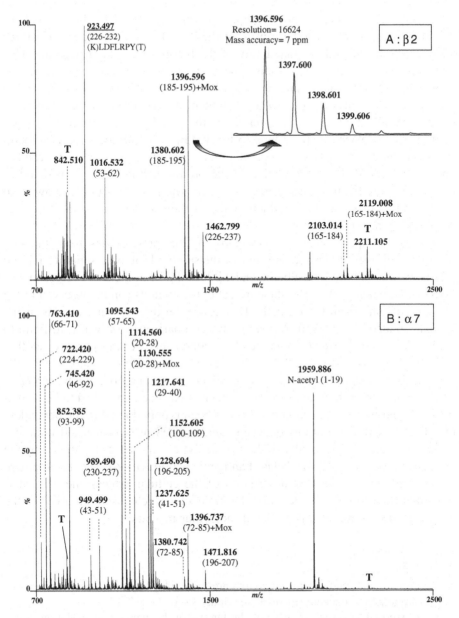

Fig. 3. MALDI-TOF mass spectra of the β2 (**A**) and α7 (**B**) subunit tryptic digests of Caco2 purified 20S proteasome. The mass spectra correspond to the analysis of the subunit digest supernatants without further peptide extraction from gel pieces. They have been internally calibrated using autodigestion peaks of trypsin indicated by T at *m/z* 842.510 and 2211.105. An inset of the isotopic peak distribution of the labeled monoisotopic peak at *m/z* 1396.596 shows a resolution of 16624 and a mass accuracy of 7 ppm. Protein sequence numbers corresponding to identified peptides are indicated

After database search, 4 and 15 peptides were attributed to the β2 and α7 subunits, respectively. The base peak of the β2 subunit digest spectrum at m/z 923.497, however, did not match any of the theoretical subunit peptides. This peak still corresponds to a β2 subunit peptide, which results from an unspecific cleavage of trypsin and is due to a chymotryptic activity (*see* **Note 30**). Even if all peaks in a spectrum cannot be attributed, it is nevertheless recommended to identify all major peaks. Similarly, in the α7 digest spectrum, the major peak at m/z 1959.886 corresponds to the modified N-terminus of α7 subunit that bears an N-acetylation. The peptide mass would not have been matched to any α7 peptide if the N-acetylation of the protein had not been considered as a possible modification during the database search. This example shows that modified peptides can be prominent and that their identification increases the protein sequence coverage.

A close comparison of the 2D maps of 20S proteasomes from the two different cell lines shows the presence of the standard catalytic subunits, β1 and β2, as well as the immunosubunits, β1i, β2i and β5i, in the 20S proteasome of U937 cells, whereas only the standard catalytic subunits can be observed in the 20S proteasome from Caco2 cells. Differences in the subunit isoforms distribution can also be observed. A quantitative comparison, however, is difficult to achieve when comparing Coomassie blue-stained gels. For this purpose, another approach has been described, which is based on an isotopic labeling of the subunits before their 2D gel separation and a relative quantification by MS *(15)*. The presence of both proteasome types in cells has been described to be constitutive *(7)*, and their proportion in cells has been reported to be tissue dependent *(8)*. The structural variations between standard 20S proteasome and immunoproteasome are known to change their proteolytic activities as illustrated by the change in processing of MHC class I antigenic peptides *(3–6)*. The development of tools allowing a detailed description of the cellular proteasome content is thus important to better understand the molecular mechanisms involved in the activity and function of this essential molecular machine.

4. Notes

1. Protease and phosphatase inhibitors are to be added just before use and can be aliquoted as concentrated stock solutions (store at –20°C).
2. It is possible to run gel strips of shorter size at the expense of resolution. Some isoforms of pI might not be observed if shorter strips are used.
3. We use either ready-made 2D gels (ExcelGel™ 2-D Homogeneous 12.5; GE Healthcare, reference: 17-6002-21) adapted to the Multiphor II Pharmacia Biotech system or we run home-made gels (0.375 M Tris–HCl, pH 8.8, 12% acrylamide, 0.03% TEMED, and 0.05% ammonium persulfate). We recommend the use of Duracryl 30% (30T, 2.6C; Genomic Solutions, reference: 0080-0148)

for higher tensile strength of the gel. The system must accept at least 17-cm-large gels to transfer the proteins from the first dimension gel strip to the second dimension SDS–PAGE gel.

4. The preparation of the solution needs the sequential addition of the different components for complete dissolution: first dissolve the ammonium sulphate in water (under agitation, sonicate if necessary), slowly add the methanol in the solution (under agitation), and sonicate if salts precipitate. Add phosphoric acid in the homogeneous solution. The staining solution containing 1% (w/v) colloidal blue is prepared in the same way: the stain is added at the end, and the solution is agitated at least for 30 min, ideally overnight. Filtrate if residual blue is not solubilized.

5. When using a 5-mL column, do not load more than 1 L of antibody solution from **step 3** to avoid saturation of the column interaction sites.

6. If the concentration of the MCP21 antibody solution is lower than 0.5 mg/mL, concentrate the solution using an ultrafiltration step with a cut-off membrane of 10 kDa.

7. Be careful to maintain the beads hydrated.

8. This quantity of covalently bound MCP21 antibodies is sufficient to purify 20S proteasomes from 0.5 to 1.0×10^9 cells.

9. Do not store longer than 1 month for best performance. It is preferable to perform the immobilization of MCP21 antibodies on sepharose beads just before proteasome purification. The beads coupled with MCP21 antibodies can be reused once or twice without significant yield loss, provided the purification procedures are performed within a short period of time. Use preferably proteasome lysis extracts from the same origin to avoid cross-contaminations.

10. At this point, you can store the pellet at –80°C.

11. Make sure that the phase is always hydrated.

12. Proteasomes can be stored in the elution buffer for a few weeks at 4°C without significant decrease in activity. However, for a longer storage period, we recommend dialyzing the proteasome solution against 100 mM Tris–HCl, pH 8.0, and storing it at –80°C after addition of 10% (v/v) glycerol.

13. This assumes that the proteasome fraction is pure: for a majority of cell lines, the proteasome preparation is at least 90–95% pure as estimated by densitometric measurement after 2D gel electrophoresis separation of proteins. However, for some cell lines, contaminants can be present in larger quantities. In this case, the proteasome concentration is overestimated, but this does not hamper identification of the subunits.

14. These perforated caps can be reused many times after rinsing with dialysis buffer.

15. Place the centrifuge tube in the centrifuge so as to easily locate the pellet at the end of the step.

16. To avoid dehydration, use mineral oil as recommended by the supplier. To avoid electro-osmotic current, place water-soaked wicks at the two ends of the gel after the passive rehydration. The wicks are present during the active rehydration step

and all the focusing steps. After isoelectrofocusing, the gel strip can be stored at −80°C.

17. For better performance in mass spectrometric analysis, we recommend reducing the development step as much as possible. A longer development time will affect protein recovery: the darker the stain, the lower the recovery. Indeed, silver-staining procedure interferes with peptide identification by MS *(17)*.

18. We recommend the use of a laminar flow hood to prevent from keratin contaminations. The use of disposable gloves, a face mask, and a hat is mandatory.

19. For higher peptide recovery, the extraction step (**step 8**) is advised. It can, however, be omitted if there is enough material in the sample.

20. Measure the desired volume of eluting solution in a centrifuge tube prior to performing the elution step because the back pressure created by the stationary phase may result in pipetting a lower volume. Take care never to dehydrate the phase except at the end of the elution step.

21. We perform the analyses using a MALDI-TOF/TOF mass spectrometer (4700 proteomics analyzer; Applied Biosystems, Foster City, CA) equipped with a delayed extraction device (delay time automatically adjusted to the focus mass at m/z 2100) and a Nd : YAG laser operating at 200 Hz.

22. The source parameters we use are 20 kV accelerating voltage and 70% grid voltage. The mass resolution of the instrument we obtain is at least 15,000 in the selected mass range. Acquiring below m/z 700 favors detection of intense matrix ions and does not significantly increase the number of detected peptides.

23. The peptide mixture we use for external calibration is composed of des-Arg1-Bradykinin ($[M + H]^+$ at m/z 904.4681), Angiotensin I ($[M + H]^+$ at m/z 1296.6853), Glu1-Fibrinopeptide B ($[M + H]^+$ at m/z 1570.6774), ACTH [1–17] ($[M + H]^+$ at m/z 2093.0867), and ACTH [18–39] ($[M + H]^+$ at m/z 2465.1989). The mass accuracy we obtain is within 10 ppm with an internal calibration and within 30 ppm with an external calibration.

24. We usually perform an automatic peak detection and peak list extraction using the instrument software. It is usually not necessary to spend time trying to consider too small intensity peaks for protein identification.

25. In our experience, the Protein Prospector software is more discriminating for protein identifications than the Mascot software when searching databases with "peptide mass fingerprint" data. Nonetheless, it is often a good idea to perform a database search using several software tools to increase the confidence in protein identifications.

26. When searching the Swiss-Prot database for proteasome subunits, be aware that known subunit variants will not be directly identified, as only one sequence per entry is used for database search. However, all known sequence modifications are listed in the protein description entry.

27. Most 20S proteasome subunits possess an acetylated N-terminus, which may or may not include the initial methionine residue.

28. The mass accuracy we routinely set for database searches is 25 ppm when an internal calibration is used and 50 ppm when an external calibration is used.

29. The sequence coverage that can be reached is dependent on a number of parameters: *(1)* the quantity of protein (the smaller the quantity, the lower the sequence coverage) and *(2)* the efficiency of tryptic digestion (both inaccessible cleavage sites and long sequence stretches without trypsin cleavage sites lower the sequence coverage).

30. A chymotryptic activity is often observed in the proteolytic peptides generated by trypsin, even though the latter enzyme is treated to inhibit this unwanted activity.

Acknowledgments

We thank Dr. Frédéric Lévy (Ludwig Institute for Cancer Research, Lausanne, Switzerland) for fruitful discussions and advice about the 20S proteasome purification. We are grateful to Dr. Jean-François Haeuw (Centre d'Immunologie Pierre Fabre, Saint-Julien en Genevois, France) for providing the Caco2 cells. This work was supported in part by the Réseau National des Génopoles, the Région Midi-Pyrénées, and the ASG program from the French Ministry of Research.

References

1. Unno, M., Mizushima, T., Morimoto, Y., Tomisugi, Y., Tanaka, K., Yasuoka, N., et al. (2002) The structure of the mammalian 20S proteasome at 2.75 Å resolution. *Structure* **10**, 609–618.

2. Orlowski, M. and Wilk, S. (2000) Catalytic activities of the 20S proteasome, a multicatalytic proteinase complex. *Arch. Biochem. Biophys.* **383**, 1–16.

3. Rock, K. L. and Goldberg, A. L. (1999) Degradation of cell proteins and the generation of MHC class I-presented peptides. *Annu. Rev. Immunol.* **17**, 739–779.

4. Morel, S., Levy, F., Burlet-Schiltz, O., Brasseur, F., Probst-Kepper, M., Peitrequin, A. L., et al. (2000) Processing of some antigens by the standard proteasome but not by the immunoproteasome results in poor presentation by dendritic cells. *Immunity* **12**, 107–117.

5. Van den Eynde, B. J., and Morel, S. (2001) Differential processing of class-I-restricted epitopes by the standard proteasome and the immunoproteasome. *Curr. Opin. Immunol.* **13**, 147–153.

6. Burlet-Schiltz, O., Claverol, S., Gairin, J. E., and Monsarrat, B. (2005) The use of mass spectrometry to identify antigens from proteasome processing. *Methods Enzymol.* **405**, 264–300.

7. Macagno, A., Gilliet, M., Sallusto, F., Lanzavecchia, A., Nestle, F. O., and Groettrup, M. (1999) Dendritic cells up-regulate immunoproteasomes and the proteasome regulator PA28 during maturation. *Eur. J. Immunol.* **29**, 4037–4042.

8. Noda, C., Tanahashi, N., Shimbara, N., Hendil, K. B., and Tanaka, K. (2000) Tissue distribution of constitutive proteasomes, immunoproteasomes, and PA28 in rats. *Biochem. Biophys. Res. Commun.* **277**, 348–354.

9. Husom, A. D., Peters, E. A., Kolling, E. A., Fugere, N. A., Thompson, L. V., and Ferrington, D. A. (2004) Altered proteasome function and subunit composition in aged muscle. *Arch. Biochem. Biophys.* **421**, 67–76.

10. Claverol, S., Burlet-Schiltz, O., Girbal-Neuhauser, E., Gairin, J. E., and Monsarrat, B. (2002) Mapping and structural dissection of human 20S proteasome using proteomic approaches. *Mol. Cell. Proteomics* **1**, 567–578.

11. Iwafune, Y., Kawasaki, H., and Hirano, H. (2002) Electrophoretic analysis of phosphorylation of the yeast 20S proteasome. *Electrophoresis* **23**, 329–338.

12. Kurucz, E., Ando, I., Sumegi, M., Holzl, H., Kapelari, B., Baumeister, W., et al. (2002) Assembly of the Drosophila 26S proteasome is accompanied by extensive subunit rearrangements. *Biochem. J*. **365**, 527–536.

13. Yang, P., Fu, H., Walker, J., Papa, C. M., Smalle, J., Ju, Y. M., et al. (2004) Purification of the Arabidopsis 26S proteasome: biochemical and molecular analyses revealed the presence of multiple isoforms. *J. Biol. Chem.* **279**, 6401–6413.

14. Huang, L. and Burlingame, A. L. (2005) Comprehensive mass spectrometric analysis of the 20S proteasome complex. *Methods Enzymol* . **405**, 187–236.

15. Froment, C., Uttenweiler-Joseph, S., Bousquet-Dubouch, M. P., Matondo, M., Borges, J. P., Esmenjaud, C., et al. (2005) A quantitative proteomic approach using two-dimensional gel electrophoresis and isotope-coded affinity tag labeling for studying human 20S proteasome heterogeneity. *Proteomics* **5**, 2351–2363.

16. Orr, A., Ivanova, V. S., Bonner, W. M. (1995) "Waterbug" dialysis. *BioTechniques* **19**, 204–206.

17. Richert, M., Luche, S., Chevallet, M., Van Dorsselaer, A., Leize-Wagner, E., and Rabilloud, T. (2004) About the mechanism of interference of silver staining with peptide mass spectrometry. *Proteomics* **4**, 909–916.

18. Hendil, K. B., Kristensen, P., and Uerkvitz, W. (1995) Human proteasomes analysed with monoclonal antibodies. *Biochem. J*. **305**, 245–252.

22

Characterization of *E. coli* Ribosomal Particles
Combined Analysis of Whole Proteins by Mass Spectrometry and of Proteolytic Digests by Liquid Chromatography–Tandem Mass Spectrometry

Isabelle Iost, Julie Charollais, Joëlle Vinh, and Delphine Pflieger

Summary

This chapter describes the purification of ribosomal particles from a mutant strain of *Escherichia coli* using sucrose gradients and the characterization of their protein composition by a combination of mass spectrometry (MS) techniques. The main objective is to identify the ribosomal proteins that are missing in an aberrant ribosomal particle corresponding to a defective large subunit. To address this question, the tryptic digests of the purified ribosomal particles are analyzed by the coupling between liquid chromatography and tandem MS. The presence or absence of a given ribosomal protein in the defective particle is determined by comparing the MS intensities of its identified tryptic peptides with that of the mature large subunit. These analyses also allow identification of proteins copurifying with the ribosomal particles. To detect low-mass proteins escaping identification by the above method, intact proteins are also analyzed by matrix-assisted laser desorption ionization time of flight (MALDI-TOF) and nano-ESI-QqTOF MS.

Key Words: *E. coli*; ribosomes; capillary liquid chromatography; tandem mass spectrometry; MALDI-TOF; nano-ESI-QqTOF.

1. Introduction

Ribosomes are large macromolecular ribonucleoprotein particles responsible for protein synthesis in the cell. Bacterial ribosomes are composed of two unequal subunits that together constitute the 70S ribosome. The small subunit (30S) contains 21 ribosomal proteins (r-proteins) and a single ribosomal RNA (rRNA) (16S), and the large subunit (50S) consists of 33 r-proteins and two

From: *Methods in Molecular Biology, vol. 432: Organelle Proteomics*
Edited by: D. Pflieger and J. Rossier © Humana Press, Totowa, NJ

rRNAs (23S and 5S). As one of the largest macromolecular complexes found in the cell, the ribosome has represented a formidable challenge for mass spectrometry (MS) approaches. Different MS methods have been developed to define the protein composition of ribosomes from different sources as well as their post-translational modifications. In particular, MS has proven successful in identifying the full complement of r-proteins present in mature ribosomes from cytoplasm or organelles (such as mitochondria or chloroplasts) of different organisms (bacteria, yeast, mammals and plants) [see for example *(1–5)*]. The analysis of precursors of yeast ribosomes has also benefited from MS. In that case, an epitope-tagged non-ribosomal protein associated with the pre-ribosome was used as affinity bait and co-purifying proteins were identified by MS *(6)*.

The aim of our study is to compare the composition of ribosomal particles from wild-type and mutant *Escherichia coli* strains *(7)*. Here, we describe the isolation and purification of ribosomal particles and several analytical methods based on MS, with the aim of identifying and quantifying the r-proteins present in these particles. Protein samples are enzymatically digested for characterization by the coupling between capillary liquid chromatography and tandem mass spectrometry (LC-MS/MS) and are also analyzed as intact proteins by matrix-assisted laser desorption ionization time of flight mass spectrometry (MALDI-TOF MS) and nano-ESI-QqTOF MS. In addition to r-proteins, the LC-MS/MS analyses allow to identify non-ribosomal proteins that are associated with ribosomal particles.

2. Materials

2.1. Purification of Ribosomal Particles

2.1.1. Growth Medium and Extract Preparation

1. Growth medium: rich LB medium.
2. Chloramphenicol (Sigma).
3. Lysosyme (Sigma).
4. Lysis buffers (*see* **Note 1**):

   a. Buffer A: 10 mM Tris–HCl, pH 7.5, 60 mM KCl and 10 mM $MgCl_2$.
   b. Buffer B: buffer A supplemented with 0.5% (w/v) polyoxyethylene 20 cetyl ether (Brij 58), 0.5% (w/v) deoxycholic acid and 0.1 unit/µL Promega RQ1 RNase-free DNase (*see* **Note 2**).

2.1.2. Sucrose Gradients

Sucrose solutions are made in buffer C and are filtered through 0.22-µm filters.

1. Buffer C: 10 mM Tris–HCl, pH 7.5, 50 mM NH_4Cl, 10 mM $MgCl_2$ and 1 mM dithiothreitol [dithiothreitol (DTT) should be added freshly to the buffer].

2. Solutions of 5 and 20% (w/w) sucrose for linear 5–20% gradients are obtained by dilution of a 66% (w/w) sucrose solution, using a sucrose dilution chart (the International Critical Table) that gives the volume of the 66% solution required to obtain the desired concentration. We also used linear 10–40% (w/v) sucrose gradients. Both types of gradients give a good separation of r-particles.
3. Centrifuge with a Beckman SW28 rotor or equivalent.
4. Centrifuge with a Kontron TFT 50.38 rotor or equivalent.
5. ISCO UA-6 detector for monitoring at 254 nm or equivalent.

2.1.3. Extraction of r-Proteins

1. 1 M $MgCl_2$.
2. Glacial acetic acid.
3. Resuspension buffer: 100 mM Tris–HCl, pH 8.5, and 4 M urea.
4. 100 mM Tris–HCl, pH 8.5.

2.2. Mass Spectrometric Analyses

2.2.1. Digestion of Ribosomal Samples, LC-MS/MS Analyses and Protein Identification

1. Digestion enzyme: sequencing grade bovine trypsin (Roche).
2. HPLC solutions are prepared with HPLC grade solvents and Milli-Q water at 18.2 MΩ. Buffer A = H_2O/acetonitrile/formic acid, 96/4/0.1 (v/v/v); buffer B = H_2O/acetonitrile/formic acid, 10/90/0.085 (v/v/v) (*see* **Note 3**).
3. Chromatographic system Famos-Switchos-UltiMate (Dionex) equipped with a reversed-phase pre-column (Pepmap C18, 3-μm particle size, 100 Å porosity, 300 μm internal diameter and 5 mm length) and a capillary column (Pepmap C18, same stationary phase characteristics as the pre-column, 75 μm internal diameter and 15 cm length).
4. Nano-ESI-QqTOF instrument (QTof2, Micromass; Waters).
5. Tip adapter for LC-ESI connection (reference ADPT-PRO, New Objective).
6. Metallized tips, distal coated (reference FS360-20-10-D, New Objective) (*see* **Note 4**).
7. Identification of peptide sequences from MS/MS spectra is provided by Mascot software (www.matrixscience.com).

2.2.2. Mass Spectrometric Analysis of Whole Proteins

1. ZipTip$_{C4}$ pipette tips (Millipore) for protein desalting before MS analysis.
2. Sinapinic acid matrix for MALDI-TOF analysis of whole proteins: saturated solution in H_2O/acetonitrile/formic acid, 66/33/1 (v/v/v) (*see* **Note 5**).
3. Stainless steel 100-spot MALDI target plate (Applied Biosystems).
4. MALDI-TOF instrument (DE-STR; PerSeptive Biosystems).
5. Metallized glass needles (Proxeon) for direct infusion of proteins in the nano-ESI-QqTOF instrument.

3. Methods

Excellent reviews on methods for isolation of ribosomes have been published [see for example *(8)*]. Most protocols employ a series of differential centrifugations, which lead to the preparation of a crude ribosomal fraction. This crude extract is then further purified using sucrose gradient centrifugation, from which the fractions containing the ribosomes are recovered. Proteins can be identified directly or after separation by two-dimensional gel electrophoresis. Different techniques for lysing the cells and diverse buffers for centrifugations have been proposed. Here, we present the techniques we used to isolate mature and incompletely assembled ribosomes that accumulate in a mutant strain deficient in ribosome biogenesis. This strain is deleted for *srmB*, which encodes an RNA helicase of the DEAD-box protein family. Ribosome analysis using centrifugation on sucrose gradients reveals a deficit in large ribosomal subunit and the accumulation of an aberrant ribosomal particle that sediments at about 40S *(7)*. This particle lacks some r-proteins of the large subunit and thus corresponds to an incomplete 50S subunit. To understand what step of assembly is affected in the mutant, it is essential to determine which ribosomal proteins are missing in that particle. Different MS approaches are implemented to address this question. This study turns out to be a real challenge: indeed, to determine whether an r-protein from the 40S fraction really originates from 40S particles rather than from contaminating 50S subunits (both particles could not be completely separated by sucrose gradient fractionation), the relative abundance of the r-proteins present in the different particles (40S and 50S) must be quantified. The particles isolated from the wild-type and mutant ΔsrmB strains are, respectively, designated as W30S, W50S, and Δ30S, 40S, Δ50S.

LC-MS/MS analysis of the proteolytic peptide mixture obtained by enzymatic digestion is a common approach for protein identification. MS/MS analysis of peptides induces fragmentation patterns that give amino acid sequence information. The ribosomal proteins identified in the Δ30S sample with this simple approach are truly constitutive of the particle. For 40S, W50S and Δ50S, the relative amounts of tryptic peptides identifying r-proteins are compared. This is performed by calculating the ratio of ESI-MS intensities detected for reliably identified peptides in the LC-MS/MS analyses. An MS/MS spectrum was considered to be reliably interpreted when a continuous series of minimally five amino acids could be read in terms of y or b fragment ions. Finally, the relative abundance of a given protein between two samples is calculated by averaging the ratios measured on its identified tryptic peptides.

LC-MS/MS analysis allows identification of most expected r-proteins, as well as some non-ribosomal proteins associated with the ribosomal particles. However, some r-proteins produce peptides that escape LC-MS/MS detection. This failure may generally occur for several reasons: the peptides either do not ionize well in ESI or do not give informative MS/MS spectra (e.g., very large or small peptides). Some r-proteins are extremely small (down to 4.4 kDa for L36) and very basic, thus generating few and small peptides by trypsin digestion. To detect these proteins, it appears more straightforward and reliable to analyze the intact proteins by MALDI-TOF or nano-ESI-QqTOF MS. Both sources favour ionization and detection of different sets of proteins and thus provide complementary insights into the protein sample compositions. The analysis by MALDI-TOF MS focuses on proteins of less than 15 kDa, because only low-molecular-weight r-proteins escape identification by LC-MS/MS. Moreover, detection of higher mass proteins (the molecular mass of r-proteins ranges between 4 and 62 kDa) would require their separation from the smaller ones, due to suppression effects in MALDI-TOF MS analysis.

The final composition of the W30S and Δ30S particles is determined using the combined results obtained from LC-MS/MS and MALDI-TOF MS analyses. The composition of the 40S and Δ50S particles is inferred from the relative quantification of the r-proteins in these mutant particles versus in the W50S and completed by MALDI-TOFMS and nano-ESI-QqTOF MS analysis for the few small proteins escaping LC-MS/MS identification.

In this chapter, we describe a gentle lysis method using lysosyme and mild detergents, followed by sucrose gradient centrifugations and a combination of MS analyses to isolate and characterize the different r-particles. The presented techniques can be applied to analyze certain ribosomal particles (such as assembly intermediates or aberrant particles) as well as mature ribosomes and also allow the identification of non-ribosomal proteins that associate with ribosomes.

3.1. Purification of Ribosomal Particles

3.1.1. Extract Preparation

1. Wild-type and mutant strains are grown in LB medium (around 200 mL) at 30°C to an absorbance at 600 nm (A_{600}) ranging from 0.5 to 0.8 (*see* **Note 6**).
2. Chloramphenicol is added to the cultures (final concentration 100 μg/mL) 5 min before harvesting to avoid polysome run-off (*see* **Note 7**).
3. Cells are rapidly cooled on ice and collected by centrifugation at 4000 g for 10 min (*see* **Note 8**).
4. After a first quick freezing and slow thawing cycle (on ice), cells are lysed by adding a volume of lysosyme (final concentration 0.5 mg/mL, freshly prepared in buffer A) equal to 1/125th of the initial cell culture volume.
5. After a second freeze-thaw cycle, a volume of buffer B equal to 1/167th of the initial culture volume is added (*see* **Note 9**). The mixture is incubated on ice for about 20 min (until the viscosity has decreased).

6. The lysate is then clarified by centrifugation in a microfuge at 18,000 g for 10 min at 4°C.

7. The extract concentration is estimated by measuring the $A_{260\ nm}$ (about 30 $A_{260\ nm}$ units can be obtained from a 50-mL culture). The extract is quickly frozen (using liquid nitrogen or a dry ice/ethanol bath) and can be stored at –80°C (for only a few days, *see* **Note 10**) or immediately loaded on sucrose gradients.

3.1.2. Purification of Ribosomal Particles

The ribosomal particles are separated according to their sedimentation rate by density gradient centrifugation. Two consecutive sucrose gradient fractionations are necessary to obtain well separated species.

1. Between 25 and 100 $A_{260\ nm}$ units of lysates are layered onto preformed 5–20% (w/w) sucrose gradients (*see* **Note 11**) in buffer C and centrifuged at 141,000 g for 7–8 h at 4°C (28,000 rpm in a Beckman SW28 rotor).

2. Gradients are then analyzed with an ISCO UA-6 detector with continuous monitoring at 254 nm. A dense solution (e.g., 60% sucrose) is used to push the gradient towards the optical unit.

3. Fractions containing the ribosomal particles are collected, combined, and pelleted by centrifugation at 148,000 g (35,000 rpm in a Kontron TFT 50.38 rotor) overnight at 4°C.

4. Ribosomal pellets are resuspended in 200 μL of buffer C.

5. Ribosomal fractions are again layered onto 5–20% sucrose gradients and centrifuged as in **step 1**.

6. Gradients are analyzed as in **step 2**. To avoid contamination, only fractions corresponding to clearly separated species are collected and pelleted (as in **step 3**).

7. Pellets are resuspended in 100 μL of buffer C.

3.1.3. Extraction of r-Proteins

1. r-proteins are extracted from the purified particles by adding, in rapid succession, 1/10th volume of 1 M $MgCl_2$ and 2 volumes of glacial acetic acid. The mixture is incubated on ice for about 45 min and then centrifuged at 18,000 g for 30 min at 4°C. rRNA is precipitated by this treatment.

2. Four to five volumes of acetone is added to the supernatant, and proteins are allowed to precipitate at –20°C for 30 min (for high-protein concentrations) or overnight (for low-protein concentrations).

3. After centrifugation at 18,000 g for 30 min, the pellets are washed twice with acetone.

4. Pellets are resuspended in 25 μL of resuspension buffer and diluted with 100 mM Tris–HCl, pH 8.5, to reach the composition of digestion buffer, 100 mM Tris–HCl, pH 8.5 and 1 M urea.

3.2. Mass Spectrometric Analyses of Ribosomal Samples

3.2.1. Digestion of Ribosomal Samples, LC-MS/MS Analyses and Protein Identification

1. r-proteins are digested by addition of trypsin at an enzyme to protein ratio of 1:50–1:20 (w/w) and overnight incubation at 37°C (*see* **Note 12**).

2. Peptide samples are then acidified by addition of 1% (v/v) formic acid in water to reach a pH around 2–3. Between 0.5 and 1 µg of digested sample is injected per LC-MS/MS analysis.

3. Peptides are first loaded onto the pre-column, with a 30-µL/min flow rate of buffer A. Salts are eliminated by a 5-min wash with buffer A.

4. After valve switching of the Switchos system, pre-column and column are brought in series.

5. Chromatographic separation of peptides is performed at 200 nL/min by running a gradient from 100% of buffer A to 45% of buffer B. The gradient is developed in 1 h 20 min and 1 h 40 min for the Δ30S and Δ50S particles, respectively (*see* **Note 13**).

6. A stable spray should be obtained by applying a capillary voltage around 2 kV and setting the cone voltage to 45 V (these values strongly depend on the design of the LC–ESI interface).

7. Create an acquisition method on the nano-ESI-QqTOF that automatically alternates between detection in MS mode of peptides eluted from the chromatographic column and fragmentation in MS/MS mode of the three species giving the most intense signals in each MS scan. For Q-Tof2, set the total duration of MS scans to 2 s and that of MS/MS scans to 4 s. Acquisitions are limited to the mass range [400; 1300] Da in MS mode and span [50; 2600] Da in MS/MS mode (*see* **Note 14**). The intensity threshold in a MS scan for a peptide to be selected for fragmentation is set to 4 c/s. The collision energy profile adapted to consecutive mass ranges is given in **Table 1** (*see* **Note 15**). The selected peptides are listed automatically in an exclusion list to prevent their reselection over the next 10 min. This time was set intentionally long because of the tailed chromatographic peaks for some species at the end of the gradient.

8. To identify more peptides, the same sample is analyzed twice while scanning in MS mode over two complementary mass ranges: [400; 700] Da and [680; 1300] Da (*see* **Note 16**).

Table 1
Collision Energy Profile Applied to Fragment Peptides

Mass range	400–500	500–600	600–700	700–800	800–900	900–1000	1000–1100	1100–1300
Energy (eV)	24	26	30	32	34	38	42	44

9. To identify peptide sequences from MS/MS spectra, perform searches using Mascot search engine (Matrix Science, www.matrixscience.com), while considering the database Swiss-Prot, two possibly missed tryptic cleavages (*see* **Note 17**), possible methionine oxidation, and specifying the use of an 'ESI-4sector' instrument (*see* **Note 18**).

10. In addition to the intrinsic r-proteins, a significant number of copurified proteins are identified in the different ribosomal particles, which are listed in **Table 2**.

11. Peptides from proteins of the large subunit, which are identified on the basis of a five amino-acid-long sequence or more, are listed and characterized by their mass to charge ratio (m/z), charge (z) and retention time (t_R).

3.2.2. LC-MS Analyses of Digested Ribosomal Samples Without Fragmentation

1. Between 0.3 and 0.5 μg of digested sample is injected per LC-MS analysis (without fragmentation of peptides).

2. Use the same gradient as for LC-MS/MS analysis.

3. Create an acquisition method on the nano-ESI-QqTOF that only scans in MS mode over the mass range [400; 1300] Da without switching to MS/MS. Thus, the species ionized in the ESI source are detected continuously.

4. After acquisition, integrate the MS signal detected for the listed peptides (m/z, z, t_R) throughout their elution peak, as represented in **Fig. 1**. This is done for the different analyzed samples and allows calculation of the intensity ratios: I(40S)/I(Δ50S), I(Δ50S)/I(W50S) and I(40S)/I(W50S) for every peptide. Finally, the relative abundance of a given protein between two samples is obtained by averaging the ratios measured on its identified tryptic peptides. This allows calculation of a mean and a standard deviation. Note that we present the manual alignment of chromatographic elution peaks; a few programs have been developed to perform automated and large-scale comparisons: see for example (*9,10*).

5. Similar protein amounts of 40S, Δ50S and W50S samples are injected per LC-MS analysis. Yet, these samples may contain different amounts of proteins copurifying with the ribosomal particles and thus different amounts of intrinsic r-proteins. To compensate for this possible difference, protein ratios are normalized with respect to a protein significantly present in all three samples, L1 [the three ratios I(40S)/I(Δ50S), I(Δ50S)/I(W50S) and I(40S)/I(W50S) are normalized to 1 for this protein].

Quantification results obtained for the mutant 40S and 50S particles are shown in **Table 3**.

3.2.3. Mass Spectrometric Analysis of Whole Proteins

3.2.3.1. MALDI-TOF ANALYSIS OF WHOLE RIBOSOMAL PARTICLES

1. Perform the cleaning of 0.5 μg of protein sample using a ZipTip$_{C4}$, pipette tip containing a bed of reversed-phase chromatographic medium. Follow instructions from the manufacturer.

Table 2
Non-Ribosomal Proteins Identified in the Ribosomal Particles

SwissProt acces n°	description	function	localization					
			W30S	Δ30S	W50S	Δ50S	40S	
A								
P02998	Translation Initiation Factor 1	translation initiation	+	+				
P02995	Translation Initiation Factor 2	translation initiation	+	+				
P02999	Translation Initiation Factor 3	translation initiation	+	+				
P09170	Ribosome Binding Factor A	16S rRNA processing	+	+				
P21504	RimM	30S maturation		+				
P22257	Chaperone Trigger Factor	folding of the nascent polypeptide			+	+		
P42641	GTP-binding protein CgtA	probably in 50S biogenesis				+		
P23304	RNA helicase DeaD	50S assembly	+	+		+	+	
P02990	Elongation Factor EF-Tu	translation				+		
B								
P37765	Ribosomal large subunit pseudouridine synthase B (rluB)	Pseudouridylation of 23S rRNA	+	+	+	+	+	
P33643	Ribosomal large subunit pseudouridine synthase D (rluD)	Pseudouridylation of 23S rRNA and 50S assembly				+		

(Continued)

Table 2
(Continued)

SwissProt acces n°	description	function	localization				
			W30S	Δ30S	W50S	Δ50S	40S
P39406	Ribosomal RNA small subunit methyltransferase C	16S rRNA methylation		+			
P37912	Ribosomal RNA small subunit methyltransferase E	16S rRNA methylation		+			
P21499	RNase R	rRNA and mRNA degradation					+
P11028	DNA binding protein Fis	activates rRNA transcription		+		+	
P33635	Hypothetical tRNA/rRNA methyltransferase YfiF	Unknown	+				+
C							
P00574	RNA polymerase α chain	Subunits of the RNA polymerase core, involved in transcription					+
P00575	RNA polymerase β chain			+			+
P00577	RNA polymerase beta chain			+			+

ID	Name	Function			
P00391	dihydrolipoamide dehydrogenase	Components of pyruvate dehydrogenase and 2-oxoglutarate dehydrogenase complexes, which convert pyruvate to acetyl-coA and 2-oxoglutatarate to succinyl-coA		+	+
P07015	2-Oxoglutarate dehydrogenase E1			+	
P07016	dihydrolipoamide succinyltransferase			+	
P17547	aldehyde-alcohol dehydrogenase			+	
P06958	pyruvate dehydrogenase E1			+	+
P06959	dihydrolipoamide acetyltransferase			+	+
D					
P06984	integration host factor α subunit	recombination, transcription,		+	
P08756	integration host factor β subunit	replication, DNA compaction		+	
P45577	ProP effector	regulator for the ProP transporter	+	+	
P02342	histone-like protein HU α	DNA supercoiling and compaction		+	+
P07028	UvrC	DNA repair			+
P04128	type-1 fimbrial protein	pilin subunit			+
P25888	RNA helicase RhlE	unknown			+
P23830	phosphatidylserine synthase	phospholipid metabolism			+
P21774	hydroxymyristoyl dehydratase	lipid synthesis			+
P00490	maltodextrin phosphorylase	carbohydrate metabolism			+

(Continued)

Table 2
(Continued)

SwissProt acces n°	description	function	localization				
			W30S	Δ30S	W50S	Δ50S	40S
P36564	hypothetical protein YibL	unknown					+
P42550	hypothetical protein YhbY	translation? (RNA binding protein with an IF3-like core fold)					+
P75864	hypothetical protein YcbY	unknown					+
Q8X9X2	hypothetical protein YncE	unknown					+
P05848	hypothetical protein YbeB	unknown				+	
P26650	hypothetical protein YjgA	unknown				+	

The identified proteins have been divided into four classes. Class A is known ribosome-associated proteins; class B, proteins whose functions are related to ribosome biogenesis; class C, proteins belonging to large complexes and class D, proteins whose functions are either unknown or unrelated to ribosome biogenesis.

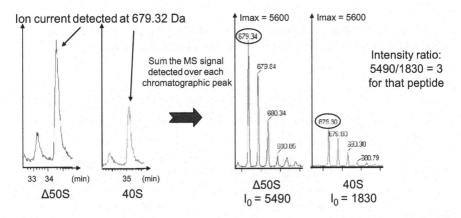

Fig. 1. Principle of relative quantification of proteins between two digests analyzed by LC-MS.

2. Concentrate the eluted proteins by Speedvac from 10 μL of elution volume to around 2 μL.

3. Deposit 1 μL of sample on the MALDI target plate and mix it with 1 μL of sinapinic acid solution. Allow sample to air-dry and matrix crystals to form.

4. Acquire mass spectra in positive linear mode over the mass range 2–15 kDa. The following parameters are given for a MALDI-TOF DE-STR (Applied Biosystems) mass spectrometer as starting conditions and should be optimized on the operating system: low-mass gate 2 kDa, accelerating voltage 20 kV, grid voltage 95%, extraction delay time 400 ns, and 500 laser shots/spectrum were averaged for each spectrum (with a N_2 laser at 4Hz).

5. Perform external calibration by spotting bovine insulin (average mass M + H = 5734.59 Da), *E. coli* thioredoxin (M + H = 11674.48 Da), and horse apomyoglobin [(M + 2H)/2 = 8476.78 Da]. Standards must have been deposited in the close vicinity of the sample spots to be calibrated.

Figure 2 shows the spectra obtained by MALDI-TOF analyses of 40S and Δ50S samples and **Table 3** summarizes the proteins of the large subunit detected at their expected masses.

3.2.3.2. NANO-ESI-QQTOF ANALYSIS OF 40S, Δ50S AND W50S SAMPLES

1. Perform the cleaning of 1 μg of protein sample using a ZipTip$_{C4}$. Instead of TFA-containing solutions, use 1% (v/v) formic acid solutions, because TFA is detrimental to ESI ionization. Elute proteins with a few microliters of solution (H_2O/acetonitrile/formic acid, 35/64/1, v/v/v).

2. Ionize the proteins directly by nano-ESI in a metallized glass capillary. A stable spray must be obtained by applying a capillary voltage between 900 and 1200 V and a cone voltage of 45 V. The analysis is performed over the mass range [400; 1850] Th.

Table 3
Identification of Ribosomal Proteins by LC-MS/MS (Tryptic Digests) and MALDI-TOF and Nano-ESI-QqTOF MS (Intact Proteins) Analyses

Technique	Particle	1	2	3	4	5	6	7/12	9	10	11	13	14	15	16	17	18	19
LC-MS/MS	Δ50S	■	■	■	■	■	■	■	■	■	■	■	■	■	■	■	■	■
	40S	■	■	■	■	■	■	■	■	■	■	□	■	■	■	■	■	■
MALDI-TOF	Δ50S	■	■	■	■	■	■	■	■	■	■	■	■	■	■	■	■	■
	40S	■	■	■	■	■	■	■	■	■	■	■	■	■	■	■	■	■
nanoESI-QqTOF	Δ50S	■	■	■	■	■	□	■	■	■	□	■	■	■	■	■	■	■
	40S	□	■	■	■	■	□	■	■	■	□	■	■	■	▩	▩	■	■

Technique	Particle	20	21	22	23	24	25	27	28	29	30	31	32	33	34	35	36
LC-MS/MS	Δ50S	■	■	■	■	■	■	■	■	■	■	□	■	■	■	■	■
	40S	■	■	■	■	■	■	□	■	■	■	■	□	■	■	■	■
MALDI-TOF	Δ50S	■	■	□	■	□	■	■	■	■	■	■	■	■	■	■	■
	40S	■	■	■	□	■	■	□	■	■	▩	■	■	■	■	■	■
nanoESI-QqTOF	Δ50S	■	■	■	■	■	■	■	□	■	■	■	■	■	■	■	■
	40S	■	■	■	■	■	▩	■	■	■	■	■	■	■	■	■	■

All proteins identified in W50S by LC-MS/MS analysis were found to be significantly present in the Δ50S mutant particle (ratios Δ50S/W50S were all superior to 0.6 after normalization with respect to L1). A protein was considered as present (black square) or absent (white square) in the 40S particle when its relative abundance compared with the same protein in the Δ50S and W50S subunits were higher than 45% or lower than 10%, respectively. Proteins detected at intermediate levels are noted with grey squares.

3. Each protein gives rise to a series of peaks corresponding to several consecutive charge states. To simplify the spectrum, we run the MaxEnt deconvolution algorithm implemented in the Masslynx software piloting the Q-Tof2 mass spectrometer. This process results in a synthetic spectrum representing all detected species at their molecular weight. The isotopic patterns of L34 and L36 proteins detected in 40S and Δ50S samples at charge states of 7+ and 10+, respectively, are represented in **Fig. 3**. The whole deconvoluted spectra obtained with these two particles are given in **Fig. 4**.

Finally, **Table 3** shows the r-proteins from 40S and Δ50S samples detected at their expected masses in the nano-ESI-QqTOF analyses.

3.3. Results

The three combined MS techniques allow identification of all r-proteins from the large and small subunits. **Table 3** summarizes the complementary data provided by each approach to characterize the 40S and Δ50S particles. Moreover, other proteins of diverse nature that are not considered to be

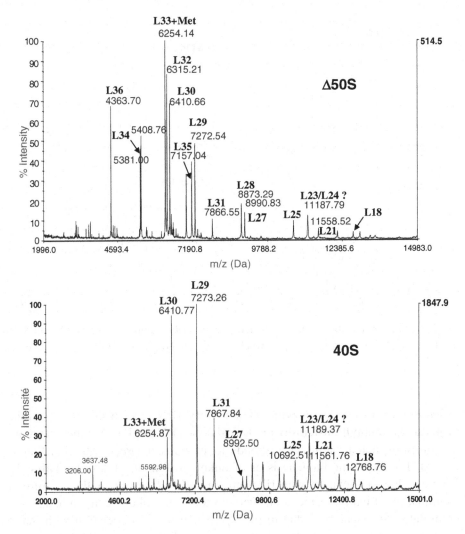

Fig. 2. MALDI-MS spectra of whole ribosomal particles: particle Δ50S and particle 40S.

ribosomal components are also identified (*see* **Table 2**). These non-ribosomal proteins co-migrate with the different ribosomal particles during gradient centrifugation, which suggests their possible association with the ribosomes. Indeed, some of them correspond to proteins (class A, **Table 2**) that are known to associate with ribosomes, such as translation initiation factors, RbfA *(11)*, trigger factor *(12)*, CgtA *(13)*, RimM *(14)* or RNA helicase DeaD *(15)*. Others correspond to proteins (class B, **Table 2**) for which a physical association with ribosomes has not been demonstrated but whose functions are related

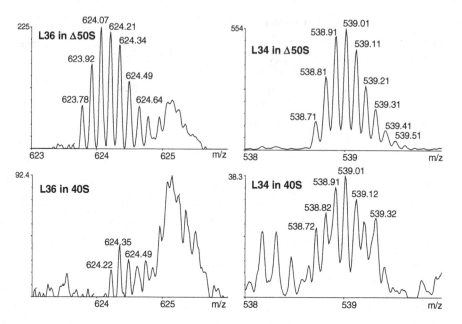

Fig. 3. Zoom on the signals detected for the ribosomal proteins L34 and L36 at charge states 10+ and 7+, respectively, in the nano-ESI-QqTOF MS analysis of particles Δ50S and 40S.

to ribosome biogenesis (enzymes that modify rRNA, such as pseudouridine synthase or methyltransferase) or degradation (RNase R). Another category of non-ribosomal proteins corresponds to proteins having a function that is unrelated to ribosome biogenesis or of unknown function (class D, **Table 2**). Whether their association with ribosomes is relevant or not will need further examination. Finally, some identified proteins are likely contaminants that comigrate with r-particles due to their presence in, or association with, large multienzyme complexes such as the pyruvate dehydrogenase and 2-oxoglutarate dehydrogenase complexes or RNA polymerase (class C, **Table 2**).

While some of these non-ribosomal proteins are unique to either small, large or 40S particles, others are found in more than one r-particle species. For example, it is interesting to note that RluB, which modifies the rRNA of the large subunit, is also found in the small subunit. One possible explanation is that a precursor of the large subunit comigrates with the 30S subunit; the association of RluB with this precursor would then suggest a role at an early step of 50S biogenesis. However, although great care has been taken for purifying the different particles on the sucrose gradients, we cannot rule out minor contamination between the species.

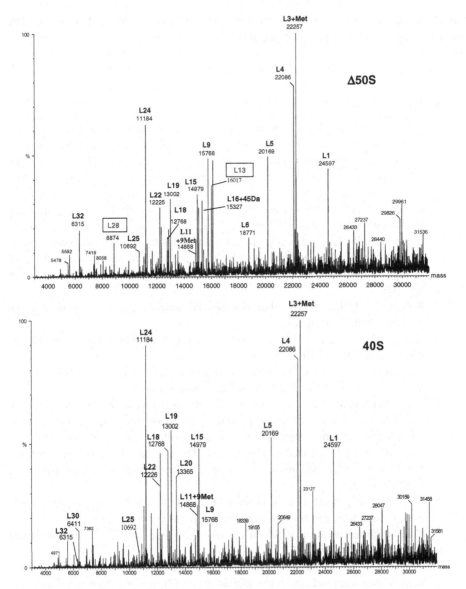

Fig. 4. Deconvoluted spectra obtained by nano-ESI-QqTOF MS analysis of the
Δ50S and 40S intact particles.

The MS analyses also allow an estimation of the relative quantity of each
r-protein present in the different r-particles. The fact that some r-proteins are
found in reduced amounts in the 40S particle may indicate that the 40S particles

are heterogeneous or that these proteins are loosely bound and thus partially lost during the isolation of the particle.

4. Notes

1. All buffers should be 'RNase-free' to prevent degradation of rRNA by ribonucleases. It is important to wear gloves throughout the preparation and analysis of ribosomes, and reagents (such as sucrose) used should be of high quality and where possible RNase-free. Some protocols recommend diethylpyrocarbonate (DEPC) treatment to inactivate ribonucleases. In our experience, the use of high-quality reagents and water and of clean glassware or plasticware is sufficient. Equipment used in these 'RNase-free' experiments should be reserved to this usage only. Buffers used for preparation and centrifugation of ribosomes should contain 10 mM Mg^{2+} and 50–100 mM salt. Low concentrations of Mg^{2+} (1 mM) or high-salt concentrations (400 mM) result in ribosome dissociation into 50S and 30S.

2. RNase-free DNase from other commercial sources can also be used (final concentration of 20 µg/mL).

3. HPLC solutions can be filtered through 0.2-µm filter (Fluoropore™ membrane filter, 0.2 µm FG Millipore). This step is optional because the liquid pathways are equipped with frits. We recommend keeping a unique pair of HPLC bottles devoted to buffers A and B to be refilled on a weekly basis.

4. We prefer 'distal coated' tips, which are externally metallized except at their tip end, where the column eluate is sprayed. In our hands, these tips provide a stable ESI spray for a longer period (up to several weeks) than the 'proximal-coated' tips, which are metallized at the tip end. The contact of the tip end with the HPLC liquid phase progressively damages the metal coating, which may result in ESI instability.

5. We prefer to prepare matrix solutions for MALDI analysis extemporaneously even though they can be kept in the fridge in darkness for 1 or 2 days.

6. This is the temperature at which the ribosome assembly defect of our mutant strain is clearly visible (7).

7. To facilitate the isolation of ribosome intermediates such as the aberrant 40S particle that accumulates in the mutant strain (7), it is recommended avoiding polysome run-off and ribosome dissociation into ribosomal subunits that would contaminate the free (newly synthesized) subunits. Chloramphenicol stabilizes the elongating ribosomes and prevents polysome run-off. Best results are obtained using a fresh solution of chloramphenicol (stock solution of 40 mg/mL in ethanol). To avoid polysome run-off, all steps should be performed at 4°C and as quickly as possible.

8. The cell pellet may be stored at –20°C (for a few days only) at this stage before continuing the protocol.

9. Because of release of DNA, the material will be very viscous at this stage and thus hard to pipet. A good resuspension is achieved by using tips that are cut at their extremities (do not vortex).

10. Degradation of the extracts is observed when stored for more than 1 week.

11. The gradients can be prepared the same day as the centrifugation or the day before; in this case, they are kept at 4°C overnight. Linear gradients are prepared using a standard two-chamber mixing device and are poured into centrifugation tubes.

12. Several groups have described the LC-MS/MS analysis of r-proteins digested by Lys-C, an endoprotease that specifically cleaves at the C-terminus of lysine, because r-proteins contain numerous lysine and arginine residues. Yet, our goal is not only to identify intrinsic r-proteins but also non-ribosomal proteins that associate with the ribosomes. Thus, trypsin, an enzyme classically used in proteomic studies, was chosen. We verified that, in our LC-MS/MS conditions, at least as many proteins constitutive of the W50S subunit could be identified from the tryptic digest as from the Lys-C digest.

13. These gradient durations were chosen by comparing the numbers of proteins identified with the LC-MS/MS analyses of digested W30S and W50S subunits when using gradients of 1 h 20min, 1 h 40 min or 2 h. Note that a longer gradient is required for the more complex sample. When optimizing HPLC gradient slope, one must keep in mind that a longer gradient provides a better peptide separation but also results in chromatographic peak broadening and a reduced analyte concentration at the peak maximum. This is detrimental to ESI-MS sensitivity, because the ESI response is directly correlated to the analyte concentration in the sprayed solution. A balance must then be found between these two constraints.

14. Preliminary LC-MS/MS experiments in which MS data were acquired on the mass range [400; 1800] Da provided a negligible number of peptide identifications associated with fragmented species at m/z above 1300 Da. We then limited the scanning of the quadrupole to the narrowest useful mass range to ensure optimal sensitivity in MS mode.

15. Efficient fragmentation of peptides requires a collision energy that depends on the primary sequence. As a general trend, peptides of larger masses and/or lower charge states have a lower fragmentation yield and may necessitate higher collision energies.

16. These mass intervals were shown to provide similar numbers of peptide identifications when analyzing mitochondrial proteins. These samples are less biased towards highly basic proteins than ribosomal subunits and thus provide larger tryptic peptides in average. For analyzing ribosomal digests, we could have shortened the low-mass range and extended the high-mass one. Yet, to favour the identification of larger, more informative peptides, it was more valuable to keep the intervals [400; 700] Da and [680; 1300] Da.

17. When analyzing protein samples that cover a large spectrum of isoelectric points, we usually perform database searches considering that trypsin may miss one

cleavage site only (K or R). Because r-proteins contain numerous lysine and arginine residues, which are sometimes closely grouped, two missed cleavages are more frequently observed for these proteins.

18. We mainly observed fragments y, b, z and a, as well as immonium ions and internal fragments (especially containing proline residues that hamper fragmentation) in the MS/MS spectra generated with our nano-ESI-QqTOF instrument. Those fragment types are best considered by the instrument description 'ESI-4sector' provided by Mascot.

Acknowledgments

We thank Dr. Marc Dreyfus for critical reading of the manuscript. This work was supported by the CNRS and l'Oréal through PhD fellowship to D. P. and by l'Agence Nationale de la Recherche through grant to Dr. Marc Dreyfus (grant NT05-1-44659).

References

1. Link, A. J., Eng, J., Schieltz, D. M., Carmack, E., Mize, G. J., Morris, D. R., et al. (1999) Direct analysis of protein complexes using mass spectrometry. *Nat. Biotechnol.* **17**, 676–682.

2. Arnold, R. J. and Reilly, J. P. (1999) Observation of Escherichia coli ribosomal proteins and their posttranslational modifications by mass spectrometry. *Anal. Biochem.* **269**, 105–112.

3. Koc, E. C., Burkhart, W., Blackburn, K., Moyer, M. B., Schlatzer, D. M., Moseley, A., et al. (2001) The large subunit of the mammalian mitochondrial ribosome. Analysis of the complement of ribosomal proteins present. *J. Biol. Chem.* **276**, 43958–43969.

4. Yamaguchi, K. and Subramanian, A. R. (2000) The plastid ribosomal proteins. Identification of all the proteins in the 50 S subunit of an organelle ribosome (chloroplast). *J. Biol. Chem.* **275**, 28466–28482.

5. Chang, I. F., Szick-Miranda, K., Pan, S., and Bailey-Serres, J. (2005) Proteomic characterization of evolutionarily conserved and variable proteins of Arabidopsis cytosolic ribosomes. *Plant Physiol.* **137,** 848–862.

6. Tschochner, H. and Hurt, E. (2003) Pre-ribosomes on the road from the nucleolus to the cytoplasm. *Trends Cell Biol.* **13**, 255–263.

7. Charollais, J., Pflieger, D., Vinh, J., Dreyfus, M., and Iost, I. (2003) The DEAD-box RNA helicase SrmB is involved in the assembly of 50S ribosomal subunits in Escherichia coli. *Mol. Microbiol.* **48,** 1253–1265.

8. Spedding, G. (ed.) (1990) *Ribosomes and Protein Synthesis. A Practical Approach.* Oxford University Press, New York.

9. Radulovic, D., Jelveh, S., Ryu, S., Hamilton, T. G., Foss, E., Mao, Y., et al. (2004) Informatics platform for global proteomic profiling and biomarker discovery

using liquid chromatography-tandem mass spectrometry. *Mol. Cell. Proteomics* **3**, 984–997.

10. Li, X. J., Yi, E. C., Kemp, C. J., Zhang, H., and Aebersold, R. (2005) A software suite for the generation and comparison of peptide arrays from sets of data collected by liquid chromatography-mass spectrometry. *Mol. Cell. Proteomics* **4**, 1328–1340.

11. Dammel, C. S. and Noller, H. F. (1995) Suppression of a cold-sensitive mutation in 16S rRNA by overexpression of a novel ribosome-binding factor, RbfA. *Genes Dev*, **9**, 626–637.

12. Lill, R., Crooke, E., Guthrie, B., and Wickner, W. (1988) The "trigger factor cycle" includes ribosomes, presecretory proteins, and the plasma membrane. *Cell* **54**, 1013–1018.

13. Wout, P., Pu, K., Sullivan, S. M., Reese, V., Zhou, S., Lin, B., et al. (2004) The Escherichia coli GTPase CgtAE cofractionates with the 50S ribosomal subunit and interacts with SpoT, a ppGpp synthetase/hydrolase. *J. Bacteriol.* **186**, 5249–5257.

14. Bylund, G. O., Persson, B. C., Lundberg, L. A., and Wikstrom, P. M. (1997) A novel ribosome-associated protein is important for efficient translation in Escherichia coli. *J. Bacteriol.* **179**, 4567–4574.

15. Charollais, J., Dreyfus, M., and Iost, I. (2004) CsdA, a cold-shock RNA helicase from Escherichia coli, is involved in the biogenesis of 50S ribosomal subunit. *Nucleic Acids Res.* **32**, 2751–2759.

II

Proteins Identified in the Organelle Sample: True Components or Contaminants?

23

Assessment of Organelle Purity Using Antibodies and Specific Assays

The Example of the Chloroplast Envelope

Daniel Salvi, Norbert Rolland, Jacques Joyard, and Myriam Ferro

Summary

Proteomics provides a powerful tool to characterize the protein content of an organelle. However, identifications obtained through mass spectrometry and database searching only make sense if the organelle sample is not heavily cross-contaminated. Besides the proteomic analysis, which gives an overview of possible cross-contamination, biochemical methods can be used to assess sample purity. These methods use specific markers that are detected and measured. Here, we describe the use of immunological, enzymatic, lipid, and pigment markers that allow the purity of chloroplast envelope fractions to be estimated.

Key Words: Sample purity; markers; western blot; chlorophyll; enzymatic markers.

1. Introduction

Sample preparation is a crucial step in the context of organelle proteomics. A thorough study of the organelle purity is essential for a precise determination of the subcellular localization of the proteins of interest. An example of a protein previously expected to be located in the plasma membrane, but actually residing to the inner envelope membrane, is given by Ferro et al. *(1)*. Thus, erroneous conclusions can be quoted if no assessment of organelle purity is performed and if no proper validation of a subcellular localization (e.g., using green fluorescent protein fusions) is undertaken. Therefore, to assess whether proteins identified from an organelle fraction are actually located in this organelle, evaluation of organelle purity must be carried out. For this purpose, many assays can be used

From: *Methods in Molecular Biology, vol. 432: Organelle Proteomics*
Edited by: D. Pflieger and J. Rossier © Humana Press, Totowa, NJ

to assess the results of subcellular fractionation experiments. First, quantitative western blotting to monitor the distribution of specific organelle-marker proteins can be used. Antibodies must be chosen to be specific of a marker protein that is known to be exclusively located in a subcellular compartment. Second, another method consists in measuring the activity of marker enzymes. Other methods are more specific to the studied organelle. For instance, pigments like chlorophyll and carotenoids are specific chloroplast markers, and their amount indicates the degree of purity of chloroplast fractions or the degree of contamination of non-plastid fractions. Besides, protein identifications related to the studied subcellular fraction are also indicative of the sample purity. For example, the excellent purity of the *Arabidopsis* chloroplasts prepared on Percoll gradients was confirmed through proteomic analysis: only 5% of the *Arabidopsis* proteins identified may correspond to non-plastid proteins *(2)*. Among them, one protein appeared to correspond to a previously characterized major plasma membrane component; four proteins may indicate contamination by major tonoplast proteins, and one by glyoxysomes. Considering the high sensitivity of current mass spectrometers, it is not surprising to detect minute amounts of these few extra-plastidial contaminants, which are major proteins in their respective subcellular compartments. It is also important to notice that none of the proteins identified in *Arabidopsis* envelope membranes by Ferro et al. *(2)* appears to derive from mitochondria, a classical contaminant of plastid preparations.

To illustrate the different methods used for assessing organelle purity, the present chapter will describe methods developed for the chloroplast envelope. On a routine basis, besides proteomics experiments, three types of markers are used to characterize the purified chloroplast envelope fractions: enzymatic markers, immunological markers, and lipid/pigments markers.

2. Materials

2.1. Enzymatic Markers

2.1.1. Assay Media for Nitrate-Sensitive ATPase Activity (EC 3.6.1.35)

1. Reaction medium: 50 mM MES, 160 mM sucrose, 0.1 mM Na_2MoO_4, 5 mM $MgSO_4$, 1 mM NaN_3, 0.02% (w/v) Brij 58, 1 mM dithiothreitol, and 50 mM KCl, adjusted to pH 8.0 with Tris (*see* **Note 1**).
2. Inhibitor: 50 mM KNO_3.
3. 23.7 mM Tris–ATP in water.
4. Fiske and Subbarow reagent (Sigma, St. Louis, MO, USA).
5. UV-visible spectrophotometer (Uvikon 810 or equivalent; Kontron, Munchen, Germany), with 1-cm (disposable, glass or UV silica) cuvettes, for enzymatic assays and/or pigment analyses.

2.1.2. Assay Media for Vanadate-Sensitive ATPase Activity (EC 3.6.1.35)

1. Reaction medium: 50 mM MES, 160 mM sucrose, 0.1 mM Na_2MoO_4, 5 mM $MgSO_4$, 1 mM NaN_3, 25 mM K_2SO_4, 0.02% (w/v) Brij 58, and 1 mM dithiothreitol, adjusted to pH 6.5 with Tris.
2. Inhibitor: 0.6 mM Na_3VO_4.
3. 23.7 mM Tris/ATP in water.
4. Fiske and Subbarow reagent (Sigma).

2.1.3. Assay Media for Cytochrome c Oxidase (EC 1.9.3.1)

1. 50 mM Na_2HPO_4/NaH_2PO_4, pH 7.5, and 0.3% (w/v) Triton X-100.
2. 5 mM cytochrome c (Sigma) (*see* **Note 2**).

2.1.4. Assay Media for Fumarase (EC 4.2.1.2)

1. 50 mM Tricine-NaOH, pH 7.5.
2. 50 mM malate.

2.1.5. Assay Media for Hydroxypyruvate Reductase (EC 1.1.1.81)

1. 50 mM Mes-NaOH, pH 6.4.
2. 200 mM NADH.
3. 1mM hydroxypyruvate.

2.2. SDS–PAGE and Protein Transfer on Nitrocellulose

1. Gel electrophoresis apparatus (Bio-Rad Protean 3 or equivalent), with the different sets of accessories for protein separation by electrophoresis (combs, plates, and casting accessories).
2. Acrylamide stocks: 30% (w/v) acrylamide—0.8% bisacrylamide: 300 g of acrylamide, 8 g of bisacrylamide, H_2O to 1 L. 60% (w/v) acrylamide—0.8% bisacrylamide: 600 g of acrylamide, 8 g of bisacrylamide, and H_2O to 1 L and store in amber bottles at 4°C.
3. SDS stock solution: 10% (w/v) SDS: 10 g of SDS, H_2O to 1 L and store at room temperature.
4. 4× Laemmli stacking gel buffer (0.5 M Tris–HCl, pH 6.8): 363 g of Tris, H_2O to 900 mL, adjust to pH 6.8 at 25°C with concentrated HCl, make up volume to 1 L and store at room temperature.
5. 8× Laemmli resolving gel buffer (3 M Tris–HCl, pH 8.8): 60.6 g of Tris, H_2O to 900 mL, adjust to pH 8.8 at 25°C with concentrated HCl, make up volume to 1 L and store at room temperature.
6. Stacking gel (5% acrylamide): 5 mL of 30% acrylamide—0.8% bisacrylamide stock solution, 7.5 mL of 4× Laemmli stacking gel buffer, 17.1 mL of H_2O, and 40 μL of TEMED, 4 mL of 10% (w/v) ammonium persulfate (10 g of ammonium

persulfate, H_2O to 100 mL, stored at 4°C, prepare fresh every month), total volume: 30 mL.

7. Single acrylamide resolving gels (10, 12 or 15% acrylamide): (1) for 10% acrylamide gel: 33.3 mL of 30% acrylamide—0.8% bisacrylamide stock solution, 12.5 mL of 8× Laemmli resolving gel buffer, 54 mL of H_2O, 20 μL of TEMED, and 0.2 mL of 10% (w/v) ammonium persulfate, total volume: 100 mL; (2) for 12% acrylamide gel: 40 mL of 30% acrylamide—0.8% bisacrylamide stock solution, 12.5 mL of 8× Laemmli resolving gel buffer, 47.3 mL of H_2O, 20 μL of TEMED, and 0.2 mL of 10% (w/v) ammonium persulfate, total volume: 100 mL; and (3) for 15% acrylamide gel: 50 mL of 30% acrylamide—0.8% bisacrylamide stock solution, 12.5 mL of 8× Laemmli resolving gel buffer, 37.3 mL of H_2O, 20 μL of TEMED, and 0.2 mL of 10% (w/v) ammonium persulfate, total volume: 100 mL.

8. 4× stock solution for protein solubilization: 200 mM Tris–HCl, pH 6.8, 40% (v/v) glycerol, 4% (w/v) SDS, 0.4% (w/v) bromophenol blue, and 100 mM dithiothreitol.

9. Gel reservoir buffer: 38 mM glycine, 50 mM Tris, and 0.1% (w/v) SDS (prepare about 400 mL for each reservoir).

10. Gel-staining medium: acetic acid/isopropanol/water, 10/25/65 (v/v/v), supplemented with 2.5 g/L of Coomassie brilliant blue R250 in clean and closed boxes.

11. Gel destaining medium: acetic acid/ethanol/water, 7/40/53 (v/v/v).

12. Protein transfer medium (for western blots): Gel reservoir buffer (see above) diluted with ethanol to obtain 20% (v/v) final ethanol concentration. Final concentration: 30.4 mM glycine, 40 mM Tris, and 0.08% (w/v) SDS (about 800 mL).

2.3. Western Blot

1. System for protein transfer on nitrocellulose membranes (central core assembly, holder cassette, nitrocellulose filter paper, fiber pads, and cooling unit).

2. Protein transfer medium: gel reservoir buffer (see **Subheading 2.2.**) diluted with ethanol to obtain 20% (v/v) final ethanol concentration. Final concentration: 30.4 mM glycine, 40 mM Tris, and 0.08% (w/v) SDS (prepare about 800 mL).

3. Nitrocellulose membranes (BA85, Schleicher & Schuell or equivalent).

4. Tris-buffered saline with Triton (TBST): 0.15 M NaCl, 50 mM Tris–HCl, pH 7.5, and 0.05% (w/v) Triton X-100.

5. Milk-containing TBST: TBST with 50 g/L of milk powder.

6. Anti-H^+-ATPase (P-type) antibody *(3)* raised against the plasma membrane H^+-ATPase of *Nicotiana plumbaginifolia* (used at a 1/250 dilution).

7. Anti-TIP antibody *(4)* raised against a tobacco (*Nicotiana tabacum*) tonoplast protein (used at a 1/2000 dilution).

8. Anti-Nad-9 antibody *(5)* raised against an extrinsic protein of the wheat (*Triticum aestivum*) mitochondrial inner membrane (used at a 1/2000 dilution).

9. Anti-TOM 40 antibody (formerly MOM42) *(6)*, which recognizes an outer membrane protein from *Vicia faba* mitochondria (used at a 1/2000 dilution).

10. Anti-T subunit of the glycine–decarboxylase complex *(7)*, which recognizes a matrix protein from pea (*Pisum sativum*) mitochondria (used at a 1/10,000 dilution).
11. Anti-E-37 antibody *(8)* raised against a protein from the inner envelope membrane of spinach (*Spinacia oleracea*) chloroplast (used at a 1/20,000 dilution).
12. Anti-ceQORH antibody *(9)* raised against a protein from the inner envelope membrane of *Arabidopsis* chloroplast (used at a 1/10,000 dilution).
13. Anti-LHCP antibody *(10)* raised against a thylakoid membrane protein from *Chlamydomonas reinhardtii* chloroplast (used at a 1/20,000 dilution).
14. Solution A: 90 mM P-coumaric acid (14 mg/mL in DMSO).
15. Solution B: 250 mM luminol (3-aminophalhydrazin) (44 mg/mL in DMSO).
16. 100 mM Tris–HCl, pH 8.5.
17. Developer and fixer solutions.
18. Chemiluminescence-adapted films (Hyperfilm ECL, Amersham).

2.4. Separation and Analysis of Lipids

1. Silica gel pre-coated thin layer chromatography (TLC) plates, chromatography chamber and oven at 110°C.
2. Chloroform, methanol, acetone, acetic acid, perchloric acid, and argon (or nitrogen).
3. Anilinonaphtalene sulfonate [0.2% (w/v) in methanol].
4. α-naphthol-sulfuric acid reagent: 8 g of 1-naphtol in 250 mL of methanol, then add 20 mL of H_2O and finally 32 mL of concentrated H_2SO_4.
5. Molybdate reagent: dissolve 1 g of ammonium molybdate in 10 mL of water, then add 30 mL of 65 % (w/v) perchloric acid and 100 mL of acetone, and finally heat at 120°C.

3. Methods

3.1. Enzymatic Markers

For the following assays, purified envelope proteins are tested for the detection of enzymatic marker activities from other cell compartments. The specific activity of these marker enzymes is measured in purified envelope extracts, in crude cell extract, and, when available, in purified other cell compartments and then compared to estimate the cross-contaminations.

1. Measure of nitrate-sensitive ATPase activity, marker for tonoplasts: this enzyme hydrolyses ATP to ADP and inorganic phosphate. This activity is measured according to the method of Briskin et al. *(11)*, measuring phosphate release resulting from the hydrolysis of ATP. The reaction is started with addition of 23.7 mM Tris/ATP in the reaction medium. Addition of ± 50 mM KNO_3 in the reaction medium allows distinguishing nitrate-sensitive ATPase activity. Release

of inorganic phosphate is measured at 660 nm after addition of 175 µL of Fiske and Subbarow reagent (Sigma).

2. Measure of vanadate-sensitive ATPase activity, marker for plasma membrane: this enzyme hydrolyses ATP to ADP and inorganic phosphate. This activity is measured according to the method of Briskin et al. *(11)*, measuring phosphate release resulting from the hydrolysis of ATP. The reaction is started with addition of 23.7 mM Tris/ATP in the reaction medium. Addition of ± 0.6 mM of Na_3VO_4 in the reaction medium allows distinguishing vanadate-sensitive ATPase activity. Release of inorganic phosphate is measured at 660 nm after addition of 175 µL of Fiske and Subbarow reagent (Sigma).

3. Measure of cytochrome c oxidase activity, marker for the inner mitochondrial membrane: this mitochondrial membrane protein catalyzes the following reaction: 4 ferrocytochrome c + O_2 = 4 ferricytochrome c + 2 H_2O. This activity is measured according to the method described in Briskin et al. *(11)* in 50 mM Na_2HPO_4/NaH_2PO_4, pH 7.5, and 0.3% (w/v) Triton X-100. After addition of 0.45 mM reduced cytochrome c, oxidation of cytochrome c is followed at 550 nm.

4. Measure of fumarase activity (EC 4.2.1.2), marker for mitochondrial matrix (*see* **Note 3**): this mitochondrial enzyme catalyzes the reversible hydratation of fumaric acid to malic acid. This activity is measured according to the method of Hill and Bradshaw *(12)* in 50 mM Tricine-NaOH pH 7.5. The reaction is started after addition of 50 mM malate in the reaction mixture and fumarate synthesis is recorded at 240 nm.

5. Measure of hydroxypyruvate reductase activity (EC 1.1.1.81), marker for peroxisomes (*see* **Note 3**): this peroxisomal enzyme catalyzes the interconversion of D-glycerate + $NAD(P)^+$ = hydroxypyruvate + $NAD(P)H$ + H^+. This activity is measured according to the method of Tolbert et al. *(13)* in 50 mM Mes-NaOH, pH 6.4, and 200 µM NADH. The reaction is started with addition of 1 mM hydroxypyruvate in the reaction mixture and oxidation of NADH is followed at 340 nm (ε_M = 6220/M/cm).

3.2. Immunological Markers: Western Blot Analyses (see Note 4)

For the following assays, purified envelope proteins are tested for the presence of marker proteins from other cell compartments. The presence of these markers is tested in purified envelope extracts, crude cell extract, and, when available, corresponding purified other cell compartments and then compared to estimate the cross-contamination. Western blots are performed after separation of membrane proteins by SDS–PAGE. After gel migration, the proteins are transferred onto a nitrocellulose membrane using a gel transfer apparatus.

1. Prepare the cassette as follows: add successively one fiber pad, three nitrocellulose filter papers, the gel, a nitrocellulose membrane, three nitrocellulose filter papers, one fibber pad, and then insert the sandwich in the holder cassette (the membrane should be placed beside the + electrode).

2. Insert the cassette in the central core assembly unit (together with the cooling unit).

3. Perform the transfer for 2 h at 80 V in protein transfer medium.
4. Recover the nitrocellulose membrane. The following incubation and washing steps (*see* **Subheading 3.2., points 5–10**) require agitation on a rocking plate.
5. Rinse the nitrocellulose membrane with TBST for 10 min.
6. Saturate the nitrocellulose membrane with milk-containing TBST. Leave it for at least 1 h at room temperature.
7. Add the primary antibody diluted in milk-containing TBST. Leave it for 3–12 h (*see* **Note 5**).
8. Wash the nitrocellulose membrane three times (10 min) with TBST.
9. Add the secondary antibody diluted at 1/10,000 in TBST (*see* **Note 6**). Leave it for 1 h 30 min at room temperature.
10. Wash the nitrocellulose membrane 3 times 10 min with TBST.
11. Mix 3 mL of 100 mM Tris–HCl, pH 8.5, with 13.3 µL of solution A.
12. Mix 3 mL of 100 mM Tris–HCl, pH 8.5, with 30 µL of solution B.
13. Mix the two above solutions (same volume of solutions prepared in **Subheading 3.2., points 11** and **12**) in a dark room.
14. Incubate the nitrocellulose membrane for 1 min in the previous mixture.
15. Expose to film for a few seconds up to several minutes depending on the detected signal.
16. Incubate the film successively in the developer solution (for 1–3 min depending on the signal to noise ratio), in water (for 10 s), and in the fixer solution (for 2 min). Rinse the film in water and dry it.

3.3. Lipids and Pigments

3.3.1. Determination of the Chlorophyll Content of a Fraction (see Note 7).

1. Add 10 µL of the extract to be analyzed to 1 mL of acetone/water, 80/20 (v/v), in a 1-mL Eppendorf tube.
2. Vortex and incubate for 15 min on ice and in the dark.
3. Centrifuge for 15 min at 16,000 *g*.
4. Pour in a 1-mL spectrophotometer glass cuvette. 5. Measure the absorbance at 652 nm against a tube containing acetone/water, 80/20 (v/v). A ratio of $OD_{652}/36 = 1$ corresponds to 1 mg of chlorophyll per mL.

3.3.2. Lipid and Pigment Extraction and Analyses

3.3.2.1. Lipid and Pcigment Extraction [Adapted from Bligh and Dyer (14)]

1. To obtain a liquid phase and subsequently extract the lipids, mix 200 µL of membrane suspension in 10 mM MOPS–NaOH, pH 7.5, with 750 µL of a methanol/chloroform (2:1, v/v) mixture. Homogenize with a vortex, and then add 250 µL of water and 250 µL of chloroform. Homogenize with a vortex.
2. Centrifuge the mixture for 10 min at 14,000 *g* to get a two-phase system. Discard the upper phase with a pipette.

3. Remove the lower phase (*see* **Note 8**) by aspiration with a Pasteur pipette. Dry it under a stream of argon (or nitrogen). The residue, containing the lipids and pigments, is dissolved in a minimal volume of chloroform or acetone/water, 80/20 (v/v).

3.3.2.2. TLC Lipid Analyses

Lipids are a heterogeneous group of compounds characterized as being sparingly soluble in water but highly soluble in non-polar solvents (e.g., chloroform, ether, hexane, etc.). A classical way to analyze the lipids contained in membrane extracts is to make use of this solubility difference to extract and separate lipid constituents from the many polar compounds by two-dimensional (2D) TLC. TLC is indeed a highly versatile and effective technique for separation of complex lipids (phospholipids, glycolipids, etc.) from membrane extracts. TLC is based on the separation of a mixture of compounds as it migrates with the help of a suitable solvent through a thin layer of adsorbent material that has been applied to an appropriate support. The most common adsorbent used is silica gel (silicic acid combined with a small amount of gypsum as a binding agent). In adsorption TLC, the sample is continually fractionated as it migrates through the adsorbent layer. Competition for active adsorbent sites between materials to be separated and the developing solvent produces continuous fractionation. As the process continues, the various components move different distances, depending on their relative affinities for the adsorbent as compared with the migrating solvent. In general, the more polar compounds are held back by the adsorbent while less polar molecules advance further. Polar solvent effects the movement of sample material over the adsorbent. The migration of a compound in a given TLC system is described by its R_f value, that is, the ratio of the distance traveled by the compound to the distance of the solvent front from the origin.

1. Prior to the experiment, heat a silica gel pre-coated TLC plate in an oven at 110°C for at least 1 h. This heat activation of the TLC plate is necessary to remove all traces of water in the silica.
2. Load, with the tip of a Pasteur pipette, the chloroform lipid extract (prepared above) onto the silica gel pre-coated TLC plate (*see* **Note 9**).
3. After spotting, place the plate in a rectangular chromatography chamber containing 100–150 mL of the following mixture: chloroform/methanol/water (65/25/4, v/v/v) for the first dimension of chromatography. Allow the solvent system to run to the front of the plate. Remove the plate, and place it in a glass chamber under a slow stream of argon (or nitrogen) for 1–2 h.
4. Turn the plate to have the lipids at the bottom, place it in a rectangular chromatography chamber containing 100–150 mL of the following mixture: chloroform/acetone/methanol/acetic acid/water (50/20/10/10/5, v/v) for the second dimension of chromatography (perpendicular to the first dimension). Remove the

plate when the solvent reaches the top of the plate, and place it in a glass chamber under a slow stream of argon (or nitrogen) until dry or lipid staining.

5. After migration, lipids can be identified using specific procedures. Polar lipids are located under UV light (360 nm) after spraying the plates with anilinonaphtalene sulfonate [0.2% (w/v) in methanol]. Spraying the plates with α-naphthol-sulfuric acid reagent followed by charring is used to detect carbohydrate residues in glycolipids (*see* **Note 10**). Phospholipids are detected with a molybdate reagent, which is specific for the phosphorus moiety in phospholipids.

6. When necessary, quantification of the amount of marker lipids in each sample is done by gas chromatography [for details, *see* references *(15,16)*].

3.3.2.3. PIGMENTS ANALYSES

1. Use in this case a lipid extract solubilized in acetone/water, 80/20, v/v (1 mL, final volume).
2. Pour the solution in a 1-mL spectrophotometer cuvette.
3. Record the absorption spectrum between 350 and 750 nm. Carotenoids are responsible for a series of peaks in the 400–500 nm region of the spectrum whereas chlorophylls show in addition a sharp peak with a maximum in the 650–700 nm region (*see* **Note 11**).

Notes

1. Reaction media ± inhibitor can be stored at –20°C.
2. As Sigma's cytochrome c products are supplied mainly in the oxidized form of the protein, the reduced form of cytochrome c can be prepared with either sodium dithionite or sodium ascorbate, followed by gel filtration *(17)*.
3. The activity of the mitochondrial marker fumarase was not detected in purified chloroplasts from *Arabidopsis* leaves (no detected cross-contamination). The specific activity of the mitochondrial marker fumarase in *Arabidopsis* mitochondria purified from cell cultures *(18)* is about 400 nmol/min/mg proteins. This corresponds to an enrichment of a factor of 10 compared to the specific activity of the same marker in a crude protoplast extract (about 40 nmol/min/mg proteins). The specific activity of the peroxisomal marker hydroxypyruvate reductase in chloroplast purified from *Arabidopsis* leaves was 6.8 nmol/min/mg proteins for an original specific activity of 256 nmol/min/mg proteins in a crude leaf extract (2–3% cross-contamination). By comparison, the specific activity of hydroxypyruvate reductase in mitochondria purified from *Arabidopsis* cell cultures was 67 nmol/min/mg proteins for an original specific activity of 103 nmol/min/mg proteins in a crude protoplast extract deriving from *Arabidopsis* cell cultures *(18)*.
4. Follow the instructions for saturation and incubation of the membrane with primary and secondary antibodies provided by the manufacturers. Purified envelope membranes were demonstrated to be enriched in envelope markers

when compared to a crude cell extract or a crude chloroplast extract. When correctly purified, chloroplast envelope membranes do not contain detectable amounts of specific markers of the mitochondrial membranes, plasma membrane and thylakoid membranes *(19)*.

5. Several dilutions of the primary antibodies should be tested to determine the best signal/noise ratio.

6. In our case, the secondary antibody is an anti-rabbit-IgG antibody coupled to the horseradish peroxidase for further detection. The secondary antibody has to be adapted to the primary antibody and the detection procedure used.

7. The chlorophyll content was 170 µg per mg of protein in chloroplasts purified from *Arabidopsis* leaves and 84 µg per mg of protein in a crude leaf extract (enrichment by a factor of 2). By comparison, chlorophyll concentration in a crude protoplast extract is about 4.5 µg chlorophyll per mg of protein *(18)*. This procedure is adapted from Arnon *(20)*.

8. The chloroformic (lower) phase contains lipids and pigments.

9. The lipids dissolved in chloroform are applied at the lower right hand corner of the plate, about 3 cm from each edge. This is done under a stream of nitrogen to accelerate solvent evaporation. 100 to 500 µg of total lipids allow detection of lipid contaminants. These figures refer to the total amount of lipids in which contamination is just a minor part. Lipid analyses are a very accurate procedure to see whether you have contaminants: phosphatidylethanolamine (PE) is absent from pure chloroplasts; traces of PE in a chloroplast fraction indicate a contamination by extraplastidial membranes. There is about 1.2 mg of lipids per mg of protein in the envelope (about 3 mg lipids/mg protein in the outer envelope membrane).

10. Galactolipids are markers for plastid membranes, whereas phospholipids such as PE are markers for extraplastidial membranes *(21)*. However, in some very specific conditions, such as phosphate deprivation, phosphate present in phospholipids is mobilized and non-phosphorous membrane lipids, such as digalactosyldiacylglycerol (DGDG), increase in non-plastidial membranes [see, for instance, reference *(22)*].

11. When correctly purified, chloroplast envelope membranes do not contain chlorophylls, but only carotenoids. Being the most likely source of membrane contamination of the purified envelope fraction, thylakoid cross-contamination needs to be precisely assessed. Chlorophyll being the most conspicuous thylakoid component, the presence of this pigment in envelope membrane preparations is specifically analyzed.

References

1. Ferro, M., Salvi, D., Rivière-Rolland, H., Vermat, T., Seigneurin-Berny, D., Grunwald, D., et al. (2002) Integral membrane proteins of the chloroplast envelope: identification and subcellular localization of new transporters. *Proc. Natl. Acad. Sci. U.S.A.* **99**, 11487–11492.

2. Ferro, M., Salvi, D., Brugière, S., Miras, S., Kowalski, S., Louwagie, M., et al. (2003) Proteomics of the chloroplast envelope membranes from *Arabidopsis thaliana. Mol. Cell Proteomics* **2**, 325–345.

3. Morsomme, P., Dambly, S., Maudoux, O., and Boutry, M. (1998) Single point mutations distributed in 10 soluble and membrane regions of the *Nicotiana plumbaginifolia* plasma membrane PMA2 H+-ATPase activate the enzyme and modify the structure of the C-terminal region. *J. Biol. Chem.* **273**, 34837–34842.

4. Gerbeau, P., Guclu, J., Ripoche, P., and Maurel, C. (1999) Aquaporin Nt-TIPa can account for the high permeability of tobacco cell vacuolar membrane to small neutral solutes. *Plant J.* **18**, 577–587.

5. Combettes, B. and Grienenberger, J. M. (1999) Analysis of wheat mitochondrial complex I purified by a one-step immunoaffinity chromatography. *Biochimie* **81**, 645–653.

6. Perryman, R. A., Mooney, B., and Harmey, M. A. (1995) Identification of a 42-kDa plant mitochondrial outer membrane protein, MOM42, involved in the import of precursor proteins into plant mitochondria. *Arch. Biochem. Biophys.* **316**, 659–664.

7. Vauclare, P., Macherel, D., Douce, R., and Bourguignon, J. (1998) The gene encoding T protein of the glycine decarboxylase complex involved in the mitochondrial step of the photorespiratory pathway in plants exhibits features of light induced genes. *Plant Mol. Biol.* **37**, 309–318.

8. Joyard, J., Grossman, A.R., Bartlett, S.G., Douce, R., and Chua, N.-H. (1982) Characterization of envelope membrane polypeptides from spinach chloroplasts. *J. Biol. Chem.* **257**, 1095–1101.

9. Miras, S., Salvi, D., Ferro, M., Grunwald, D., Garin, J., Joyard, J., et al. (2002) Non-canonical transit peptide for import into the chloroplast. *J. Biol. Chem.* **277**, 47770–47778.

10. Vallon, O., Wollman, F. A., and Olive, J. (1986) Lateral distribution of the main protein complexes of the photosynthetic apparatus in *Chlamydomonas reinhardtii* and in spinach: an immunocytochemical study using intact thylakoid membranes and a PSII enriched membrane. *Photobiochem. Photobiophys.* **12**, 203–220.

11. Briskin, D. P., Leonard, R. T., and Hodges, T. K. (1987) Isolation of the plasma membrane: membrane markers and general principles. *Methods Enzymol.* **142**, 542–558.

12. Hill, R. L., and Bradshaw, R. A. (1969) Fumarase. *Methods Enzymol.* **13**, 91–99.

13. Tolbert, N. E., Yamazaki, R. K., and Oeser, A. (1970) Localization and properties of hydroxypyruvate and glyoxylate reducteses. *J. Biol. Chem.* **245**, 5129–5136.

14. Bligh, E. G. and Dyer, W. J. (1959) A rapid method of total lipid extraction and purification. *Can. J. Med. Sci.* **37**, 911–917.

15. Kates, M. (1986) Separation of lipid mixtures. In *Laboratory Techniques in Biochemistry and Molecular Biology. Techniques of Lipidology* (Burdon, R. H. and van Knippenberg, P. H., eds), Elsevier, Amsterdam, pp.186–278.

16. Douce, R., Joyard, J., Block, M.A, and Dorne, A.-J. (1990) Glycolipids analyses and synthesis in plastids. In *Methods in Plant Biochemistry, Vol. 4, Lipids, Membranes and Aspects of Photobiology* (Harwood, J.L. and Bowyer, J.R., eds.), Academic Press, London, pp. 71–103.

17. Dixon, H. B. and McIntosh, R. (1967) Reduction of methaemoglobin in haemoglobin samples using gel filtration for continuous removal of reaction products. *Nature* **213**, 399–400.

18. Brugière, S., Kowalski, S., Ferro, M., Seigneurin-Berny, D., Miras, S., Salvi, D., et al. (2004) The hydrophobic proteome of mitochondrial membranes from *Arabidopsis* cell suspensions. *Phytochemistry* **23**, 1693–1707.

19. Seigneurin-Berny, D., Gravot, A., Auroy, P., Mazard, C., Kraut, A., Finazzi, G., et al. (2006) HMA1, a new Cu-ATPase of the chloroplast envelope, is essential for growth under adverse light conditions. *J. Biol. Chem.* **281**, 2882–2892.

20. Arnon, D. L. (1949) Copper enzymes in isolated chloroplasts. Polyphenoloxidase in *Beta vulgaris. Plant Physiol.* **24**, 1–15.

21. Douce, R. and Joyard, J. (1982) Purification of the chloroplast envelope. In *Methods in Chloroplast Molecular Biology* (Edelman, M., Hallick, R., and Chua, N.H., eds.), Elsevier/North-Holland, Amsterdam, pp. 139–256.

22. Jouhet, J., Marechal, E., Baldan, B., Bligny, R., Joyard, J., and Block, MA. (2004) Phosphate deprivation induces transfer of DGDG galactolipid from chloroplast to mitochondria. *J. Cell Biol.* **167**, 863–874.

24

Identifying Bona Fide Components of an Organelle by Isotope-Coded Labeling of Subcellular Fractions

An Example in Peroxisomes

Marcello Marelli, Alexey I. Nesvizhskii, and John D. Aitchison

Summary

Organelles are biochemically distinct subcellular compartments that perform specific functions within a cell. These roles are regulated by the complement of proteins associated with each organelle. Thus, a comprehensive understanding of an organelle's proteome is necessary to elucidate the diverse roles of each organelle. Mass spectrometry combined with biochemical fractionation methods has enabled the proteomic characterization of several organelles. However, due to the poorly quantitative nature of mass spectrometry and the inability to generate pure fractions of an organelle, distinguishing bona fide components of the organelle from contaminants of the fraction is a significant challenge. We have addressed this limitation by adopting quantitative mass spectrometric approaches to identify proteins that enrich in a purified fraction of organelles relative to a crude or contaminating fraction. The methods for the analyses of the yeast peroxisome are described in detail; however, these concepts are generally applicable to the study of other organelles.

Key Words: ICAT; organelle proteomics; quantitative mass spectrometry; peroxisome; organelle proteome; organelle isolation.

1. Introduction

In eukaryotes, organelles constitute distinct organizations of proteins that contribute to cellular metabolism. An in-depth understanding of their metabolic roles requires a comprehensive knowledge of their components, the changes in their composition under different conditions, and the dynamic interplay between

From: *Methods in Molecular Biology, vol. 432: Organelle Proteomics*
Edited by: D. Pflieger and J. Rossier © Humana Press, Totowa, NJ

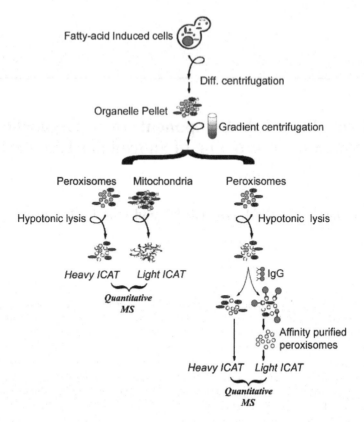

Fig. 1. Outline of quantitative mass spectrometry approaches used to identify bona fide peroxisomal proteins. Cells induced to proliferate peroxisomes are lysed, subcellular organelles are isolated by a series of differential centrifugation steps. Organelles are separated by density gradient centrifugation. Peroxisomal and mitochondrial fractions are collected, the organelles are hypotonically lysed, and the membrane fraction is collected by high-speed centrifugation. Proteins in each sample are labeled with either heavy or light isotope-coded affinity tags (ICAT) reagents and the relative abundance of ICAT-labeled peptide pairs is determined by mass spectrometry. In a second approach, peroxisomes are affinity purified from a peroxisomal membrane fraction derived from a yeast strain expressing a peroxisomal membrane protein tagged with the IgG-binding domains of protein A. Affinity-purified peroxisomes and peroxisomal membranes are differentially labeled with ICAT reagents and analysed by mass spectrometry.

subcellular structures. Proteomic approaches aimed at defining organellar constituents have been enabled by advances in the sensitivity and throughput of mass spectrometry. However, the comprehensive characterization of an organelle by mass spectrometry remains a formidable undertaking, due in part to the large number of proteins associated with organelles and dynamic range of their components. These studies are also hindered by the limited resolving power of subcellular fractionation procedures and the inevitable copurification of contaminating organelles. The issue of sample complexity can be partially addressed by pre-fractionation of the samples [reviewed in *(1)*]. In parallel, a number of biochemical approaches have been developed to generate samples of high purity [reviewed in *(2)*]. These studies have yielded extensive lists of proteins derived from an organelle fraction but are unable to distinguish proteins of the organelle from residual contaminants. Constituents of a fraction have been classically defined as proteins that enrich with that fraction relative to other fractions, a classification that requires relative quantification of protein abundance in the fractions *(3)*. Thus, we developed methods based on the use of isotope-coded affinity tags (ICAT) and quantitative mass spectrometry to identify proteins that enrich with peroxisomes during their purification *(4)*. In one experiment, we compared the abundance of proteins in the peroxisome versus a fraction containing mitochondria, the most common contaminant of the peroxisomal fraction. A second complementary approach was designed to compare a crude peroxisomal fraction against a highly purified fraction of immunopurified peroxisomal membranes (*see* **Fig. 1**). While these studies did not provide a complete catalogue of the peroxisomal proteome, an in-depth prioritized list of the components of this organelle was assembled that included the identification of a number of proteins not previously known to associate with peroxisomes *(4)*.

The methods below describe procedures for the induction of peroxisomes in the yeast *Saccharomyces cerevisisae*, the isolation of peroxisomal membranes [based largely on Smith et al. *(5)*, with modifications], the affinity purification of peroxisomal membranes, and the labeling of sample pairs with the ICAT reagents. In addition, a method for the analysis of these data is described.

2. Materials

2.1. Yeast Growth and Peroxisome Isolation

1. YEPD: 2% (w/v) peptone, 1% (w/v) yeast extract, and autoclave to sterilize. Supplement with glucose [to 2% (w/v)] before use. Autoclave with 20g/L of agar for YEPD plates.
2. *S. cerevisiae* induction medium (SCIM): 0.7% (w/v) yeast nitrogen base, 0.5% (w/v) yeast extract, 0.5% (w/v) peptone, 0.5% (v/v) Tween-40, 0.79 g/L CM medium (Q-Biogene, Irvine, CA, USA), and 0.5% (w/v) ammonium sulphate. Medium can be stored at room temperature after autoclaving. Oleic acid is added to a final concentration of 0.1% (v/v) prior to the addition of cells.
3. Oleic acid: store in the dark at –20°C in 10-mL aliquots (MP Biomedicals, Solon, OH, USA).

4. TSD reduction buffer: 0.1 M Tris–HCl, pH 9.4 and 10 mM dithiothreitol (DTT). Store a 1 M Tris–HCl, pH 9.4, stock solution at room temperature and dilute to the working concentration when needed. Add DTT prior to use from a 1 M stock stored at –20°C.

5. 50 mM potassium phosphate buffer: combine 40 mL of 1 M K_2 HPO, 10 mL of 1 M KH_2 PO, and 2 mL of 0.5 M ethylenediaminetetraacetic acid (EDTA), and add water up to 1 L to obtain a phosphate buffer at pH 7.5.

6. 1.2 M sorbitol-K: 218.60 g of sorbitol is dissolved in 1 L of 50 mM potassium phosphate buffer. Store at 4°C.

7. Zymolyase 100T (MP Biomedicals).

8. MES buffer: 5 mM 2-(N-morpholino)ethanesulfonic acid, pH 5.5, 1 mM EDTA, and 1 mM KCl.

9. 1.2 M sorbitol/MES: 218.60 g of sorbitol dissolved in 1 L of MES buffer. Store at 4°C.

10. Disruption buffer: 0.6 M sorbitol in MES buffer. Dissolve 109.3 g of sorbitol in 1 L of MES buffer. Store at 4°C.

11. P-solution (1000×): 90 mg of phenylmethylsulfonylfluoride (PMSF), 2 mg of pepstatin, and 2 mg of leupeptin in 5 mL of 100% ethanol. Store at –20°C.

12. Nycodenz gradient solutions: dissolve 50 g of Nycodenz (Axis-Shield, Oslo, Norway) in 100 mL of disruption buffer overnight with constant mixing. Dilute 50% stock solution to 35, 25, and 17% Nycodenz with disruption buffer. Using a 10-mL syringe fitted with a 6-inch pipetting needle, dispense 2.2 mL of 17% Nycodenz solution into an Optiseal centrifugation tube (11 mL Beckman Optiseal tubes). Using the syringe, slowly underlay 5.5 mL of 25% Nycodenz solution ensuring that the gradient solutions do not mix. In a similar manner, underlay 1.5 mL of 35% and 1 mL of 50% Nycodenz solution into the sample tube (*see* **Fig. 2**). Gradients can be stored at 4°C overnight; however, best results are obtained when gradients are formed within 4 h of use.

2.2. Extraction of Organelle Membranes

1. Ti8 buffer: 20 mM Tris–HCl, pH 8.0, and 1 mM EDTA.

2.3. ICAT Labeling of Peroxisomal and Mitochondrial Membrane Fractions

1. Light ($^1H_8$ or $^{12}C_9$) and heavy ($^2H_8$ or $^{13}C_9$) ICAT reagents (Applied Biosystems, Foster City, CA, USA). Cleavable ICAT reagent kit from Applied Biosystems contains 10 vials of light and heavy versions of the reagent. Each vial can be used to label 100 μg of protein. $^1H_8$ and $^2H_8$ ICAT reagents contain 175 nmol of the reagent. Resuspend vial contents in 10 μL of 100% methanol prior to use. Each kit also contains all the necessary reagents, buffers (load, wash 1, wash 2, elute, and storage buffers), an avidin affinity column, and an assembly for sample labeling and purification.

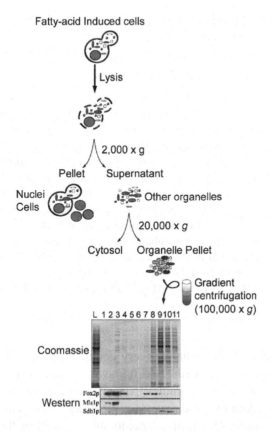

Fig. 2. Subcellular fractionation to isolate yeast peroxisomes. Yeast cells induced to proliferate peroxisomes are converted to spheroplasts and lysed by homogenization. The homogenate is subjected to 2000 *g* to separate unbroken cells and nuclei (Pellet) from the remainder of the cytoplasmic components (Supernatant). The organelles in the supernatant are collected by centrifugation at 20,000 *g* and are named 20kgP. The individual components of the organelle pellet are separated by density centrifugation using a Nycodenz step gradient. Twelve equal fractions are collected from the bottom of the gradient. Equal portions of each fraction *(1–11)* are resolved by SDS–PAGE. Western blotting using antibodies directed against Fox2p and Mls1p are used to identify peak peroxisomal (P) and antibodies to Sdh1p used to identify peak mitochondrial (M) fractions.

2. Ti8-SDS buffer: 20 mM Tris–HCl, pH 8.0, 1 mM EDTA, and 0.05% (w/v) SDS. Store at room temperature.
3. 500 mM Tris (2-carboxyethyl)-phosphine-HCl (Pierce, Rockford, IL, USA).
4. Proteomics grade porcine trypsin (Sigma, St. Louis, MO, USA).
5. Phosphoric acid/water, 85/15 (w/v).
6. Polysulphoethyl A (strong cation exchange) column (2.1 × 200 mm; PolyLC Inc., Columbia, MD, USA).

7. Buffer B: 5 mM KH PO , pH 3.0, 600 mM KCl, and 25% (v/v) CH_2 CN.
8. 0.1% (v/v) formic acid in water.

2.4. Affinity Purification of Peroxisomes

1. M_{280} Tosylactivated beads (Invitrogen, Carlsbad, CA, USA), pre-equilibrated in Ti8 buffer (*see* **Note 1**): IgG-conjugated beads are generated by an extended incubation of affinity-purified rabbit IgG to Tosylactivated beads at 37°C as described by the manufacturer. Briefly, 10^7 beads are washed in 0.1 M borate buffer, pH 9.5. Three to eight micrograms of affinity-purified IgGs in 0.1 M borate buffer, pH 9.5, is added to the beads and incubated overnight at 37°C. After this incubation, the bound material is collected using a magnet, washed extensively, and unbound sites are blocked by incubation in 0.2 M Tris-HCl, pH 8.5, and 0.1% (w/v) BSA and incubated for 4 h at 37°C. The beads are then washed, resuspended in phosphate-buffered saline (PBS) solution at $\sim 2 \times 10^9$ beads/mL, and stored in PBS containing 0.02% (w/v) sodium azide.
2. Affinity-purified rabbit IgG (MP Biochemicals).
3. Phosphate buffer: 47 µL of 1 M K_2 HPO_4 and 3 µL of 1 M KH_2 PO_4.

3. Methods

The efficiency of organelle recovery is dependent on effective induction of peroxisomes, lysis of the cells, and homogenization of the sample. However, unlike other more robust organelles, peroxisomes are susceptible to lysis. Thus, care should be taken to avoid excessive pipetting, homogenization, or handling throughout the procedure (critical points are outlined in the methods below). The method described here has been optimized to yield high-purity peroxisomes at the expense of yield (*see* **Fig. 2**).

3.1. Yeast Growth and Peroxisome Isolation

1. Yeast strains are streaked to single colonies in YEPD agar plates. Pick a single colony, seed in 5 mL of YEPD, and grow overnight at 30°C to an optical density exceeding OD 600_{nm} of 1.0. Use 100 µL of this starter culture to seed 50 mL of YEPD and grow overnight at 30°C.
2. Transfer the culture to 1 L of YEPD and grow for an additional 6-8 h at 30°C. The culture should approach an OD 600_{nm} between 2 and 2.5.
3. Collect cells by centrifugation at 2000 *g* in a swinging bucket rotor (e.g., Beckman JS 4.3 at 3500 rpm).
4. Resuspend cells in a small amount of SCIM medium and transfer the cells to 4 L of SCIM. Induce peroxisome proliferation by incubating in this medium for 8–12 h at 30°C with constant agitation. To ensure adequate aeration and proper emulsification of the oleic acid in the media use only 20–25% of the capacity of the Erlenmeyer flasks (e.g., 1 L of medium in a 4-L flask) (*see* **Note 2**)

5. Collect cells by centrifugation at 2000 g (e.g., Beckman JA-10 at 3400 rpm). Decant media and resuspend cells in 500 mL of water. Harvest cells as described above. Using pre-weighed centrifuge bottles to determine the wet weight of the cell pellet. Expect 10–20 g of cells.

6. Resuspend the cells in 8 mL of TSD reduction buffer for each gram of cells and incubate at 30°C for 30 min with gentle agitation. This incubation increases the efficacy of the subsequent Zymolyase treatment by compromising the yeast cell wall.

7. Harvest cells by centrifugation at 2000 g for 10 min (e.g., JA-10 rotor at 3400 rpm) and wash pellet with 100 mL of 1.2 M sorbitol-K solution.

8. Collect cells by centrifugation at 2000 g for 10 min as above and resuspend cells in 8 mL of 1.2 M sorbitol-K per gram of cells.

9. Add 1 mg of Zymolyase 100T per gram of cells and incubate at 30°C for 30 min with gentle agitation. At the end of this incubation, the degree to which cells have been converted to spheroplasts is monitored by diluting 10 μL of cells suspension in 50 μL of water and observing the cells in a phase-contrast microscope. No intact (phase light) cells should be observed. If necessary, extend the incubation until 100% of the cells are converted to spheroplasts.

10. Harvest cells by centrifugation at 1000 g for 10 min in a swinging bucket rotor (e.g., Beckman JS 13.1 at 2500 rpm).

11. Decant supernatant and wash the sides of the tubes with 1.2 M sorbitol/MES without disrupting cell pellet. Decant again.

12. Add 3 mL of disruption buffer containing P-solution per gram of cells to the pellet and gently resuspend the pellet by tapping the side of the tube until the pellet is loosened.

13. Transfer the spheroplasts to an appropriate glass tube and homogenize using a motorized Potter–Elvehjem PTFE homogenizer. 15–20 strokes should be performed and the sample must be kept on ice (*see* **Note 3**).

14. Examine the homogenate with a phase contrast microscope to ensure efficient cell lysis (expect ~60–75% of cells to lyse). From this point on, ensure that fresh P-solution (diluted 1000 fold in each buffer) is added to each buffer and that samples are kept on ice to prevent excessive proteolytic degradation.

15. Transfer the homogenate to fresh tubes and centrifuge at 2000 g for 10 min at 4°C in a swinging bucket rotor (e.g., a JS13.1 rotor at 3600 rpm) to collect unbroken cells and nuclei. Decant the supernatant to fresh tubes and repeat the centrifugation. Repeat centrifugation several times until the pellet is significantly reduced in size and the initially yellow supernatant appears whiter in color. Generally, five spins are required to remove the majority of nuclei and unbroken cells.

16. Transfer the supernatant to fresh tubes and collect organelles by centrifugation at 20,000 g at 4°C in a swinging bucket rotor (e.g., JS13.1 rotor, 11,300 rpm).

17. Decant the supernatant and save for analysis. Add 1 mL of disruption buffer to the pellet and resuspend gently by tapping the tube or with the aid of a glass rod until no particulates are visible. Avoid vortexing and excessive pipetting of the sample to prevent lysis of the organelles.

18. Transfer the homogenate to a 1.5-mL microcentrifuge tube and centrifuge at 2000 g to remove any residual nuclei and cells. Collect the supernatant and transfer to a fresh tube. Repeat this centrifugation step until the obtained pellet has a volume smaller than 50 μL. It is important to efficiently clear the sample as residual nuclei will result in their contamination of peak peroxisomal fractions following isopycnic gradient centrifugation.

19. Determine the protein concentration of supernatant using a Bradford assay. Each gram of cells will produce ~0.5–1 mg of organelles.

20. Load 2–4 mg of protein on each Nycodenz gradient. Top up the tube with disruption buffer. Do not overload the gradient as this will result in poor resolution of organelles.

21. Centrifuge at 100,000 g_{avg} for 1 h at 4°C in a near vertical rotor (e.g., Beckman NVT65 at 35,000 rpm or equivalent). If a near vertical rotor is not available, use a vertical rotor (e.g., Beckman VTi65).

22. Peroxisomes should be visible as a white cotton-like material at the 35–50% Nycodenz interface. Using a peristaltic pump collect twelve 1-mL fractions from the bottom of the gradient into 1.5-mL microcentrifuge tubes (*see* **Note 4**).

23. Determine the protein concentration of each fraction using a Bradford assay. Peak protein amounts in fractions ~2–4 and ~8–10 correspond to the peroxisomal and mitochondrial fractions, respectively. Resolve 5 μL of each fraction by SDS–PAGE and examine by Coomassie blue staining and/or western blotting to identify peak peroxisomal fractions and peak mitochondrial fractions (*see* **Fig. 2**).

3.2. Extraction of Organelle Membranes

1. Combine the peroxisomal fractions (fractions 2–4) and mitochondrial fractions (fractions 8–10). Concentrate the organelles by diluting each sample with 6 volumes of disruption buffer followed by centrifugation at 20,000 g in a swinging bucket rotor (e.g., JS13.1 at 11,300 rpm) for 1 h at 4°C. The organelles will be visible as a white pellet at the bottom of the tube. Decant the supernatant being careful not to disrupt the pellet.

2. Hypotonically lyse the organelles by resuspending the pellets in 1 mL of Ti8 buffer containing P-solution and incubate at 4°C for 2 h. Ensure that the pellets are well resuspended by repeated pipetting or using a hand-held Potter–Elvehjem homogenizer (*see* **Note 5 and Fig. 3**).

3. Collect the membrane fractions by centrifugation at 160,000 g_{avg} (65,000 rpm in a Beckman TLA120.1) for 1 h at 4°C. Decant supernatants and save for analysis. Resuspend pellets in 100 μL of Ti8 buffer containing P-solution and sonicate if necessary to obtain a homogeneous suspension. Determine the protein concentration of this sample by Bradford assay. Each peroxisome-derived fraction may yield ~125–250 μg of protein; mitochondrial fractions can yield 750 μg. Several peroxisome isolations and membrane extractions may be necessary to obtain enough material for ICAT labeling and analyses. Aim to collect ~0.5–1 mg total protein for mass spectrometric analyses.

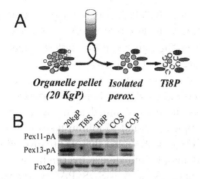

Fig. 3. Extraction of organelles reduces sample complexity. (**A**) Peroxisomes isolated from an organelle pellet (20kgP) can be lysed by incubation in a hypotonic buffer to release its matrix components and membranes pelleted by centrifugation to generate a Ti8P. (**B**) The differential extraction of peroxisomal proteins. An organelle pellet was hypotonically lysed and subjected to centrifugation to generate a supernatant (Ti8S) enriched for matrix enzymes and a pellet (Ti8P) enriched for membranes. Ti8P was further extracted with sodium carbonate and centrifuged to obtain a supernatant (CO_3 S) containing membrane-associated proteins and a pellet (CO_3P) containing integral membrane proteins. Cellular equivalents of each fraction were analysed by western blot to observe the distribution of a membrane-associated protein (Pex11-pA), an integral membrane protein (Pex13-pA), and a matrix protein (Fox2p). Note the reduction of the matrix protein, Fox2p, in Ti8P fractions relative to the 20kgP fraction.

3.3. ICAT Labeling of Peroxisomal and Mitochondrial Membrane Fractions

1. Resuspend membrane pellets in 100 μL of Ti8-SDS buffer.
2. Sonicate samples to homogeneity.
3. Collect a volume of mitochondrion-derived sample and dilute with Ti8-SDS. Because of the complexity of the mitochondrial proteome and the wide range of protein abundances of its components, the optimal amount of the mitochondrion-derived sample must be determined empirically to maximize the number of peroxisome identifications. Initially, use an equal quantity of protein as that of peroxisomal fractions, in the same sample volume.
4. Denature proteins by adding urea to a final concentration of 6 M to each sample followed by incubation at 37°C for 30 min.
5. Reduce samples in 5 mM Tris–(2-carboxyethyl)-phosphine-HCl for 45 min at 23°C.
6. Calculate the amount of ICAT reagent needed to label the cysteines of the sample. To do this, estimate the molar amount of cysteines in the sample and use a fivefold molar excess of ICAT reagent while maintaining at least a 1.2 mM concentration of the reagent (*see* **Note 6**).

7. Add heavy ($^2H_8$ or $^{13}C_9$) ICAT reagent to the peroxisomal fraction and light ($^1H_8$ or $^{12}C_9$) ICAT to the mitochondrial fraction and incubate at room temperature for 2 h protected from the light. At the end of the incubation quench the reaction by the addition of 12 mM DTT to each tube.

8. Save 10 μL of sample from each reaction for analyses, such as SDS–PAGE.

9. Combine the peroxisomal and mitochondrial samples and dilute with 5 volumes of Ti8 buffer. This dilution is aimed at reducing the concentrations of SDS and urea in the sample so as not to inhibit trypsin digestion (aim to reduce the concentration of SDS to 0.01% and that of urea to below 2 M).

10. Add trypsin to 10 ng/mL and digest overnight at 37°C. Save 10 μL of the tryptic digests for analyses, such as SDS–PAGE, to check completion of protein digestion.

11. Add phosphoric acid/water, 85/15 (w/v) drop-wise to the samples while monitoring their pH using pH strips until they reach a pH of ~3.

12. Fractionate the sample by strong cation exchange chromatography using a polysulphoethyl A column following this elution protocol: add 0–25% buffer B over 30 min, followed by 25–100% buffer B over 20 min at 0.44 mL/min. Collect sixty 0.4-mL fractions. Obtain a chromatographic trace (OD_{280nm}) to visualize the elution profile of the peptides.

13. Reduce the acetonitrile concentration of the strong cation exchange fractions containing the bulk of the peptides (usually 30–35 fractions) by vacuum centrifugation (e.g., Speedvac) for 30 min.

14. Capture the ICAT-derivatized peptides in each fraction using a monomeric avidin column (kits for the purification of ICAT-labeled peptides are available from Applied Biosystems) following the manufacturer's instructions. Prepare the avidin column by washing with 2 mL of elution buffer followed by 1 mL of load buffer.

15. Apply the sample to the column collecting the flow-through and reapplying it to the column. Repeat this three times to ensure efficient retention of the biotin-labeled peptides on the column. Save the final flow-through for further analysis if necessary.

16. Wash the column with sequential 1-mL washes of load, wash 1, and wash 2 buffers. Elute captured peptides with 1 mL of elution buffer collecting the sample into a fresh tube.

17. Regenerate the column by a further wash with 1 mL of elution buffer (do not save) followed by a wash with 1 mL of load buffer.

18. Apply and isolate peptides from the next sample (**steps 15–17**). Repeat this process until all fractions have been purified.

19. Dry samples to completion using a vacuum centrifuge (e.g., Speedvac).

20. Resuspend each fraction in 0.1% (v/v) formic acid in water and analyse each fraction by reverse-phase liquid chromatography coupled to electrospray tandem mass spectrometry (LC-ESI-MS/MS).

21. Analyse the data using SEQUEST *(6)* for peptide identification, and INTERACT software for user interface *(7)*. Relative quantification of differentially labeled

peptide pairs can be determined using ASAPRatio software *(8)*. Data analysis tools for the statistical validation of identified peptides (PeptideProphet) and proteins (ProteinProphet) are available and recommended *(9,10)*.

3.4. Affinity Purification of Peroxisomal Membranes

1. Peroxisomal membranes can be further enriched by affinity purification. This can be achieved with the use of antibodies directed against a membrane component of the peroxisome or alternatively, by isolating peroxisomes from a strain expressing a peroxisomal membrane protein tagged with one of the many available affinity tags. The procedure used for the isolation of protein A tagged Pex11 (Pex11-A) using IgG-coupled paramagnetic beads is described here (*see* **Note 7**).
2. Dilute the extracted peroxisomal membranes (described in **Subheading 3.2.**) in 3 mL of Ti8 buffer and add 50 μL of phosphate buffer and 3 μL of P-solution.
3. Add 150 μL of pre-equilibrated IgG beads (10^9 IgG coated beads/ mL) and incubate overnight at 4°C with end-over-end mixing (*see* **Note 1**).
4. Collect the beads using a magnet and carefully remove unbound material (*see* **Note 8**).
5. Wash beads three times with 1 mL of Ti8 buffer. Resuspend the beads in 50 μL of 0.05% SDS and incubate for 5 min at room temperature to elute the bound material. Centrifuge the samples at 10,000 *g* for 30 s in a microcentrifuge and collect the supernatant. Repeat this elution three times and combine the eluates.
6. Analyse unbound and bound fractions by SDS–PAGE and western blotting to define the enrichment factor of marker proteins. Determine the protein concentration of the fractions using a Bradford assay.

3.5. ICAT Labeling of Affinity-Purified Peroxisomal Proteins

1. Perform ICAT labeling of affinity-purified peroxisomal membranes (*see* **Subheading 3.4.**) isolated from oleate-induced cells using 500 μg of sample. Several affinity purifications may be necessary to obtain enough material for mass spectrometric analyses.
2. Generate 500 μg of extracted peroxisomal membranes as described in **Subheadings 3.1** and **3.2.**
3. Equalize sample volumes of the two fractions, peroxisomal membranes and affinity-purified sample, with Ti8-SDS buffer.
4. Differentially label each fraction with the appropriate heavy or light ICAT reagent and analyse samples as described in **Subheading 3.3**.

3.6. Analysis of Data Sets

1. For each protein, the relative quantitation between two samples is obtained. In many studies, proteins whose ICAT ratio is greater than some arbitrary selected threshold, for example, 1.2 or 1.5, are considered to be significantly enriched in one sample over the other. However, because of the variability of sample

preparation, such an approach may not be accurate. Instead, we applied a statistical method to convert ICAT ratios into probabilities that proteins are enriched in one fraction over the other, as described below.

2. For each experiment, the ICAT ratios (r = heavy/light) for all the quantified proteins is plotted as a histogram. The resulting distributions had a shape that could be described as a normal distribution with an extended shoulder, indicating the presence of proteins with high ICAT ratios. Assume that each quantified protein can be divided into proteins that enrich (E) or do not enrich (U) with the peroxisomal fraction.

3. The distribution of ICAT ratios is then modeled using overlapping Gaussian distributions (*see* **Fig. 4**) and the mixture model fitted to the data using the expectation maximization algorithms (*11,12*). One curve (higher ratios) describes those proteins that are over-represented in the purified peroxisomal fraction and the other those that are over-represented in the crude or contaminant fraction (*4*). Mathematically, the model can be expressed as $f(r) = p(E)p(r|E) + p(U)p(r|U)$,

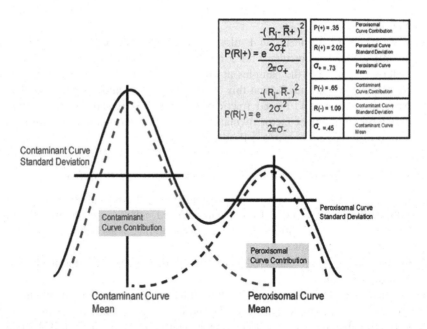

Fig. 4. Probabilistic analysis of quantitative mass spectrometry data to identify proteins enriched in peroxisomal fractions. A plot of quantified identifications shows a normal distribution with an extended shoulder in the direction of high isotope-coded affinity tags (ICAT) ratios (heavy/light). We assume that every identified protein must be either enriched or not enriched with peroxisomes. Two overlapping Gaussian curves representing peroxisomal and contaminant distributions are modeled and the mixture model fitted to the data using expectation maximization algorithms. For every ICAT ratio, the probability of being in the peroxisomal curve is determined.

where $p(r|E)$ and $p(r|U)$ are the distributions of ICAT ratios among enriched and not enriched proteins, respectively, and $p(E)$ and $p(U)$ are the mixing proportions and represent the overall distribution of the enriched and not enriched proteins in the data set.

4. The mixture model is initialized with distributions $p(r|E)$ and $p(r|U)$ with parameters obtained using a k-mean clustering algorithm. For each protein in the data set, the probability of it belonging to the enriched or not enriched subpopulations is calculated as $p(E|r) = p(E)p(r|E)/(p(E)p(r|E) + p(U)p(r|U))$ and $p(U|r) = 1 + p(E|r)$, respectively (E-step). The parameter estimates are updated using all the proteins in the data set, with the contribution from each of the proteins going toward the distributions $p(r|E)$ and $p(r|U)$, and proportions $p(E)$ and $p(U)$, being weighted according to its current probabilistic labels, $p(E|r)$ and $p(U|r)$ (M-step).

5. At convergence, the learned distributions of ICAT ratios among enriched and not enriched proteins is used to compute, for each protein in the data set, the probability of being enriched in the peroxisomal fraction as a function of its ICAT ratio r, $P_E = p(E|r)$.

4. Notes

1. Affinity-purified rabbit IgG is conjugated to Tosylactivated paramagnetic beads following the manufacturer's (Invitrogen) recommendations. These beads react with the amino terminus of proteins allowing the Fc region, the domain that binds protein A, to face outwards. Chemistry reactive to the carboxy terminus is necessary when using antibodies directed against a specific target of the organelle.

2. Ensure that the yeast strain being used does not have a peroxisome deficiency prior to commencing this procedure. Because yeast does not require peroxisomes for growth in glucose-containing media (a standard growth condition for the maintenance of yeast strains), several common laboratory strains of yeast have acquired mutations in genes essential for the biogenesis of peroxisomes. The presence of peroxisomes can be easily tested for by their ability to proliferate and by tracking the abundance of a peroxisomal marker protein tagged with Green Fluorescent Protein (e.g., Pex11-GFP or Pex3-GFP) by epifluorescence microscopy or western blotting. Cells grown in SCIM should contain numerous punctuate structures.

3. The optimal speed of the homogenizer must be determined empirically to maximize cell lysis, but at the same time minimize peroxisome lysis. In general, we recommend ~1000 rpm with a tight-fitting pestle (~50–80% of maximum speed). To reduce proteolysis, maintain the vessel on ice during the homogenization.

4. Samples can be stored at 4°C overnight if necessary. Ensure that samples are not frozen as this will result in premature lysis of the organelles.

5. Hypotonic lysis of the organelles is performed to reduce the dynamic range of protein levels in the organelles. Like many organelles, the peroxisome contains abundant matrix proteins that dominate the MS analyses. Depletion of these

abundant proteins allows for greater coverage of the organelle proteome and the identification of proteins of lower abundance. Carbonate extractions of the membrane fractions can also be performed to fractionate integral membrane proteins from those that are membrane-associated and thus further simplify the fraction (*see* **Fig. 3**).

6. Calculations: in 125 µg of protein sample, estimate 2.5 nmol of protein assuming an average molecular weight of 50 kDa. For each protein assume six cysteines, thus 15 nmol of cysteines in 125 µg of protein. Use a fivefold excess (75 nmol) of ICAT reagent.

7. The approach described in **Subheading 3.3** is better at distinguishing mitochondrial contaminants from peroxisomal proteins than the method described in **Subheading 3.5** (affinity-purified peroxisomes versus crude peroxisomes). Mitochondrial proteins are common and abundant contaminants of the peroxisomal fractions and are easily recognized as such based on their ICAT ratio. However, in this approach, the sample complexity is increased when mitochondria are added to peroxisomes, and this hinders the ability to identify lower abundance proteins in the highly complex sample. Lower abundance proteins can generally be identified in samples of lower complexity. Thus, more peroxisomal proteins can be expected to be identified using the method described in **Subheading 3.5**. Using this approach, the complexity of the sample mixture should not be increased above that of the unenriched peak peroxisomal fraction. However, one should keep in mind that proteins with affinity to the resin used for the immuno-isolation procedure may be artificially enriched in these fractions. For further details refer to reference *(4)*.

8. We have observed, on occasion, the formation of membranous strings during the binding step with the IgG beads. Magnetic separation of the beads from the supernatant, however, has proved efficient at partitioning these unbound membranes from the bound substrates. Alternatively, a small amount of non-ionic detergents (like Tween-20 used at a final concentration of 0.05%) added to the binding reaction buffer can prevent the formation of these membranes but may not be amenable to every use.

References

1. Aebersold, R. and Mann, M. (2003) Mass spectrometry-based proteomics. *Nature* **422,** 198–207.
2. Brunet, S., Thibault, P., Gagnon, E., Kearney, P., Bergeron, J. J., and Desjardins, M. (2003) Organelle proteomics: looking at less to see more. *Trends Cell Biol.* **13,** 629–638.
3. de Duve, C. (1992) Exploring cells with a centrifuge. In *Nobel Lectures in Physiology 1971–1980* (Lindsten, J., ed.), World Scientific Publishing Co., London, pp. 152–172.

4. Marelli, M., Smith, J. J., Jung, S., Yi, E., Nesvizhskii, A. I., Christmas, R. H., et al. (2004) Quantitative mass spectrometry reveals a role for the GTPase Rho1p in actin organization on the peroxisome membrane. *J. Cell Biol.* **167,** 1099–1112.

5. Smith, J. J., Marelli, M., Christmas, R. H., Vizeacoumar, F. J., Dilworth, D. J., Ideker, T., et al. (2002) Transcriptome profiling to identify genes involved in peroxisome assembly and function. *J. Cell Biol.* **158,** 259–271.

6. Eng, J. K., McCormack, A. L., and Yates, J. R. III. (1994) An approach to correlate tandem mass spectral data of peptides with amino acid sequences in a protein database. *J. Am. Soc. Mass Spectrom.* **5,** 976–989.

7. Han, D. K., Eng, J., Zhou, H., and Aebersold, R. (2001) Quantitative profiling of differentiation-induced microsomal proteins using isotope-coded affinity tags and mass spectrometry. *Nat. Biotechnol.* **19,** 946–951.

8. Li, X. J., Zhang, H., Ranish, J. A., and Aebersold, R. (2003) Automated statistical analysis of protein abundance ratios from data generated by stable-isotope dilution and tandem mass spectrometry. *Anal. Chem.* **75,** 6648–6657.

9. Keller, A., Nesvizhskii, A. I., Kolker, E., and Aebersold, R. (2002) Empirical statistical model to estimate the accuracy of peptide identifications made by MS/MS and database search. *Anal. Chem.* **74,** 5383–5392.

10. Nesvizhskii, A. I., Keller, A., Kolker, E., and Aebersold, R. (2003) A statistical model for identifying proteins by tandem mass spectrometry. *Anal. Chem.* **75,** 4646–4658.

11. Lee, R. T. (2000) Use of microarrays to identify targets in cardiovascular disease. *Drug News Perspect.* **13,** 403–406.

12. Pan, W., Lin, J., and Le, C. T. (2002) How many replicates of arrays are required to detect gene expression changes in microarray experiments? A mixture model approach. *Genome Biol.* **3,** 22.

25

Determination of Genuine Residents of Plant Endomembrane Organelles using Isotope Tagging and Multivariate Statistics

Kathryn S. Lilley and Tom P. J. Dunkley

Summary

The knowledge of the localization of proteins to a particular subcellular structure or organelle is an important step towards assigning function to proteins predicted by genome-sequencing projects that have yet to be characterized. Moreover, the localization of novel proteins to organelles also enhances our understanding of the functions of organelles. Many organelles cannot be purified. In several cases where the degree of contamination by organelles with similar physical parameters to the organelle being studied has gone unchecked, this has lead to the mis-localization of proteins. Recently, several techniques have emerged, which depend on characterization of the distribution pattern of organelles partially separated using density centrifugation by quantitative proteomics approaches. Here, we discuss one of these approaches, the localization of organelle proteins by isotope tagging (LOPIT) where the distribution patterns of organelles are assessed by measuring the relative abundance of proteins between fractions along the length of density gradients using stable isotope-coded tags. The subcellular localizations of proteins can be determined by comparing their distributions to those of previously localized proteins by assuming that proteins that belong to the same organelle will cofractionate in density gradients. Analysis of distribution patterns can be achieved by employing multivariate statistical methods such as principal component analysis and partial least squares discriminate analysis. In this chapter, we focus on the use of the LOPIT technique in the assignment of membrane proteins to the plant Golgi apparatus and endoplasmic reticulum.

Key Words: Golgi; ER; LOPIT; density gradient centrifugation; PCA; PLS-DA.

From: *Methods in Molecular Biology, vol. 432: Organelle Proteomics*
Edited by: D. Pflieger and J. Rossier © Humana Press, Totowa, NJ

1. Introduction

Proteins are spatially organized according to their function within eukaryote cells and thus, knowledge of their location is an important step towards assigning function to the thousands of proteins predicted by genome-sequencing projects that have yet to be characterized. Moreover, the localization of novel proteins to organelles will also improve our understanding of the functions of organelles. Proteomics approaches can provide powerful tools for characterizing the proteins resident within organelles. To enable confident protein localization, it is prerequisite that organelle preparations are free of contaminants or that techniques are utilized which discriminate between genuine organelle residents and contaminating proteins *(1)*. In the case of some organelles, such as mitochondria, it is relatively straightforward to produce a largely homogeneous preparation of the organelle. The components of the endomembrane system such as the Golgi apparatus and endoplasmic reticulum are largely impossible to produce in a purified form as the organelles within the endomembrane system have similar sizes and densities, making them difficult to separate *(2,3)*. Another confounding issue with the study of this class of organelle is that the proteins that reside within them are in a constant state of flux owing to the fact that proteins traffic through the system en route to their final destination. Furthermore, proteins within the endomembrane system cycle between compartments, for example, ER residents continuously escape to the Golgi and are retrieved in COPI vesicles *(4)*. These factors necessitate the measurement of the steady state distributions of proteins within the whole endomembrane system if a realistic insight into the subcellular localization of endomembrane proteins is to be achieved.

Recently, a couple of techniques have emerged which depend on characterization of the distribution pattern of organelles partially separated using density centrifugation by quantitative proteomics approaches *(1,5–7)*. Here we discuss one of these methods, localization of organelle proteins by isotope tagging (LOPIT), which is a proteomic technique for protein localization that is not dependent on the preparation of pure organelles *(5)*. In this approach, organelles are first partially separated using equilibrium centrifugation through density gradients. The distribution patterns of proteins within the gradient are then determined by the measurement of the relative abundance of proteins amongst the fractions along the length of the gradient. This can be readily and reliably achieved using stable isotope tagging procedures where protein quantitation is determined by mass spectrometric (MS) techniques. There are several methods that can be used to incorporate stable isotope labels into the proteins within fractions in a differential manner, here we describe the use of iTRAQ (stable-isotope-tagged amine-reactive reagents) tagging, which tags peptides generated from proteins within fractions from along the length of a gradient *(8)*. The sheer number of peptides generated then necessitates the use of multi-dimensional peptide chromatography before mass spectrometric analysis of the resultant-tagged peptides.

The subcellular localization of proteins can then be determined by comparing their distributions to those of previously localized proteins, as proteins that belong to the same organelle will cofractionate in the density gradients. An effective method to assess such distribution patterns on a large scale is to employ multivariate statistical methods such as principal component analysis and partial least squares discriminate analysis. The combination of these two approaches allows the confident assignment of proteins to specific subcellular structures with a low rate of mis-assignment as determined by the application of orthogonal methodologies.

In this chapter we focus on the use of the LOPIT technique in the assignment of membrane proteins to the plant endomembrane system, in particular the Golgi apparatus from *Arabidopsis thaliana* callus material.

2. Materials

2.1. Membrane preparation and fractionation

1. Homogenization medium (HM): 250 mM sucrose, 10 mM HEPES–NaOH, pH 7.4, 1 mM ethylenediaminetetraacetic acid (EDTA), and 1 mM dithiothreitol (DTT). This solution must be prepared freshly before use and kept at 4°C.
2. Iodixanol stock solution: 60% (v/v) solution of iodixanol (Optiprep) in water (Axis Shield, Huntingdon, UK).
3. Iodixanol working solution buffer: 60 mM HEPES–NaOH, pH 7.4, 6 mM EDTA, and 6 mM DTT. This solution must be prepared freshly before use and kept at 4°C.
4. Iodixanol working solution: prepared by diluting five parts of Optiprep stock solution with one part of iodixanol working solution buffer. This solution must be prepared freshly before use and kept at 4°C. This 50% iodixanol stock solution is further diluted with iodixanol working solution buffer to obtain 18 or 16% solutions.
5. Refractometer (Leica Solms, Wetzlar Germany).
6. 60 g of Arabidopsis callus material (wet weight).
7. Polytron homgenizer plus 94 PTA 10EC head attachment (Kinematica, Littav, Switzerland).
8. SW28 centrifuge tubes and SW28 rotor (Beckman Coulter, Fullerton, CA).
9. Optima L-XP ultracentrifuge (Beckman Coulter).
10. Vti65.1 polycarbonate Optiseal tubes (Beckman Coulter).
11. Vti65.1 rotor (Beckman Coulter).
12. Auto Densiflow collection device (Labconco Corporation, Kansas City, MO).
13. For carbonate washing of membranes: 162.5 mM Na_2CO_3.

14. TLA 100.3 rotor (Beckman Coulter).
15. Optima TLX ultracentrifuge.

2.2. Sodium Dodecyl Sulphate–Polyacrylamide Gel Electrophoresis

1. Sodium dodecyl sulphate (SDS) loading buffer: 100 mM Tris–HCl, pH 6.8, 8% (w/v) SDS, 30% (v/v) glycerol, 200 mM DTT, and 5 μL/mL saturated bromophenol blue solution.
2. 10× SDS running buffer: 25 mM Tris–HCl, pH 8.3, 192 mM glycine, and 0.1% (w/v) SDS.

2.3. Western Blotting

1. Hybond-ECL nitrocellulose membranes (GE Healthcare, Little Chalfont, UK).
2. Mini Trans-Blot Cell (Bio-Rad, Hercules, CA).
3. PBS-T buffer: 137 mM NaCl, 2.7 mM KCl, 10 mM KH_2PO_4, pH 7.4, and 0.1% (v/v) Tween 20.
4. Anti-rabbit secondary antibody conjugated to horseradish peroxidase (Bio-Rad).
5. Luminol enhancer solution: 100 mM Tris–HCl, pH 6.8, 1.25 mM luminol (Sigma, St. Louis, MO), 2.7 mM H_2O_2, and 6.7 mM trans-4-hydroxycinnamic acid in dimethyl sulphoxide (enhancer).
6. X-OMAT film (Kodak, Harrow, UK).

2.4. Tryptic digestion and iTRAQ labeling

1. iTRAQ-labeling buffer: 25 mM triethylammonium bicarbonate (TEAB), 8 M urea, 2% (v/v) Triton X-100, and 0.1% (w/v) SDS.
2. BCA protein assay kit (Pierce, Rockford, IL).
3. iTRAQ reagent (Applied Biosystems, Framingham, MA).
4. 200 mM methyl methanethiosulphonate (MMTS).
5. 50 mM Tris(2-carboxyethyl)phosphine hydrochloride (TCEP).
6. Sequencing grade trypsin (Promega, Madison, WI).
7. Speed Vac centrifugal vacuum concentrator (Thermo Electron, Waltham, MA).

2.5. Cation-exchange chromatography

1. BioLC HPLC system (Dionex, Amsterdam, Netherlands).
2. Polysulphoethyl A column, 2.1 mm internal diameter and 200 mm length, particles of 5 μm diameter and 300 Å porosity (PolyLC, Columbia, MD, USA).
3. Speed Vac centrifugal vacuum concentrator (Thermo Electron).
4. Cation-exchange buffer A: 20% (v/v) acetonitrile and 10 mM KH_2PO_4–H_3PO_4, pH 2.7, in HPLC grade water.
5. Cation-exchange buffer B: 1 M KCl, 20% (v/v) acetonitrile and 10 mM KH_2PO_4–H_3PO_4, pH 2.7, in HPLC grade water. The pH of the buffers should be checked prior to the addition of acetonitrile.
6. LC-MS/MS resuspension buffer: acetonitrile/TFA/water, 2/0.1/98 (v/v/v).

2.6. LC-MS/MS

1. Ultimate nano-LC system (Dionex) with a FAMOS autosampler (Dionex), Acclaim PA C_{16} pre-column (5 mm length and 300 µm internal diameter), and PepMap C_{18} column (15 cm length and 75 µm internal diameter, Dionex).
2. HPLC buffers: buffer A is formic acid/HPLC water/acetonitrile, 0.1/95/5 (v/v/v) and buffer B is formic acid/HPLC water/acetonitrile, 0.1/5/95 (v/v/v).
3. QSTAR XL QqTof mass spectrometer (Applied Biosystems) with a 10 µm internal diameter PicoTip electrospray needle (New Objective, Woburn, MA).

2.7. Peptide identification and quantitation

1. MASCOT version 2.0.01 (Matrix Science, London, UK).
2. MIPS *Arabidopsis* protein database (ftp://ftpmips/gsf.de/cress/arabiprot/arabi_all_ proteins_v090704.gz).
3. GAPP software (www.ccbit.org/gapp).
4. MySQL database (MySQL version 4.0, MySQL AB, Uppsala, Sweden).
5. i-Tracker software (www.dasi.org.uk/download/itracker.html).

2.8. Multivariate analysis of iTRAQ results

1. SIMCA 10.0 (Umetrics, Umea, Sweden).

3. Methods

The LOPIT technique is designed to enable the confident assignment of membrane proteins to organelles and is based upon analytical centrifugation and hence is not dependent on the production of pure organelles. The technique involves partial separation of organelles by density gradient centrifugation followed by the analysis of protein distributions in the gradient by stable isotope labeling strategies coupled with mass spectrometry. Multivariate data analysis techniques are then used to group proteins according to their distributions and hence localizations.

The first stage of the LOPIT procedure involves the collection of membranes from *Arabidopsis callus* and subsequent equilibrium density centrifugation using iodixanol self-forming gradients to produce a partial separation of organelles (*see* **Fig. 1** for LOPIT schema). The extent of this separation and identification of fractions enriched in specific organelles is then achieved by use of 1D SDS–polyacrylamide gel electrophoresis (PAGE) of fractions collected from the density gradient and western blotting utilizing antibodies raised against marker proteins whose organelle location has previously been verified by an orthogonal technique. The degree of separation of organelles can be enhanced by manipulation of the iodixanol gradient. Membranes from fractions that show

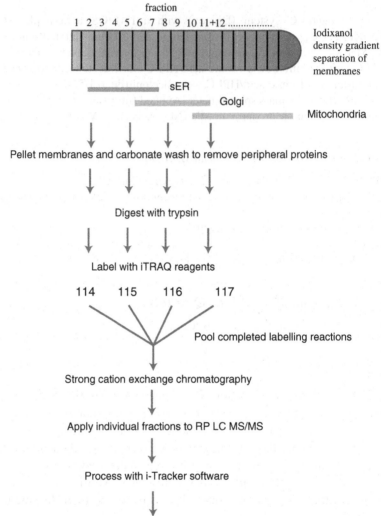

Fig. 1. The LOPIT schema. The LOPIT schema utilizes iTRAQ labeling and 2D peptide separation. The colored bars represent a hypothetical distribution of organelles along the gradient. Proteins from carbonate-washed membrane samples from for example, fractions 2, 5, 8, and 11+12 (pooled) are digested to peptides and labeled with the 114,115,116, and 117 iTRAQ tags, respectively. These are then pooled and the peptides separated over two dimensions of chromatography and analysed by MS/MS. The output of MS/MS is then further processed by i-Tracker software to identify and quantitate peptides and group them into proteins. The organelle assignment of proteins is then determined using multivariate statistical packages.

peak enrichment of the organelles of choice are then subjected to a high pH (carbonate) wash resulting in soluble and peripheral proteins being stripped from the membranes. The next stage involves isotope tagging of peptides generated by trypsinolysis of the pelleted membrane proteins. In the example described here, the iTRAQ tagging system is employed, which allows labeling of up to four fractions showing peak enrichment per experiment. The subsequent stages involve, the separation of peptides by cation-exchange chromatography, and further separation of peptides within fractions eluted from cation-exchange columns by reverse-phase chromatography coupled to MS/MS. The final stages of the process involve the identification and quantitation based on peptide fragmentation data generated by MS/MS using a bespoke open access suite of programmes, i-Tracker *(9)*. The i-Tracker software can be used to extract reporter ion peak ratios from non-centroided tandem MS peak lists in a format that is easily associated with the results of protein identification tools such as Mascot. Following any purity correction, the peak areas are normalized. These normalized areas are used to calculate the quantitative ratios between each reporter ion, and the ratios for each corresponding protein are thus calculated taking an average value where several peptides per protein have been quantified. Finally, analysis of the resulting data sets is carried out using unsupervised and supervised multivariate statistical approaches, principal components analysis (PCA), partial least squares discriminant analysis, PLS-DA (*see* **Fig. 2**).

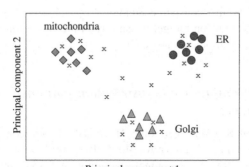

Fig. 2. PCA plot. Schematic representation of a PCA scores plot showing clustering of proteins according to their density gradient distributions and therefore localizations. Each point on the plot represents a protein, the colored shapes representing known organelle markers, and the crosses proteins of unknown localization. Proteins that share similar distributions in the gradient cluster together on the plot, and consequently each cluster represent a subcellular structure. For classification, a regression extension of PCA and partial least squares-discriminant analysis (PLS-DA) can be used to compare proteins of unknown location to training sets consisting of 20–30 marker proteins.

3.1. Membrane preparation and fractionation

1. Approximately 60 g of c is homogenized in 60 mL of homogenization buffer at 4°C, using a polytron with a 94 PTA 10EC head attachment for two pulses of 7 s (*see* **Note 1**).
2. The homogenate is then centrifuged twice for 5 min at 2200 *g* before loading the supernatant onto a 6-mL cushion of 18% iodixanol and centrifuging it at 100,000 *g* for 2 h in a SW28 rotor (*see* **Note 2**).
3. Crude membranes are then collected from the interface and adjusted to 16% iodixanol. These membranes are then fractionated according to their density by centrifuging at 350,000 *g* for 3 h in a VTi65.1 rotor, during which time an iodixanol gradient is generated.
4. Fractions of 0.5 mL are harvested from the top of the gradient using an Auto Densi-flow collection device (*see* **Note 3**).
5. The shape of the resulting gradient is then determined using a refractometer as per the manufacturer's instructions (see **Note 4**).

3.2. SDS–Polyacrylamide Gel Electrophoresis

1. Take 1/100 of each 0.5-mL fraction, add an equal volume (5 µL) of SDS loading buffer and heat at 70°C for 5 min.
2. Load the protein samples onto a 1-mm thick, 7-cm long 12% SDS–polyacrylamide mini gel with a 5% SDS–polyacrylamide stacking gel, using a Mini-Protean 3 cell (Bio-Rad).
3. Prepare the running buffer by diluting 10× SDS running buffer with deionized water.
4. Run the gel at 60 mA constant current.
5. To visualize proteins on the gel, wash three times for 5 min in deionized water, stain using SimplyBlue coomassie (Invitrogen, Carlsbad, CA) for 1 h and destain in water for 1 h.

3.3. Western blotting to assess distribution patterns of organelle markers

1. Blot protein SDS–PAGE gels onto Hybond-ECL nitrocellulose membranes in transfer buffer using a Mini Trans-Blot Cell at 100 V for one hour.
2. Incubate nitrocellulose membranes overnight in PBS-T buffer containing 5% (w/v) powdered milk.
3. Incubate nitrocellulose membranes with primary antibodies for one hour in PBS-T containing 5% (w/v) powdered milk at an appropriate dilution (*see* **Note 5**).
4. Wash nitrocellulose membranes four times for 10 min in PBS-T buffer containing 5% (w/v) powdered milk and then incubate with an appropriate secondary antibody conjugated to horseradish peroxidase at a dilution of 1:1000 in PBS-T buffer containing 5% (w/v) powdered milk for 1 h.
5. Wash the nitrocellulose once for 10 min with PBS-T buffer and then three times for 10 min in PBS buffer.

6. Incubate the nitrocellulose in 4 mL of a 100:1 luminol-enhancer solution for 2 min and expose to X-OMAT film for ~30 s.

3.4. Carbonate washing of membranes

1. To carbonate wash the membranes collected from the density gradient, add 800 µL of 162.5 mM Na_2CO_3 at 4°C to each density gradient fraction to give a final Na_2CO_3 concentration of 100 mM and incubate on ice for 30 min.
2. Pellet the washed membranes by centrifugation at 100,000 g for 25 min at 4°C in a TLA 100.3 rotor using an Optima TLX ultracentrifuge.
3. Discard the supernatant and wash the membrane pellets with 1 mL of deionized water at 4°C (*see* **Note 6**).
4. Repellet the carbonate-washed membranes by centrifugation at 100,000 g for 10 min at 4°C and discard the supernatants.

3.5. iTRAQ labeling

Choose up to four carbonate-washed membranes from fractions along the length of the density gradient. The fractions chosen should exhibit different distributions of a specific organelle based on distribution of markers as visualized by western blotting. For example, the peak Golgi apparatus and peak endoplasmic reticulum may be chosen along with two other fractions from areas within the equilibrium density gradient that do not have a great abundance of these organelles. **Figure 1** shows a typical LOPIT schema and highlights fractions chosen based on the distribution that will give maximal resolution between the Golgi and endoplasmic reticulum.

1. Solubilize the carbonate-washed membrane from each of the chosen fractions in 100 µL of iTRAQ-labeling buffer (*see* **Note 7**).
2. Determine the protein concentration of each fraction using the BCA Protein Assay kit according to the manufacturer's instructions using a BSA standard in iTRAQ-labeling buffer.
3. Reduce 100 µg of protein by adding 2 µL of 50 mM TCEP and incubate at 20°C for 1 h.
4. Block cysteine sulphydryls by the addition of 1 µL of 200 mM methyl methanethiosulfonate (MMTS) followed by incubation at 20°C for 10 min.
5. Dilute each of the four protein samples after reduction and blocking with 50 mM TEAB, such that the urea concentration is below 1 M before adding 2.5 µg of trypsin and incubating for 1 h at 37°C.
6. Add a further 2.5 µg of trypsin to each sample followed by incubation overnight at 37°C.
7. Lyophilize the resulting tryptic digests using a Speed Vac centrifugal vacuum concentrator without heating (*see* **Note 8**).

8. Resuspend the peptides in 100 µL of 250 mM TEAB in water/ethanol, 25/75 (v/v), and then add to one unit of the corresponding iTRAQ reagent followed by incubation for 1 h at 20°C. Of the four sets of peptides from the four starting fractions, label the least dense fraction with iTRAQ reagent 114; the second least dense fraction with 115; the third least dense fraction with 116; and the densest fraction with 117.

9. Hydrolyse the residual iTRAQ reagents by adding 100 µL of deionized water and incubating for 15 min at 20°C.

10. Pool all four samples together and lyophilize in a Speed Vac centrifugal vacuum concentrator (*see* **Note 9**).

3.6. Cation-exchange chromatography

The next stage in the LOPIT process is to fractionate pooled iTRAQ-tagged peptides by cation-exchange chromatography.

1. Resuspend the iTRAQ-labeled samples in 3 mL of cation-exchange buffer A ensuring that the resulting solubilized samples are below pH 3. If the pH is higher than 3 at this stage, further additions of cation-exchange buffer A can be made. Load the peptides onto the cation-exchange column as per the manufacturer's instructions.

2. After loading of the peptides onto the column, firstly wash the column with cation-exchange buffer A for 60 min at 200 µL/min.

3. Following the loading and washing steps, elute peptides from the column using a 70-min linear gradient of 30–125 mM KCl in 20% (v/v) acetonitrile, and 10 mM KH_2PO_4–H_3PO_4, pH 2.7, at 200 µL/min. Wash the column after the gradient is complete by applying the following gradients sequentially; 125–500 mM KCl over 30 min, 500–1000 mM KCl over 5 min, hold at 1000 mM KCl for 10 min, and 1000–0 mM KCl over 2 minutes. Re-equilibrate for at least 10 minutes at 0 mM KCl before the next sample injection.

4. Collect fractions at 2-min intervals and lyophilize using a Speed Vac centrifugal vacuum concentrator.

5. Resuspend the peptides in 130 µL of acetonitrile/TFA/water, 2/0.1/98 (v/v/v).

3.7. LC-MS/MS analysis of the cation-exchange fractions

Choose for further analysis fractions from the cation-exchange chromatography which contain the majority of the peptides as judged by absorbance at 214 nm. The next step in the process is to analyse each fraction by reversed-phase LC-MS/MS. The example given here makes use of an Ultimate nano-LC system coupled to a QSTAR XL QqTOF mass spectrometer.

1. Load 60 µL (approximately half) of the peptide samples onto an Acclaim PA C_{16} pre-column at 20 µL/min using a FAMOS autosampler.

2. Wash the pre-column at 20 µL/min for 25 min with HPLC water containing 0.1% (v/v) formic acid to remove the KCl.

3. Elute peptides onto a PepMap C_{18} column at 150 nL/min and perform their separation using a 165-min gradient of 5–32% acetonitrile in HPLC water containing 0.1% (v/v) formic acid.

4. The reverse-phase eluent is interfaced with the QSTAR XL mass spectrometer through a distal-coated PicoTip electrospray needle to which a voltage of 2000 V is typically applied.

5. Operate the QSTAR XL in information-dependent acquisition mode (IDA), in which 1-s precursor ion scans from 400 to 1600 *m/z* are performed, followed by 3-s product ion scans (100–1580 *m/z*) on the two most intense doubly or triply charged ions. Following the acquisition of each product ion spectrum, ions with the same *m/z* value should be excluded from further analysis for 5 min. To ensure sufficient peptide and iTRAQ tag fragmentation, the collision energy used during the acquisition of product ion spectra is increased by 8 eV for LC-MS/MS analysis of iTRAQ-labeled samples (*see* **Note 10**).

3.8. Peptide identification and quantitation

3.8.1. Peptide identification

1. MS data files resulting from LC-MS/MS analysis of iTRAQ-labeled samples are processed using the wiff2DTA software (http://sourceforge.net/projects/protms/) to generate centroided and uncentroided peak lists *(10)*.

2. Use Mascot to search the centroided peak lists against the MIPS *Arabidopsis* protein database by applying the following fixed modifications – iTRAQ (K), iTRAQ (N-term), MMTS (C) – and variable modifications: oxidation (M) and iTRAQ (Y) (*see* **Note 11**). Apply a tolerance on mass measurement of 0.2 Da in MS mode and 0.5 Da in MS/MS mode.

3. To assess the rate of false protein identifications, also search the data against a reversed version of the database. Database reversal can be achieved by using the fasta_flip.pl perl script, which can be obtained from Matrix Science (see Matrix Science website for details).

4. The GAPP software can be used to parse into a MySQL database peptide identifications and scoring information from the Mascot output files, resulting from both the normal and reversed-database searches (*see* **Note 12**). The GAPP software uses the results of the reversed database searches to determine a Mascot score threshold that results in a false protein identification rate of less than 1%.

3.8.2. Peptide quantitation

The i-Tracker software can be used to calculate iTRAQ reporter ion areas using the uncentroided peak lists *(9)*. This software calculates iTRAQ reporter ion areas using the trapezium approximation for measuring the area under a peak (*see* **Fig. 3**).

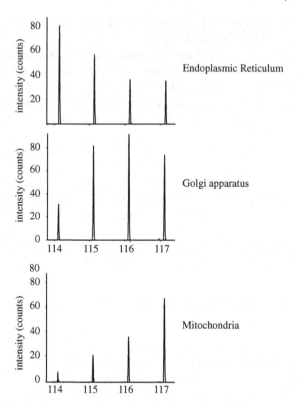

Fig. 3. Typical set of iTRAQ reporter ion intensities. Reporter ion profiles for known residents of the ER (calnexin, At5g61790), Golgi (gtl6, At2g22900), and mitochondria (AT5, At5g14040). The mean normalized peak area for each reporter ion is calculated.

The i-Tracker software then prompts the user to input the following information: the name of the folder containing the uncentroided peaklists, the purity correction values specified on the quality control certificate that accompanies every batch of iTRAQ reagents, and the reporter ion intensity threshold. The i-Tracker software then calculates the normalized iTRAQ reporter ion intensities for each product ion spectrum using the following formula:

normalized area A = area A/(area A + area B + area C + area D),

    where A, B, C, and D being the iTRAQ reporter ions.

    i-Tracker then enters the quantitation results into the MySQL database that contains the Mascot identification results.

3.9. Multivariate analysis of the iTRAQ results

The iTRAQ MS data analysis described in the previous section results in a Microsoft Excel spreadsheet containing a list of the proteins identified in addition to their iTRAQ ratios. This data can be analysed using the multivariate statistical methods: PCA and partial least squares-discriminant analysis (PLS-DA). PCA results in a 2D plot of the data in which proteins with similar density gradient profiles are clustered. PLS-DA enables proteins to be assigned to organelle classes based on their density gradient profiles.

1. The iTRAQ ratios for the proteins identified are imported into SIMCA 10.0.
2. The ratios are then logged and pre-processed with unit-variance scaling.
3. PCA is performed using the SIMCA 10 package according to the manufacturer's instructions.
4. A 2D PCA scores plot is then generated to cluster proteins according to their density gradient profiles.
5. To assign proteins to organelles, a training set of proteins with known organelle localizations is compiled. Proteins belonging to each organelle are assigned to a separate class. SIMCA 10 is then used to build a PLS-DA model based on this training set of proteins, according to the manufacturer's instructions. This PLS-DA model is then used to rank the previously unlocalized proteins according to how well they fit each of the training set classes, each of which corresponds to an organelle. In each case, the lowest ranked training set protein is used as the threshold for class, and hence organelle, membership. An ideal training set should consist of 20–30 proteins.

Notes

1. 60 g of callus wet weight – grown according to Sherrier et al. *(11)*.
2. Following the 2,200 *g* centrifugation steps, the homogenate is transferred to four clean SW28 centrifuge tubes. The 6 mL of iodixanol cushion is then set up by carefully underlayering the homogenate using a 10 mL-syringe with a blunt-ended needle.
3. The Labconco comprises a liquid-sensing probe connected to a peristaltic pump. The probe automatically detects the meniscus at the top of the gradient. As the peristaltic pump removes the gradient solution, the probe moves downwards with the meniscus. This minimizes mixing of the gradient during fraction collection.
4. The gradient can be shifted to higher density by increasing the starting iodixanol concentration or shifted to lower density by decreasing the iodixanol starting concentration. Increasing the centrifugation time, or centrifugal force, results in a steeper gradient whilst reducing the centrifugation time, or centrifugal force, results in a shallower gradient *(12,13)*.
5. The antibodies used should be specific for proteins with known localizations to the organelles of interest. The results of the western blot can then be used to determine whether the organelles of interest exhibit distinct distributions within the density gradient used. If this is not the case, further manipulation of the density gradient conditions should be performed at this stage (*see* **Note 4**).
6. The supernatant will contain cytoplasmic proteins and peripheral membrane proteins. The high pH of the carbonate buffer disrupts electrostatic interactions

between peripheral proteins and the membrane. In addition, the high pH results in closed vesicles being transformed into membrane sheets. Consequently, soluble proteins contained within organelles are released *(14)*.

7. To solubilize the carbonate-washed membrane pellets, add the iTRAQ-labeling buffer and vortex at 20°C until the pellets are dissolved. This may take up to 30 min.

8. The iTRAQ reagents are supplied as *N*-hydroxysuccinimide esters that react with primary amine moieties. It is therefore important that the samples are not heated above 30°C as the iTRAQ-labeling buffer contains urea and heating may result in carbamylation of amino groups. These groups would then not be available to react with the iTRAQ reagents.

9. To maximize the amount of separation of organelles in resulting analysis (*see* **Subheading 3.9.**), it might be necessary to repeat another four-plex analysis of fractions from the equilibrium density gradient using four interpolated different fractions. See figure 1 for an example of this arrangement.

10. Generation of the iTRAQ reporter ions is dependent on fragmentation of the iTRAQ tag. Intense reporter ions are essential for accurate quantitation. Consequently the collision energy used to fragment iTRAQ-labeled peptides during collision-induced dissociation (CID) should be increased relative to the values used during analysis of unlabeled peptides. For the QSTAR XL instrument used in our experiments the collision energy is increased by 8 eV when analysing iTRAQ-labeled peptides. This value should be optimized for every mass spectrometer by analysing the same iTRAQ-labeled sample multiple times and increasing the collision energy in consecutive experiments. If the collision energy is set too low then intense parent ions and weak reporter ions will be observed in the product ion spectra. If the collision energy is set too high then weak b and y ions will be observed resulting in low scoring peptide identifications.

11. It is expedient to repeat the search with iTRAQ (K) and iTRAQ (N-term) as variable modifications as this will indicate whether the reaction has gone to completion. Incomplete modification can arise from using suboptimal pH of labeling buffer, presence of exogenous amine (for example in Tris buffer), and the use of more than 100 μg of protein per iTRAQ vial, which results in saturation of the iTRAQ reagents. Variable labeling efficiency between the four parallel labeling reactions will compromise the accuracy of quantitation.

12. All MySQL scripts were run from MySQL using the following command: source [name of MySQL script.sql]; The version of the GAPP software used was a perl scipt "mprot7bCam.pl." The perl script was opened using Microsoft Notepad and the target database was specified in the following line: $driver = "dbi:mysql:[name of MySQL database]".

References

1. Dunkley, T. P. J., Hester, S., Shadforth, I., Runnions, J., Weimar, T., Hanton, S., et al. (2006) Mapping the *Arabidopsis* organelle proteome. *Proc. Natl. Acad. Sci. U.S.A.* **103**, 6518–6523.

2. Gabaldon, T. and Huynen, M. A. (2004) Shaping the mitochondrial proteome. *Biochim. Biophys. Acta* **1659,** 212–220.
3. Peck, S. C. (2005) Update on proteomics in Arabidopsis. Where do we go from here? *Plant Physiol.* **138,** 591–599.
4. Hanton, S. L., Bortolotti, L. E., Renna, L., Stefano, G., and Brandizzi, F. (2005) Crossing the divide-transport between the endoplasmic reticulum and Golgi apparatus in plants. *Traffic* **6,** 267–277.
5. Dunkley, T. P. J., Watson, R., Griffin, J. L., Dupree, P., and Lilley, K. S. (2004) Localization of organelle proteins by isotope tagging (LOPIT). *Mol. Cell. Proteomics* **3,** 1128–1134.
6. Andersen, J. S, Lam, Y. W., Leung, A. K., Ong S. E., Lyon, C. E., Lamond, A. I., et al. (2005) Nucleolar proteome dynamics. *Nature* **433,** 77–83.
7. Foster, L. J., de Hoog, C. L., Zhang, Y., Zhang, Y., Xie, X., Mootha, V. K., et al. (2006) A mammalian organelle map by protein correlation profiling. *Cell* **125,** 187–199.
8. Ross, P. L., Huang, Y. L. N., Marchese, J. N., Williamson, B., Parker, K., Hattan, S., et al. (2004) Multiplexed Protein Quantitation in *Saccharomyces cerevisiae* Using Amine-reactive Isobaric Tagging Reagents. *Mol. Cell. Proteomics* **3,** 1154–1169.
9. Shadforth, I., Dunkley, T., Lilley, K., and Bessant, C. (2005) i-Tracker: For quantitative proteomics using iTRAQ™. *BMC Genomics* **6,** 145.
10. Boehm, A., Galvin, R., and Sickmann, A. (2004) Extractor for ESI quadrupole TOF tandem MS data enabled for high throughput batch processing. *BMC Bioinformatics* **5,** 162.
11. Sherrier, D. J., Prime, T. A., and Dupree, P. (1999) Glycosylphosphatidylinositol-anchored cell-surface proteins from Arabidopsis. *Electrophoresis* **20,** 2027–2035.
12. Ford, T., Graham, J., and Rickwood, D. (1994) Iodixanol - a nonionic isosmotic centrifugation medium for the formation of self-generated gradients. *Anal. Biochem.* **220,** 360–366.
13. Graham, J., Ford, T., and Rickwood, D. (1994) The preparation of subcellular organelles from mouse-liver in self-generated gradients of iodixanol. *Anal. Biochem.* **220,** 367–373.
14. Fujiki, Y., Hubbard, A. L., Fowler, S., and Lazarow, P. B. (1982) Isolation of Intracellular membranes by means of sodium carbonate treatment: application to endoplasmic reticulum. *J. Cell Biol.* **93,** 97–102.

26

Quantitative Proteomic Analysis to Profile Dynamic Changes in the Spatial Distribution of Cellular Proteins

Wei Yan, Daehee Hwang, and Ruedi Aebersold

Summary

Organelle protein profiles have traditionally been analyzed by subcellular fractionation of a specific organelle followed by the identification of the protein components of specific fractions containing the target organelle(s) using mass spectrometry (MS). However, because of limited resolution of the available fractionation methods, it is often difficult to isolate and thus profile pure organelles. Furthermore, many proteins (e.g., secretory proteins) are often observed to dynamically shuttle between organelles. Therefore, the determination of their true cellular localization requires the concurrent analysis of multiple organelles from the same cell lysate. Here, we report an integrated experimental approach that simultaneously profiles multiple organelles. It is based on the subcellular fractionation of cell lysates by density gradient centrifugation, iTRAQ labeling, and MS analysis of the proteins in selected fractions and principal component analysis (PCA) of the resulting quantitative proteomic data. Quantitative signature patterns of several organelles, including the ribosome, mitochondria, proteasome, lysosome, endoplasmic reticulum (ER), and Golgi apparatus, have been acquired from a single multiplexed proteomic assay using iTRAQ reagents. Through comparison PCA, we compare organelle profiles from cells under different physiological conditions to investigate changes in organelle profiles between control and perturbed samples. Such quantitative proteomics-based subcellular profiling methods thus provide useful tools to dissect the organization of cellular proteins into functional units and to detect dynamic changes in their protein composition.

Key Words: Quantitative proteomics; organelle profiling; multiplexed proteomic analysis; iTRAQ; mass spectrometry; principal component analysis; subcellular distribution; spatial regulation.

From: *Methods in Molecular Biology, vol. 432: Organelle Proteomics*
Edited by: D. Pflieger and J. Rossier © Humana Press, Totowa, NJ

1. Introduction

The components of the cell are often organized in organelles and other subcellular structures. Comprehensive analyses of the organelle constituents, in particular the protein components, have generated useful structural and functional insights into complex cellular processes. Using mass spectrometry (MS)-based proteomics *(1)*, protein components of many "purified" organelles have been identified. These include the mitochondria, Golgi, lysosome, phagosome, exosome, nucleus, clathrin-coated vesicles, spliceosome, etc. [see reviews *(2–4)* for details]. However, preparing a pure organelle is often limited using the currently available fractionation techniques. Furthermore, many proteins dynamically shuttle between organelles, suggesting necessity of profiling organelle proteins systematically in a time- and context-dependent manner.

To circumvent the first technical difficulty on organelle profiling, several research groups have attempted to generate organelle-specific patterns using quantitative proteomic approaches in partially fractionated samples without the necessity of complete purification of a specific organelle. Using isotope-coded affinity tag (ICAT)-based quantitative proteomics to compare the composition of samples at different stages of enrichment, Marelli et al. *(5)* discriminated the protein composition of yeast peroxisomes from the "contaminant" proteins. By measuring peptides ion intensities detected in MS across sequential sucrose density fractions, Mann and colleagues *(6,7)* found that patterns of proteins in a specific organelle could be differentiated from those in other "contaminant" organelles. In a similar manner but using stable isotope-labeling techniques to achieve accurate quantification, Dunkley et al. associated plant proteins to different organelles based on their quantitative patterns across partially separated fractions from density gradient centrifugation *(8,9)*. The various quantitative proteomics methods applied in these studies led to lists of proteins likely associated with a specific organelle. Such analyses are similar to the traditional characterization of density gradient fractions by western blotting, except that MS supports an unbiased analysis of the fractions.

The second but more complicated challenge to organelle protein profiling lies in the fact that many proteins are observed in multiple subcellular compartments in a time- and condition-dependent manner, suggesting that they may dynamically shuttle among different cellular compartments. This dynamic process of protein localization is often biologically interesting. We have developed a pattern-specific organelle profiling approach to study changes in subcellular localization of protein components in cells at different states. Cells under control and drug treatment conditions were cultured and collected separately. The two harvested samples were processed in parallel under identical conditions through the following steps: (1) partial separation using iodixanol density gradient fractionation; (2) iTRAQ reagent labeling *(10)* of selected subcellular fractions;

(3) protein quantification and identification by tandem MS; and (4) principle component analysis of the generated fractionation patterns of abundance ratios, revealing specific distributions of the proteins associated with several organelles including mitochondria, ribosomes, ER/Golgi, lysosomes, and proteasomes. The simultaneous profiling of multiple organelles in a single set of iTRAQ proteomics assay has allowed us to subsequently compare the acquired organelle profiles from lysates of cells at different cellular states. As a proof of concept, we applied this method to investigate changes in organelle composition and protein subcellular localization for the unfolded protein response (UPR), a cellular process that has been reported to significantly affect the protein composition of a number of organelles *(11,12)*.

2. Materials

2.1. Cell Culture and Drug Treatment

1. Mouse macrophage RAW264.7 cells (ATCC, Manassas, VA).
2. RPMI medium 1640 (Invitrogen) containing L-glutamine and 25 mM HEPES buffer. The following components are added to the purchased medium before use: 10% (v/v) fetal bovine serum (Invitrogen, Carlsbad, CA, USA), 1% (v/v) of Penicillin–Streptomycin (100× liquid containing 10,000 units of penicillin and 10,000 μg of streptomycin) (Invitrogen) and 1% (v/v) of 200 mM L-glutamine (Invitrogen).
3. Trypsin/EDTA solution (Invitrogen): it contains 0.25% (w/v) trypsin with 1 mM ethylenediaminetetraacetic acid (EDTA).
4. Thapsigargin (ICN Biomedicals, Solon, OH, USA) is dissolved at 1 mM (5000× solution) in dimethylsulfoxide (DMSO, Sigma, St. Louis, MO, USA) and stored at –80°C until use.
5. EDTA–PBS solution: 5 mM EDTA (Thermo Fisher Scientific, Waltham, MA, USA) in 1× phosphate-buffered saline (PBS, Thermo Fisher Scientific).
6. Cell lifter (Corning Incorporated, Corning, NY).

2.2. Cell Lysis and Fractionation

1. Homogenization buffer: 10 mM HEPES (prepared from 1 M HEPES buffer solution, pH 7.5; Invitrogen), 1 mM EDTA, and 0.25 M sucrose, with 1× protease inhibitor cocktail. A 50× protease inhibitor cocktail stock solution is made by dissolving 1 tablet of Protease Inhibitor Cocktail (Roche Applied Science, Indianapolis, IN, USA) into 1 mL of the buffer and is stored at –20°C until use.
2. Dounce homogenizer.
3. 25-gauge needles.
4. Density gradient solution: 10 mM HEPES (pH 7.5, no pH adjustment), 1 mM EDTA, and 8 or 28% (v/v) iodixanol (OptiPrep™; Axis-Shield, Oslo, Norway).
5. Gradient Maker SG-30 (Hoffer Scientific Instruments, San Francisco, CA).

6. Beckman Ultra-Clear centrifuge tube (Beckman, Coulter, Fullerton, CA, USA).
7. Auto Densi-Flow Density Gradient Fractionator (Labconco, Kansas city, MO).

2.3. Immunoblot

1. SDS–PAGE: 4–15% Criterion™ Tris–HCl gel (Bio-Rad, Hercules, CA, USA).
2. Nitrocellulose membrane (Pierce, Rockford, IL, USA).
3. PBS-Tween buffer: 1× PBS supplemented with 0.5% (v/v) Tween-20 (USB Co., Cleveland, OH, USA).
4. Blocking solution: PBS-Tween buffer supplemented with 10% (w/v) nonfat dry milk.
5. Primary antibodies against individual organelle markers recognize calnexin (Stressgen, Victoria, BC, Canada), Na-K-ATPase (Abcam), Bcl-2 (BD Biosciences, San Jose, CA, USA), and Golgi matrix protein GM130 (BD Biosciences).
6. The secondary antibodies against rabbit and mouse IgGs are purchased from Jackson ImmunoResearch Laboratories (West Grove, PA, USA).
7. ECL detection reagents are purchased from Amersham Biosciences (Piscataway, NJ, USA).

2.4. iTRAQ™ Labeling and Strong Cation Exchange Fractionation

1. Detergent-containing dissolution buffer: 50 mM triethylammonium bicarbonate (TEAB) (Fluka Chemika, Switzerland), pH 8.5, supplemented with 0.2% (w/v) RapiGest™ SF (Waters, Milford, MA, USA).
2. BCA Protein Assay Reagent Kit (Pierce).
3. Reducing reagent: Tris[2-carboxyethyl] phosphine hydrochloride (TCEP–HCl) (Pierce). A stock solution of 250 mM is made and stored at $-20°C$.
4. Cysteine blocking reagent: methyl methanethiosulfonate (MMTS) (Pierce). A stock solution of 200 mM is made in isopropanol and can be stored at 4°C for months.
5. Sequencing-grade modified trypsin (Promega, Madison, WI, USA) is used for protein proteolysis.
6. iTRAQ™ reagents 4-plex kit from Applied Biosystems (Foster City, CA, USA).
7. Strong cation exchange (SCX) load buffer: 5 mM KH_2PO_4, pH 3.0, in acetonitrile (ACN)/water, 25/75 (v/v).
8. SCX cartridge: ICAT™ Cartridge Cation Exchange (Applied Biosystems).
9. MacroSpin™ C_{18} column (The Nest Group, Southboro, MA).

2.5. Liquid Chromatography-Mass Spectrometry

1. LC: solvent A is 0.1% (v/v) formic acid in ACN/water, 5/95 (v/v) and solvent B is 100% ACN.
2. Famos autosampler (Dionex Co., Sunnyvale, CA, USA).
3. HP1100 system (Agilent Technologies, Santa Clara, CA, USA) is used for solvent delivery.

4. 100 µm internal diameter fused-silica capillary pre-column packed with 2 cm of 200 Å pore-size Magic C18AQ™ materials (Michrom Bioresources, Auburn, CA).
5. 75 µm × 10 cm fused-silica capillary column packed with 100 Å pore-size Magic C18AQ™ materials (Michrom Bioresources).
6. QSTAR Pulsar i quadrupole-TOF hybrid mass spectrometer (Applied Biosystems) is used with an in-house fabricated microelectrospray source.
7. MS data processing: SEQUEST™ software (Thermo Finnigan) and Trans-Proteomic Pipeline (TPP, ISB, Seattle, WA, http://tools.proteomecenter.org/TPP.php).

3. Methods

3.1. Cell Preparation and Drug Treatment

1. Mouse macrophage RAW264.7 cells are maintained in the RPMI medium on T-75 flasks. Cells are passed with trypsin/EDTA. Experimental cultures are carried out on six 150-mm dishes for either mock- or drug-treatment experiment. About 2×10^7 cells per dish are required to provide adequately confluent cultures after 48 h (*see* **Note 1**).
2. Cells with ~80% confluence are changed with fresh RPMI medium containing either DMSO for mock- or 200 nM thapsigargin for drug-treatment. Cells are incubated for 4 h at 37°C in the incubator under these conditions. Thapsigargin is an inhibitor of ER Ca^{2+}-ATPase, and treatment with this drug induces unfolded protein response in the cell (*13*).
3. For harvesting, cells are rinsed twice with 1× PBS and incubated with EDTA–PBS solution (10 mL per dish) for 5 min at 37°C before being lifted with a cell lifter. The dishes are washed with additional 10 mL of 1× PBS per dish and the combined cell-containing solutions are centrifuged at 1500 rpm (430 *g*) for 5 min to pellet the cells. The two obtained cell pellets (~0.4 mL each) are washed once with PBS, and the repelleted cells are processed to lysis or snap-frozen by liquid nitrogen before being stored at –80°C.

3.2. Cell Lysis and Fractionation

1. The cells are resuspended in ice-cold homogenization buffer (1 mL) and lysed by dounce homogenizer (20 strokes) and passage of 25-gauge needle (five times). Since cell lysis, protein samples should be kept at 4°C or on ice until the proteins are denatured for digestion (*see* **Subheading 3.4.2.**). The cell lysates are centrifuged at 3000 *g* for 10 min to generate post-nuclear fractions from the supernatant.
2. To fractionate each post-nuclear lysate, a continuous density gradient is generated by mixing 8 and 28% iodixanol solutions (5.3 mL each) using a Gradient Maker SG-30 and is then loaded into an Ultra-Clear centrifuge tube (Beckman) using the Auto Densi-Flow Density Gradient Fractionator with the heavy solution (28%) at the bottom of the tube (*see* **Note 2**).

3. The post-nuclear fractions are loaded on the top of the continuous density gradients and centrifuged at 28,000 rpm (~100,000 *g*) for 3 h at 4°C in Beckman SW41 rotor. After centrifugation, 23 fractions (~0.5 mL each) are collected using the Auto Densi-Flow Density Gradient Fractionator.

3.3. Immunoblot

1. To examine separation of subcellular compartments, 50 μL of the fractions obtained in **Subheading 3.2, step 3**, is subjected to SDS-PAGE analysis. The gel is then transferred to nitrocellulose membrane and subjected to immunoblot analysis against individual organelle marker proteins following the protocol of ECL western blotting (*see* **Fig. 1**).

3.4. iTRAQ™ Labeling and SCX Fractionation

1. Four representative fractions from **Subheading 3.2, step 3** (#4, 8, 12, and 16, ~0.45 mL each) (*see* **Note 3**), are mixed with 0.8 mL of homogenization buffer and centrifuged at 100,000 *g* (50,000 rpm) for 1 h in TLA100.2 rotor to remove the OptiPrep™ reagent. The pellets are resuspended in 100 μL of detergent-containing dissolution buffer and boiled for 5 min to obtain their dissolution (*see* **Note 4**). Take 5 μL to determine protein amount by BCA protein assay.

2. The proteins in detergent-containing dissolution solution (95 μL) are reduced by addition of 1.9 μL of TCEP–HCl stock solution (5 mM final concentration) and incubation at 37°C for 0.5–1 h. Then 4.75 μL of MMTS is added (final concentration of 10 mM), and the samples are incubated at room temperature for

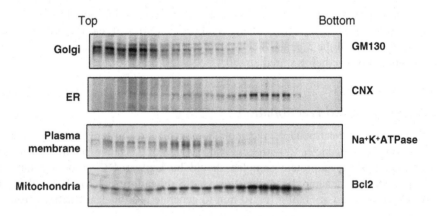

Fig. 1. Subcellular fractionation analysis. Mouse macrophage post-nuclear lysates were subjected to centrifugation in iodixanol density gradient (8–28%). The 23 collected fractions were analyzed by SDS–PAGE and immunoblot using antibodies against each organelle marker protein as indicated. CNX designates calnexin.

10 min to block cysteine residues. Trypsin is then added to a final protease: protein ratio of 1:20 (w/w). In case the pH is lowered by the added TCEP–HCl or trypsin solution, drops of 1 M TEAB are added to adjust pH to ≥ 8.0. Digestion is carried out at 37°C overnight.

3. The proteolyzed samples are dried out using Savant Speed Vac and resuspended in 30 μL of dissolution buffer supplied in the iTRAQ™ reagents 4-plex kit. Add 70 μL of ethanol (provided in the iTRAQ™ kit) to each iTRAQ™ reagent vial (114–117) and immediately transfer the dissolved iTRAQ™ reagent to each digested sample. The labeling is performed at room temperature for 1 h.

4. The four iTRAQ-labeled samples are then combined and loaded on an SCX cartridge for clean up and further fractionation. Prior to loading on the SCX cartridge, the combined sample is diluted to 4 mL (1:10) with SCX load buffer to reduce buffer salts and organic concentration. The sample pH is adjusted to 2.5–3.3 with concentrated H_3PO_4 (~85%) if needed.

5. The SCX cartridge is washed by 1 mL of SCX load buffer supplemented with 1 M KCl and then conditioned by injection of 2 mL of SCX load buffer. Then the diluted 4-mL iTRAQ-labeled sample is loaded on the SCX cartridge by injecting it slowly (about 1 drop/s). After washing the column with 1 mL of SCX load buffer, peptides are eluted from the cartridge stepwise by SCX load buffer supplemented with 80, 120, 160, and 350 mM KCl (0.4 mL per fraction) (*see* **Note 5**). The cartridge is then washed with 1 mL of SCX load buffer supplemented with 1 M KCl and finally equilibrated with 2 mL of 1 mM EDTA, pH 4.0, prior to storage at 4°C for future reuse.

6. The four eluted SCX fractions (0.4 mL each) are diluted to 2 mL (1:5 dilution) with 0.1% (v/v) trifluoroacetic acid (TFA) to reduce ACN concentration below 5%. The eluted samples are desalted on MacroSpin™ C_{18} columns following the manufacturer's protocol. The peptides eluted from the spin columns by ACN/water/TFA, 80/20/0.1 (v/v/v), are dried out by Savant Speed Vac and resuspended in 15 μL of 0.1% (v/v) formic acid in water for subsequent liquid chromatography-mass spectrometry (LC-MS) analysis.

3.5. Liquid Chromatography–Mass Spectrometry

1. Samples are automatically delivered by a FAMOS autosampler to a 100-μm-internal-diameter fused-silica capillary pre-column packed with 2 cm of 200-Å-pore-size Magic C18AQ™ material. An HP1100 system is used for solvent delivery. The loaded samples are washed with solvent A on the pre-column and then eluted with a gradient of 10–35% solvent B over 60 min to a 75 μm × 10 cm fused-silica capillary column packed with 100-Å-pore-size Magic C18AQ™ material and are then sprayed into the mass spectrometer at a constant column-tip flow rate of ~200 nL/min.

2. A QSTAR mass spectrometer is used with an in-house fabricated microelectrospray source. The eluting peptides are analyzed using the information-dependent acquisition (IDA) function of the Analyst software with the two most abundant ions detected in MS being selected for MS/MS fragmentation. Once an ion has

been selected for fragmentation, it is excluded from selection for 120 s. The scanned MS mass range is from 350–1300 Da, and the scanned MS/MS mass range is from 60–1800 Da.

3. The observed MS/MS spectra data are subjected to search the International Protein Index (IPI) database using the SEQUEST™ software with the following parameters: 144.1035 Da for iTRAQ-based static modification on N-termini and lysine residues, 45.9877 Da for MMTS-mediated cysteine modification, 15.99 Da for possible modification of oxidized methionine, mass tolerance 1 Da, and single-end tryptic restriction applied. The searched data are processed using TPP including protein identification curation through Interact, identification validation and probability assignment through PeptideProphet and ProteinProphet, and quantification through Libra (*see* **Note 6**).

3.6. Bioinformatics Data Analysis

1. Protein annotation for organelle association: the identified and quantified proteins are searched in the Database for Annotation, Visualization, and Integrated Discovery (DAVID 2006, http://niaid.abcc.ncifcrf.gov/) to gain annotation for proteins with known subcellular association. Proteins with their Refseq accession number are input for search and their corresponding organelle annotation information is output based on the SwissProt_keyword (*see* **Note 7**).

2. Principal component analysis (PCA): PCA is a popular statistical procedure that transforms, in an unsupervised manner, a number of correlated variables into a smaller number of uncorrelated variables called principal components (i.e., PC1, PC2, PC3, etc) while retaining most of its information *(14,15)*. The three ratios for each quantified protein in the iTRAQ assays are transformed into pairs of variables (PC1 and PC2) that are then plotted in a 2D graph (*see* **Fig 2 and 3**). The location of each protein in such a 2D graph is determined by the iTRAQ ratios that in turn reflect the distribution of the proteins over the four analyzed subcellular fractions. Therefore, proteins associated with the same organelle group, that have similar subcellular distribution over the fractions, are expected to cluster at similar locations in the 2D graph. PCA is conducted using pcacov.m in MATLAB (v7) (The MathWorks). Specifically, the iTRAQ-based protein abundance information is presented in a data matrix X ($n \times m$) where n and m are the numbers of quantified proteins and the number of iTRAQ channels corresponding to four density gradient fractions, respectively (i.e., $m = 4$ in this study). Prior to PCA modeling, the data matrix of $\mathbf{X}$ is scaled by the following normalization: (1) divide each row in data matrix by its sum to represent the relative protein abundances in the four fractions; (2) auto-scale $\mathbf{X}$ such that each column has zero mean and unit variance: $X_s = (X - 1\bar{\mathbf{x}}^T)\mathbf{S}^{-1}$, where $\bar{\mathbf{X}}$ is $m \times 1$ column mean vector and S is $m \times m$ standard deviation matrix. Then, a PCA model is generated with two principal components (PCs) capturing about 90% variance to represent $\mathbf{X}$: $\mathbf{X}_s = \mathbf{T}\mathbf{V}^T + \mathbf{E}$, where V is $m \times 2$ PCA loadings, T is $n \times 2$ score matrix ($\mathbf{T} = \mathbf{X}_s\mathbf{V}$), $\mathbf{E}$ is

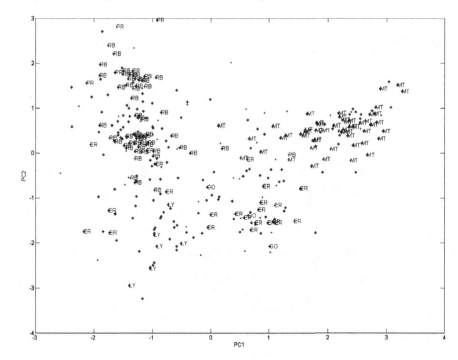

Fig. 2. The iTRAQ-quantified proteins in the mock experiment (M_1) were subjected to PCA and displayed as crosses in the PCA-plot with the newly generated PC1 and PC2 as X- and Y-axes, respectively. Proteins with organelle annotations are labeled as follows: ER (endoplasmic reticulum), MT (mitochondria), RB (ribosome), PR (proteasome), LY (lysosome), and GO (Golgi). The crosses with diamond edges indicate the proteins that are also identified in the repeated mock experiment (M_2) and thapsigargin-treatment experiment (T) and are used for comparison (see Fig. 3).

residual matrix, and the number of significant PCs (=2) is determined by *PRESS (16)*. PCA projections in the two PCs (PC1 and PC2) are displayed in a 2D graph (*see* **Figs 2 and 3**) with PC1 and PC2 as X-axis and Y-axis, respectively (*see* **Note 8**).

3. Comparison PCA: to compare organellar protein profiles under different conditions such as the drug treatment performed in **Subheadings 3.1, step 2**, two data matrices are generated, $\mathbf{X}_M$ and $\mathbf{X}_t$, corresponding to the mock- (M) and thapsigargin- (T) treatment experiments, respectively. After a PCA model for $\mathbf{X}_M$ is generated as described above (*see* **Subheading 3.6., step 2**), the scaled matrix of $\mathbf{X}_t$ is normalized by auto-scaling using the mean and the standard deviation matrix of *M*. Then the score matrix $\mathbf{T}_t$ is computed by projecting the corresponding scaled matrices onto *V* of *M*. The comparative result is displayed in the same 2D figure with PC1 and PC2 as X- and Y-axes, respectively (*see* **Fig. 3A**) (*see* **Note 9**).

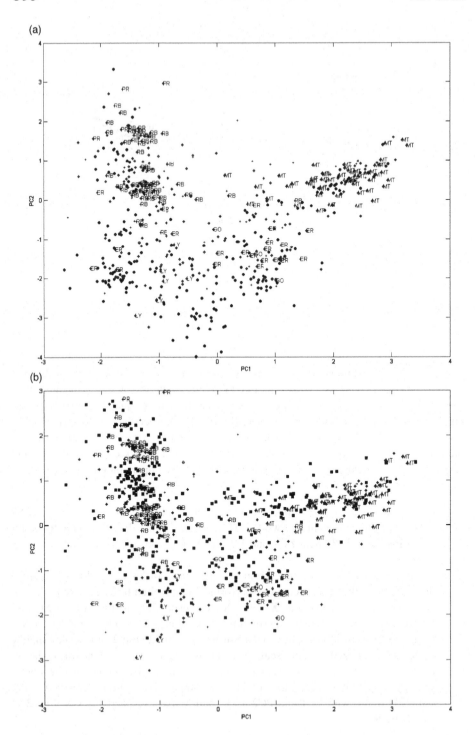

4. Notes

1. Cell type is not limited to the mouse macrophage. Any type of cells including those from tissue samples can be used as well. And perturbations can be expanded to other types, such as disease state, drug treatment, pathogen infection, etc.
2. The gradient condition is dependent on the specific cellular compartments being studied and needs to be experimentally optimized. In addition to the continuous pre-formed iodixanol gradient, as described here, other types of gradients such as discontinuous and/or self-generated gradients may offer adequate organelle separations.
3. The four representative fractions are selected based on the western blot data obtained on the organelle marker proteins. Those fractions containing the target organelles to be studied, dependent on their separation in the gradient, are labeled with the iTRAQ reagents. Ideally, the more fractions analyzed, the better the resolution of profiling different organelles. A combination of two sets of iTRAQ assays to monitor seven fractions was reported *(9)*. However, the use of multiple MS analyses resulted in a decreased total number of quantified proteins from both assays. The second generation of iTRAQ methodology, allowing analysis of up to eight fractions simultaneously, will be commercially available soon (Applied Biosystems) and is expected to provide improved resolution of organelle profiling.
4. It is suggested to dissolve the protein pellet in a small volume. If less than 30 μL can be applied to dissolve the pellet, the drying and re-suspending procedures at **step 3** of **Subheading 3.4** can be skipped to prevent potential sample loss.
5. The condition of step elution is sample specific. For samples with high complexity and protein amount, more fractions or even a gradient elution may be needed to provide adequate separation of the peptides.

Fig. 3. (**A**) Comparison PCA analysis between mock- (M_1) and thapsigargin- (T) treated experiments: the proteins from the T data set (shown by dots) were superimposed to the M_1 matrices for comparison with proteins from the M_1 data set (shown by crosses). (**B**) Comparison PCA analysis between the two mock-treated experiments (M_1 and M_2): The proteins from the M_2 data set (shown by squares) were superimposed to the M_1 data matrix for comparison with the proteins from the M_1 data set (shown by crosses). In **Fig. 3B**, a complete overlap between crosses (M1) and squares (M2) is not observed because of various experimental variations (from MS analysis, labeling, fractionation, etc). However, for those proteins affected by the UPR, the difference between M1 and M2 (as shown by the distance between cross and square in **Fig. 3B**) is much smaller than the difference between M1 and T (as shown by the distance between crosses and dots in **Fig. 3A**). For those proteins not expected to be affected by the UPR (e.g., mitochondrial proteins), a similar difference on each pair of comparison is observed.

6. The signal intensities of the four iTRAQ channels for each labeled peptide are quantified by our in-house developed Libra program (http://tools.proteomecenter.org/Libra.php) that consists of one module of the TPP and specifically quantifies multi-channel labeled peptides.

7. Information from other databases, such as that in the gene ontology (cellular component), can be applied for organelle annotation of the identified proteins as well.

8. The percentage of variance covered by PC1 and PC2 is important for displaying adequate information in the 2D figure. If a combination of PC1 and PC2 captures less than 80% of variance, a 3D figure using all the three PCs may be applied.

9. A control assay by comparison of two independent mock experiments (i.e., M_1 and M_2) is suggested before applying comparison of $\mathbf{X}_M$ and $\mathbf{X}_t$. Comparison of $\mathbf{X}_{M1}$ and $\mathbf{X}_{M2}$ will generate a baseline for accurate comparison between $\mathbf{X}_M$ and $\mathbf{X}_t$ (*see* **Fig. 3B**).

References

1. Aebersold, R. and Mann, M. (2003) Mass spectrometry-based proteomics. *Nature* **422**, 198–207.

2. Yates, J. R., III, Gilchrist, A., Howell, K. E., and Bergeron, J. J. (2005) Proteomics of organelles and large cellular structures. *Nat. Rev. Mol. Cell Biol.* **6**, 702–714.

3. Brunet, S., Thibault, P., Gagnon, E., Kearney, P., Bergeron, J. J., and Desjardins, M. (2003) Organelle proteomics: looking at less to see more. *Trends Cell Biol.* **13**, 629–638.

4. Warnock, D. E., Fahy, E., and Taylor, S. W. (2004) Identification of protein associations in organelles, using mass spectrometry-based proteomics. *Mass Spectrom. Rev.* **23**, 259–280.

5. Marelli, M., Smith, J. J., Jung, S., Yi, E., Nesvizhskii, A. I., Christmas, R. H., et al. (2004) Quantitative mass spectrometry reveals a role for the GTPase Rho1p in actin organization on the peroxisome membrane. *J. Cell Biol.* **167**, 1099–1112.

6. Andersen, J. S., Wilkinson, C. J., Mayor, T., Mortensen, P., Nigg, E. A., and Mann, M. (2003) Proteomic characterization of the human centrosome by protein correlation profiling. *Nature* **426**, 570–574.

7. Foster, L. J., de Hoog, C. L., Zhang, Y., Zhang, Y., Xie, X., Mootha, V. K., et al. (2006) A mammalian organelle map by protein correlation profiling. *Cell* **125**, 187–199.

8. Dunkley, T. P. J., Watson, R., Griffin, J. L., Dupree, P., and Lilley, K. S. (2004) Localization of Organelle Proteins by Isotope Tagging (LOPIT). *Mol. Cell Proteomics* **3**, 1128–1134.

9. Dunkley, T. P., Hester, S., Shadforth, I. P., Runions, J., Weimar, T., Hanton, S. L., et al. (2006) Mapping the Arabidopsis organelle proteome. *Proc. Natl. Acad. Sci. U.S.A.* **103**, 6518–6523.

10. Ross, P. L., Huang, Y. N., Marchese, J. N., Williamson, B., Parker, K., Hattan, S., et al. (2004) Multiplexed protein quantitation in Saccharomyces cerevisiae using amine-reactive isobaric tagging reagents. *Mol. Cell Proteomics* **3**, 1154–1169.

11. Schroder, M. and Kaufman, R. J. (2005) The mammalian unfolded protein response. *Annu. Rev. Biochem.* **74**, 739–789.

12. Federovitch, C. M., Ron, D., and Hampton, R. Y. (2005) The dynamic ER: experimental approaches and current questions. *Curr. Opin. Cell Biol.* **17**, 409–414.

13. Yan, W., Frank, C. L., Korth, M. J., Sopher, B. L., Novoa, I., Ron, D., et al. (2002) Control of PERK eIF2alpha kinase activity by the endoplasmic reticulum stress-induced molecular chaperone P58IPK. *Proc. Natl. Acad. Sci. U. S. A.* **99**, 15920–15925.

14. Holter, N. S., Mitra, M., Maritan, A., Cieplak, M., Banavar, J. R., and Fedoroff, N. V. (2000) Fundamental patterns underlying gene expression profiles: simplicity from complexity. *Proc. Natl. Acad. Sci. U.S.A.* **97**, 8409–8414.

15. Alter, O., Brown, P. O., and Botstein, D. (2000) Singular value decomposition for genome-wide expression data processing and modeling. *Proc. Natl. Acad. Sci. U.S.A.* **97**, 10101–10106.

16. Chan, C., Hwang, D., Stephanopoulos, G. N., Yarmush, M. L., and Stephanopoulos, G. (2003) Application of multivariate analysis to optimize function of cultured hepatocytes. *Biotechnol. Prog.* **19**, 580–598.

27

Use of Transcriptomic Data to Support Organelle Proteomic Analysis

Wallace F. Marshall

Summary

Genome-wide transcriptional profiling provides a rich source of data for the validation and annotation of organelle proteomic data. Organelle biogenesis is in most cases accompanied by upregulation of genes encoding organelle-specific proteins. Consequently, identification of genes whose expression correlates with organelle assembly leads to a candidate list that can be cross-checked with a preliminary organelle proteome. When proteins are found in the proteome and the corresponding genes are found to have organelle assembly-correlated expression, this greatly increases our confidence that those proteins are true components of the organelle and not contamination. Such an approach can be used to narrow down a preliminary proteomic data set and help us to focus on a smaller sub-set of proteins that are supported by transcriptomic cross-validation.

Key Words: Cilia; centrioles; meiosis; synaptonemal complex; unfolded protein response; endoplasmic reticulum; synexpression group; spermiogenesis; Daf-19; X-box.

1. Introduction

Proteomic analysis of organelles can be a difficult method, requiring careful development of an optimized purification procedure, careful evaluation of stepwise enrichment, and mastery of advanced technologies needed for protein identification. In contrast, the tools used to study gene expression are technically much simpler and faster. Although we are ultimately interested in the proteins that make up an organelle, in some cases, studies of gene expression can be used to obtain similar information by enumerating the set of genes

From: *Methods in Molecular Biology, vol. 432: Organelle Proteomics*
Edited by: D. Pflieger and J. Rossier © Humana Press Inc., Totowa, NJ

whose expression correlates with assembly of an organelle of interest. Such an analysis can provide valuable confirmatory evidence to help annotate a proteomic data set and can also provide a rapid, if indirect, set of candidate genes, which in some cases may closely approximate the actual proteomic parts list but at far less cost of time and effort. Usually, the set of genes whose expression correlates with organelle assembly will include genes that encode components of the proteome of the organelle, as well as genes whose products are involved in organelle assembly but whose products are not included in the final structure. In this sense, the approach described here is only partially overlapping with proteomics and should thus be viewed as a complementary technique. In this chapter, I describe several related approaches for obtaining lists of genes potentially involved in formation of an organelle of interest using transcriptional profiling.

2. Transcriptional Profiling for Proteomics

Organelle biogenesis is a highly regulated process, and this regulation can be exploited as the basis for a transcriptomic approach to proteomics. The general strategy is to correlate expression of genes with the formation of the organelle. The resulting organelle-related gene lists can then be used in several ways. First, it can be used to validate a proteomic data set, by asking what fractions of the putative proteome components have an expression pattern that suggests they are part of the organelle. Such an analysis can increase our confidence in a proteomic data set. Second, even in the absence of proteomic data, an organelle assembly-related gene list can suggest interesting candidate molecules related to the organelle's assembly, which can serve as a starting point for functional studies. In this way, transcriptomic approaches can provide much of the same information as proteomic approaches, albeit less directly.

This approach is rendered feasible because of the prevalence of "synexpression groups" in functional biology. A synexpression group has been defined as a set of coexpressed genes that participate in a common biological function (1). When the biological function in question refers to the biogenesis of an organelle, the corresponding synexpression group reveals genes involved in building that organelle and is likely to include genes encoding structural elements of the organelle.

2.1. Inducible Organelle Biosynthesis

2.1.1. Strategy

Growth of organelles is often accompanied by dramatic and specific increases in mRNA levels for genes encoding protein components of the organelle in question. For example, when endoplasmic reticulum is induced to proliferate by the unfolded protein response pathway, a whole array of genes are upregulated, whose products encode proteins related either to endoplasmic reticulum (ER) function or formation (*2*). A similar induction of organelle-specific genes is seen during induced proliferation of peroxisomes (*3*) and mitochondria (*4*).

To use transcriptional profiling to identify genes related to the assembly of a particular organelle, it is highly desirable to induce the synchronous expression of these genes. In the absence of prior information about upstream regulatory pathways, the most straightforward way to induce expression of organelle-related genes is to induce organelle biogenesis and then see which genes become upregulated as a result.

2.1.2. Example

During the course of our proteomic analysis of isolated centrioles (*5*), we knew that our final sample was still contaminated with proteins from other cellular components and were therefore faced with the challenge of distinguishing the real centriole proteins in our mass spectrometry data from contaminants. To help with this process, we noted that several centriole-localized proteins had previously been shown to be encoded by genes upregulated during flagellar regeneration (*6*), possibly indicating a role for such genes in the nucleation of cilia and flagella by the centriole when it becomes a basal body. We therefore compared our candidate proteome list with results of our microarray analysis of gene expression during flagellar regeneration (*7*) and found that many of the putative centriole proteins in our list were encoded by upregulated genes. On this basis, we asserted that these were particularly promising candidates. Thus, analysis of flagellar assembly-related gene expression not only helped with the annotation of the flagellar proteome (*8*) but also with validation of the proteome of centrioles.

2.1.3. Requirements to Apply to a New Organelle

The primary requirement is a model system in which the organelle of interest can be induced to form rapidly and synchronously, leading to a large discrete pulse of organelle-related gene expression. Induction is usually achieved by taking advantage of intrinsic biological regulatory pathways but in cases where the upstream regulatory mechanisms driving organelle assembly are understood, one could imagine engineering inducible control over organelle biogenesis.

A second requirement is a technology for genome-wide transcriptional profiling in the model system in question. This is most generally accomplished using microarrays. In cases where commercial oligonucleotide- or cDNA-based microarrays are not available, the burden of microarray construction will fall upon the individual researcher. Fortunately, for simple transcriptomic analyses of organelle assembly, the number of array experiments is usually rather small. For example, to study gene expression levels at three time points before, during, and after organelle assembly, with duplicate sets of data at each time point, six arrays would be needed. In such cases, it is more economical to generate each array individually using modern methods such as optical patterning with digital light processors *(9)*. These methods provide a cost-effective way to generate the limited number of transcriptomic data sets required for support of an organelle proteome project.

To obtain maximally useful transcriptomic data from a minimal number of arrays, it is important to perform pilot experiments using a few known organelle-related genes, analyzing their expression levels using RT–PCR or Northern blotting, to determine the time point following induction at which organelle-related gene expression is maximum. This time point should then be used for genome-wide analysis by microarrays.

An alternative way to employ transcriptional profiling for proteomics is to start with a proteomic data set and then measure expression of each gene encoding proteome components to see whether its expression is correlated with organelle assembly. This can be done without the investment in developing a microarray using quantitative RT–PCR with primers specific for candidate genes.

2.1.4. Potential Pitfalls

As with any proteomic study, the generation of candidate gene lists is just a first step to be followed by localization and functional studies that are generally much more costly, in terms of time and resources, than the initial generation of the list. Consequently, the main goal is to minimize the false positive rate, so as to limit the subsequent waste of resources spent on secondary analysis of false hits. This always represents a tradeoff between missing some true candidates, but at least to the first approximation, we will be most concerned about pitfalls that could lead to the erroneous inclusion of large numbers of incorrect genes in the candidate list. The most serious experimental pitfall of a transcriptomic approach is the possibility that the conditions used to induce organelle assembly might simultaneously induce a stress response within the cell. In our study of genes induced during flagellar regeneration following pH shock in *Chlamydomonas (10)*, we obtained a large number of heat shock proteins and other chaperones among the list of induced genes.

One strategy for removing stress response from the gene list is to identify all putative stress response genes based on homology with known chaperone proteins and remove them from the list. This approach would of course miss novel stress response proteins, which would represent the most serious type of contamination to a potential gene list as they would not be apparent. Moreover, the fact that a protein can act as a chaperone is not necessarily, by itself, an indicator that these genes are false positives. One could certainly imagine that assembly of large and complex cellular structures involving the synthesis of hundreds of proteins might place a stress on the cell protein folding machinery and that production of chaperones may play a critical role in assembly of the structure even in the normal biological context. Indeed, a major chaperone system (GroEL) was first identified because it is required for bacteriophage assembly. In the specific case of flagellar assembly, many chaperones were identified on the basis of expression upregulated during flagellar assembly *(7)*, but it was not clear a priori whether the large number of chaperones reflected a real role in flagellar assembly or not. At least one Heat Shock Protein (HSP) is a bona fide component of the cilium *(11)* and thus should be considered a legitimate flagella-associated gene. This result highlights the real challenge: to discriminate possible indirect stress response genes (i.e., genes whose expression is induced as a byproduct of organelle assembly and which therefore are potentially involved in assembling the organelle proteome) from the direct stress response (i.e., a response triggered by the inducing stimulus directly and not requiring organelle assembly).

One strategy is to knock down gene levels of candidate genes and ask whether the organelle can still assemble at the usual rate. However, such use of reverse genetics might be misleading, as one could imagine that loss of a stress response system might block organelle assembly following experimental induction simply because the cells become sick. Moreover, this approach requires a gene-by-gene analysis of function and is therefore potentially costly in terms of labor and resources. A better way is needed.

In the specific case of flagellar assembly, we have developed a strategy to distinguish between stress response induced by the pH shock itself from gene induction that is actually correlated with flagellar assembly. To do so, we perform a standard pH shock experiment under conditions in which flagellar assembly is prevented. This can be done either by blocking flagellar growth using microtubule inhibitors such as colchicines or nocodazole or using mutants in which regeneration cannot occur following pH shock as previously described by others *(12,13)*. We have found that in such cells, genes encoding flagella-localized proteins such as tubulin, dynein arms, or radial spoke proteins are no longer induced while at least some heat shock proteins continue to show high levels of induction. Genome-wide analysis of such cells, currently underway, should provide a clear discrimination of direct stress response versus flagellar growth-associated response. This example illustrates one potentially generalizable strategy to discriminate direct from indirect stress response: induce organelle assembly by the standard method but using additional treatments that prevent actual assembly.

An alternative strategy is to induce organelle assembly using a different type of inducing stimulus that is unlikely to cause the same direct stress response. For the case of flagellar assembly, addition of lithium to the medium causes flagella to elongate, providing a stress-free way to trigger flagellar assembly. Analysis of transcriptional profiling during lithium-induced growth may thus provide an interesting new viewpoint onto the true flagellar growth-associated gene set.

2.2. Comparison of Cell Types

2.2.1. Strategy

The strategy outlined above requires a method to induce organelle assembly. This will only be possible in some cases, for other organelles, no inducing stimulus may be known at the outset of the experiment. However, a potential work-around can be obtained by realizing that the ability of organelle assembly-specific genes to be induced did not evolve for our convenience but rather to enable cells of the appropriate cell type to form the organelle in question. We can thus attempt to use the inherent tendency of particular cell types to express these genes, by comparing gene expression patterns in cells containing the organelle of interest with those in cells lacking the organelle. As most organelles are dynamic and undergo continuous turnover at some rate throughout their lifetime, cells that contain a given organelle may express organelle-specific genes at a high enough level to allow their detection over background.

2.2.2. Example

During meiotic cell division, the cell assembles a remarkable machine onto each chromosome, called the synaptonemal complex (SC). Induction of SC by an external stimulus has not yet been achieved. However, a comparison of gene expression patterns in meiotic versus non-meiotic yeast cells *(14)* revealed a list that contained many known SC proteins.

2.2.3. Requirements

The principle requirement for this method is a way to generate pure samples of cells with or without the organelle in question. One approach would be to use cell culture to grow cell lines derived from different tissue types such that one line has the organelle of interest in large abundance while the organelle is relatively depleted in the other cell line. Alternatively, one can analyze gene expression in whole tissues. This can be problematic as most tissues contain a mixture of cell types, only some of which contain the organelle of interest. This problem can be overcome by using cell sorting methods to separate cell types with the organelle from types without. This has been done, for example in the testis, where separation of cells undergoing spermiogenesis from the non-spermiogenic sertoli cells *(15,16)* allows production of subtractive gene lists highly enriched in proteins localized in the sperm tail flagellum (W.F.M., unpublished results).

2.2.4. Potential Pitfalls

This approach is potentially limited in its applicability depending on the intrinsic rate of protein turnover for the organelle of interest. An organelle that is relatively static and does not turn over might not require a measurable rate of protein synthesis or gene expression under static conditions. The more an organelle turns over, the higher a level of gene expression one might expect. Conversely, organelles that turn over slowly might not require much gene expression over background. As a result, comparison of gene expression patterns between cells that have the organelle and cells that do not might not reveal any significant changes in expression levels. In such a case, the method would fail completely. Because of this concern, we strongly recommend that workers interested in any particular organelle invest the extra time needed to discover a reliable and rapid method to induce organelle assembly, so as to produce a large discrete burst of gene expression that can be reliably detected over background fluctuations.

A second pitfall is more subtle and involves the possibility that gene expression can vary as a result of organelle function. For instance, cells containing chloroplasts will have significant differences in metabolism from cells without them and might therefore express a range of different genes that have nothing to do with chloroplast assembly per se but instead are induced as a result of the metabolic conditions. Again, this type of concern can be largely overcome using the previously described method in which organelle assembly is induced at a time of our choosing using artificial means, because in this case, we can show that gene expression temporally correlates with organelle assembly rather than simply the presence or absence of the organelle and any consequent downstream effects on cell physiology.

A third limitation of this approach is that different cell types usually differ in many aspects of biology, some of which may have nothing to do with the organelle of interest. For example, in the case of the meiotic SC, expression profiling of meiotic yeast cells versus non-meiotic revealed many proteins involved in other aspects of meiosis having nothing to do with SC assembly and even genes whose primary function was unrelated to meiosis but that were instead involved in the formation of spores after meiosis. Thus, comparisons based on cell types are likely to contain a majority of genes that are NOT related to the organelle of interest. Because of these three large pitfalls, cell type comparison should best be viewed as a method of last resort. Direct comparisons of proteomic and transcriptomic differences between different tissues have indicated that only a fraction (10–20%) of the proteomic differences are detectable at the level of transcriptomic data *(17,18)*, further indicating the level of caution that must be applied in such studies.

2.3. Identification of Organelle-Specific Promoter Sequences

2.3.1. Strategy

Given that gene expression is often determined by the promoter and its associated enhancer elements, analysis of consensus sequences upstream of genes known to be involved with a particular organelle can often provide a simple way of obtaining organelle-related gene lists.

2.3.2. Example

One of the best examples of this approach is a strategy developed by Swoboda and coworkers to identify cilia-related genes in the nematode, *Caenorhabditis elegans*. In this organism, cilia are only found in a set of ciliated sensory neurons whose developmental fate is specified by the selector gene *Daf-19*. The binding site of *Daf-19* is a well characterized sequence called the "X-box". Analysis of the complete set of genes in the *C. elegans* genome containing X boxes revealed a large number of known cilia-related proteins *(19)*.

2.3.3. Requirements

This method requires extensive prior knowledge about the regulation of organelle-specific genes. In cases where a clear consensus sequence has been identified upstream of a set of known organelle-related genes, the strategy can be applied.

2.3.4. Potential Pitfalls

The primary pitfall for this method is the same as one raised for cell-type-specific gene expression methods, specifically the danger that promoters are likely to indicate cell-type-specific expression rather than organelle-specific expression. For instance, the X-box is really specific for ciliated sensory neurons, and thus X-box-regulated genes will include not only cilia-related genes but also sensory neuron-specific genes.

We also note that in some cases, it is possible to recognize a protein targeting sequence specific for a given organelle. For instance, one can in principle obtain a set of mitochondrial proteins by doing a genome-wide search for ORFs encoding a sequence that has a mitochondrial-targeting signal as a subsequence. This approach is generally quite limited due to the degeneracy in such localization signals when they are known, and the tremendous difficulty in determining consensus sequences when they are not already known. In most cases, we suspect that identification of putative-targeting sequences will be consigned to play a primary role in gene confirmation rather than gene discovery.

3. Comparative Genomics for Proteomics

3.1. Strategy

We have described above a strategy in which cell types containing an organelle of interest are compared with cell types from the same species in which the organelle is lacking, and the differences in gene expression pattern are used to infer organelle-related genes. A similar strategy can be applied at the level of genomes, by comparing the complete set of predicted gene models in species that have the organelle of interest versus species that lack the organelle. For instance, a comparison of green plant genomes versus animal genomes will certainly include most genes specifically involved in formation of chloroplasts.

3.2. Example

Two studies have compared genes conserved in species that have cilia with species that do not form cilia, such as higher plants and yeast *(20,21)*. Both studies successfully identified a large set of known cilia-related genes, as well as several novel genes whose relation to cilia was subsequently verified by direct experiments. One of these studies gave the unfortunate name "flagellar and basal body proteome" (FABP) to their data set, a misleading term that suggests, somewhat deceptively, that a proteomic analysis was performed. In fact, the so-called FABP is not at all a proteome but instead a set of conserved genes.

3.3. Requirements

First, one requires a detailed knowledge of organelle cell biology in a variety of species spanning eukaryotic biodiversity. Detailed information about organelle structure in different species may suggest interesting comparisons that would otherwise never be considered. For instance, comparing genes in species with motile cilia versus species with only non-motile cilia can be used to identify genes related to assembly of the dynein arms that drive ciliary motility.

A second important requirement is a reliable way to determine orthologs between species. Published methods use relatively crude mutual best-hit BLAST statistics as a criterion, but this is greatly complicated by the presence of multiple paralogs within some species. More advanced bioinformatics methods should yield far more reliable and thorough genome comparisons.

3.4. Potential Pitfalls

One pitfall can arise if workers are too hasty to force the absence of genes from a species that they assume to lack an organelle that it actually contains. For instance, a number of species that appear to not perform meiosis actually contain in their genomes many meiosis-related genes, and it is quite likely that they do in fact perform meiosis under the right conditions, but these conditions have not yet been discovered in the laboratory. With the rapid sequencing of ever more obscure organisms, it becomes increasingly important to pay attention to the detailed ultrastructural literature on each organism that one uses in the comparison. Mainstream molecular biologists have historically treated the careful documentation of phylogeny and biodiversity with an appalling degree of contempt, but that sort of attitude needs to be corrected if one can ever hope to exploit the rich source of information that these different genomes provide.

References

1. Niehrs, C. (2004). Synexpression groups: genetic modules and embryonic development. In *Modularity in Development and Evolution*, G. Schlosser and G. Wagner, eds. University of Chicago Press, Chicago, 187–201.
2. Cox, J. S., Chapman, R. E., and Walter, P. (1997). The unfolded protein response coordinates the production of endoplasmic reticulum protein and endoplasmic reticulum membrane. *Mol. Biol. Cell* **8**, 1805–1814.
3. Smith, J. J., Marelli, M., Christmas, R. H, Vizeacoumar, F. J., Dilworth, D. J., Ideker, T., et al. (2002) Transcriptome profiling to identify genes involved in peroxisome assembly and function. *J. Cell Biol.* **158**, 259–271.
4. Li, F., Wang, Y., Zeller, K. I., Potter, J. J, Wonsey, D. R., O'Donnell, K. A., et al. (2005) Myc stimulates nuclearly encoded mitochondrial genes and mitochondrial biogenesis. *Mol. Cell Biol.* **25**, 6225–3624.

5. Keller, L. C., Romijn, E. P., Zamora, I. Yates, J. R., and Marshall, W. F. (2005) Proteomic analysis of isolated Chlamydomonas centrioles reveals orthologs of ciliary disease genes. *Curr. Biol.* **15**, 1090–1098.

6. Pfannenschmid, F., Wimmer, V. C., Rios, R. M., Geimer, S., Krockel, U., Leiherer, A., et al. (2003) Chlamydomonas DIP13 and human NA14: a new class of proteins associated with microtubule structures is involved in cell division. *J. Cell Sci.* **116**, 1449–1462.

7. Stolc, V., Samanta, M. P., Tongprasit, W., and Marshall, W. F. (2005). Genome-wide transcriptional analysis of flagellar regeneration in *Chlamydomonas reinhardtii* identifies orthologs of ciliary disease genes. *Proc. Natl. Acad. Sci. U.S.A.* **102**, 3703–3707.

8. Pazour, G. J., Agrin, N., Leszyk, J., and Witman, G. B. (2005) Proteomic analysis of a eukaryotic cilium. *J. Cell Biol.* **170**, 103–113.

9. Nuwaysir, E. F., Huang, W., Albert, T. J., Singh, J., Nuwaysir, K., Pitas, A., et al. (2002) Gene expression analysis using oligonucleotide arrays produced by maskless photolithography. *Genome Res.* **12**, 1749–1755.

10. Lefebvre, P. A., Silflow, C. D., Wieben, E. D., and Rosenbaum, J. L. (1980) Increased levels of mRNAs for tubulin and other flagellar proteins after amputation or shortening of Chlamydomonas flagella. *Cell* **20**, 469–477.

11. Shapiro, J., Ingram, J., and Johnson, K. A. (2005) Characterization of a molecular chaperone present in the eukaryotic flagellum. *Euk. Cell.* **4**, 1591–1594.

12. Lefebvre, P. A., Barsel, S. E., and Wexler, D. E. (1988) Isolation and characterization of Chlamydomonas reinhardtii mutants with defects in the induction of protein synthesis after deflagellation. *J. Protozool.* **35**, 559–564.

13. Larkin, J. C., Lefebvre, P. A., and Silflow, C. D. (1989) A gene essential for viability and flagellar regeneration maps to the uni linkage group of Chlamydomonas reinhardtii. *Curr. Genet.* **15**, 377–384.

14. Chu, S., DeRisi, J., Eisen, M., Mulholland, J., Botstein, D., Brown, P. O., et al. (1998) The transcriptional program of sporulation in budding yeast. *Science* **282**, 699–705.

15. Schultz, N., Hamra, F. K., and Garbers, D. L. (2003) A multitude of genes expressed solely in meiotic or postmeiotic spermatogenic cells offers a myriad of contraceptive targets. *Proc. Natl. Acad. Sci. U.S.A.* **100**, 12201–12206.

16. Divina, P., Vlcek, C., Strnad, P, Paces, V, Forejt, J. (2005) Global transcriptomic analysis of the C57BL/6J mouse testis by SAGE: evidence for nonrandom gene order. *BMC Genomics* **6**, 29.

17. Lorenz, P., Ruschpler, P., Koczan, D, Stiehl, P, and Thiesen, H. J. (2002) From transcriptome to proteome: differentially expressed proteins identified in synovial tissue of patients suffering from rheumatoid arthritis and osteoarthritis by an initial screen with a panel of 791 antibodies. *Proteomics* **3**, 991–1002.

18. Conrads, K. A., Yi, M., Simpson, K. A., Lucas, D. A., Camalier, C. E., Yu, L. R., et al. (2005) A combined proteomic and microarray investigation of inorganic phosphate induced pre-osteoblast cells. *Mol. Cell Proteomics* **4**, 1284–1296.

19. Efimenko, E., Bubb, K., Mak, H. Y., Holzman, T, Leroux, M. R., Ruvkun, G., et al. (2005). Analysis of xbx genes in C. elegans. *Development* **132**, 1923–1934.

20. Li, J. B., Gerdes, J. M, Haycraft, C. J., Fan, Y., Teslovich, T. M., May-Simera, H., et al. (2004) Comparative genomics identifies a flagellar and basal body proteome that includes the BBS5 human disease gene. *Cell* **117**, 541–552.

21. Avidor-Reiss, T., Maer, A. M., Koundakjian, E., Polyanovsky, A, Keil, T., Subramanian, S., et al. (2004) Decoding cilia function: defining specialized genes required for compartmentalized cilia biogenesis. *Cell* **117**, 527–539.

Index

415